Sustainable Development Goals Series

The **Sustainable Development Goals Series** is Springer Nature's inaugural cross-imprint book series that addresses and supports the United Nations' seventeen Sustainable Development Goals. The series fosters comprehensive research focused on these global targets and endeavours to address some of society's greatest grand challenges. The SDGs are inherently multidisciplinary, and they bring people working across different fields together and working towards a common goal. In this spirit, the Sustainable Development Goals series is the first at Springer Nature to publish books under both the Springer and Palgrave Macmillan imprints, bringing the strengths of our imprints together.

The Sustainable Development Goals Series is organized into eighteen subseries: one subseries based around each of the seventeen respective Sustainable Development Goals, and an eighteenth subseries, "Connecting the Goals", which serves as a home for volumes addressing multiple goals or studying the SDGs as a whole. Each subseries is guided by an expert Subseries Advisor with years or decades of experience studying and addressing core components of their respective Goal.

The SDG Series has a remit as broad as the SDGs themselves, and contributions are welcome from scientists, academics, policymakers, and researchers working in fields related to any of the seventeen goals. If you are interested in contributing a monograph or curated volume to the series, please contact the Publishers: Zachary Romano [Springer; zachary.romano@springer.com] and Rachael Ballard [Palgrave Macmillan; rachael.ballard@palgrave.com].

Uday Chatterjee
Abdulaziz I. Almulhim
Editors

Urban Waterscapes

Challenges, Resilience for Future Sustainable Cities

Springer

Editors
Uday Chatterjee
Department of Geography
Bhatter College Dantan (Autonomous)
Kharagpur, West Bengal, India

Abdulaziz I. Almulhim
Department of Urban and Regional Planning, College of Architecture and Planning
Imam Abdulrahman Bin Faisal University
Dammam, Saudi Arabia

ISSN 2523-3084 ISSN 2523-3092 (electronic)
Sustainable Development Goals Series
ISBN 978-3-032-20648-0 ISBN 978-3-032-20649-7 (eBook)
https://doi.org/10.1007/978-3-032-20649-7

This Springer imprint is published by the registered company Springer Nature Switzerland AG
The registered company address is: Gewerbestrasse 11, 6330 Cham, Switzerland

Young scholars and practitioners in urban studies, environmental science, water resources management, sustainability, geospatial science, and geography, as well as policymakers striving for equitable and resilient urban water systems.

Preface

As cities grow and climate pressures intensify, the management of urban water systems has become one of the defining challenges of our time. *Urban Waterscapes: Challenges, Resilience for Future Sustainable Cities* responds to this urgent context, offering a multidimensional examination of how water governance, equity, technology, and resilience can be integrated to support sustainable urban futures. Cities are both vulnerable to and responsible for the shifting dynamics of water. From flooding and water scarcity to pollution and ecosystem degradation, the impacts of climate change and rapid urbanization are increasingly visible in our streets, rivers, and coasts. Yet, cities also hold the innovation capacity, governance frameworks, and community engagement needed to transform water challenges into opportunities for regeneration. Bringing together contributors from diverse disciplines including urban planning, hydrology, environmental science, geospatial analysis, and climate policy, this volume bridges technical expertise with social insight. Through case studies from Ghana, Namibia, Bangladesh, India, South Africa, and beyond, the chapters reveal both shared global concerns and the unique, placespecific strategies that can be leveraged to achieve resilient and equitable water systems. Organized into four thematic sections, the book moves from foundational governance and climate-related water challenges to equity and resilience in planning, to the role of smart technologies and innovative policy pathways, and finally to global case studies that offer transferable lessons. Topics explored include governance strategies for water equity, climate change adaptation in floodprone cities, wetland integration into urban planning, participatory vulnerability assessments, AI-powered water quality monitoring, blue-green infrastructure for mental well-being, wastewater recycling for sustainability, and global case studies offering transferable lessons. This is not simply a collection of studies, it is a call for integrated, cross-sectoral action. It urges policymakers, practitioners, researchers, and community leaders to move beyond fragmented approaches and embrace holistic, context-sensitive strategies that recognize the interconnected nature of water, climate resilience, and urban well-being. We hope this volume will serve as both a reference and a source of inspiration, encouraging the design and governance of urban waterscapes that are sustainable, equitable, and resilient in the face of future uncertainty.

Kharagpur, West Bengal, India — Uday Chatterjee
Dammam, Saudi Arabia — Abdulaziz I. Almulhim

Acknowledgements

This book would not have been possible without the generous support, collaboration, and guidance of many individuals and organizations. We extend our heartfelt thanks to all the contributors whose scholarly insights, dedication, and expertise have enriched each chapter, adding depth and diversity to the perspectives presented in this volume. We are grateful to our colleagues, researchers, and reviewers who played a vital role in shaping the book Urban Waterscapes: Challenges, Resilience for Future Sustainable Cities. Special thanks go to the anonymous reviewers whose constructive feedback and valuable comments greatly strengthened the quality of the work. Our sincere appreciation goes to friends and family for their patience, encouragement, and understanding during the demanding process of compiling and editing this book. Their unwavering support has been a constant source of motivation. Finally, we acknowledge the Springer publishing team for their professionalism, guidance, and commitment to excellence. Their meticulous work in editing, design, and production has been instrumental in transforming this manuscript into a comprehensive and accessible reference. To everyone who contributed in any way to the creation of this volume — we thank you for helping bring this work to life and for supporting its mission to advance knowledge and inspire change.

About This Book

This book examines the increasing challenges confronting urban water systems in the context of climate change and rapid urban growth, emphasizing the need for more effective and inclusive water management strategies. It brings together diverse global perspectives, combining theoretical insights with practical applications and real world case studies. The chapters explore how approaches in governance, planning, innovation, and community engagement can help address issues such as water scarcity, urban flooding, pollution, and social inequality. Drawing on examples from different regions, the book highlights both context specific solutions and lessons that can be adapted elsewhere. It also addresses key themes including equitable access to water, the integration of nature based solutions into urban environments, participatory assessment approaches, and the use of advanced technologies such as artificial intelligence for monitoring and prediction. Intended for urban planners, environmental scientists, hydrologists, policymakers, and researchers, this book provides practical insights to support the development of urban waterscapes that are sustainable, technologically advanced, and socially inclusive, contributing to more resilient cities in the face of climate uncertainty.

Contents

About the Editors and Contributors

About the Editors

Uday Chatterjee is an Assistant Professor at the Department of Geography, Bhatter College, Dantan, affiliated with Vidyasagar University, West Bengal, India. He earned his Doctoral Degree in Applied Geography from Ravenshaw University, Cuttack, Odisha. His research interests encompass Urban Planning, Social and Human Geography, Applied Geomorphology, Indigenous knowledge and Hazards and Disasters, Risk Assessment and Resilience, Community Response and Environmental Issues. Dr. Chatterjee has been actively involved in editing scholarly works, including the Special Issue on Urbanism, Smart Cities, and Modelling for GeoJournal, Springer, Special issue in Global Social Welfare, Springer and Frontiers Environmental Science journal, Geological journal Wiley, Editorial Board Member of Scientific Report (Nature Portfolio) journal and serving as a Book Series Editor for Developments in Environmental Science, Elsevier (https://www.elsevier.com/books-and-journals/book-series/developments-in-environmental-science) and Sustainability Solutions—An Interdisciplinary and Transdisciplinary Nexus (https://www.springer.com/series/59999), Springer Nature. His extensive research and editorial contributions highlight his commitment to advancing geographical knowledge and its applications. His research work has been funded by the West Bengal Pollution Control Board (WBPCB), Govt. of West Bengal, India, Indian Council of Social Science Research (ICSSR), Govt. of India. He has published research papers in reputed journals, edited books with reputed publishers like Springer, Elsevier, CRC Press, and author books, book chapters, and conference proceedings. Currently, Dr. Chatterjee completed an international project, in collaboration with Indonesia, Malaysia, and Japan funded by APN (Asia Pacific- Global Change Research, Grant DOI https://doi.org/10.30852/p.22128), and visited Saudi Arabia for International collaboration and conference session chair and Keio University, Tokyo, Japan as a Visiting Professor.

Abdulaziz I. Almulhim is an Associate Professor of Urban Sustainability at Imam Abdulrahman Bin Faisal University (IAU), in the College of Architecture and Planning. He earned his Ph.D. from the University of Liverpool, a Master's degree from the University of Colorado Denver, and a Bachelor's degree from IAU (formerly Dammam University). Dr. Almulhim's research focuses on urban sustainability, urbanization, sustainable cities, climate change, environmental planning, the circular economy, and resource management. He has published over 50 high-impact research papers in leading international peer-reviewed journals and edited books with reputed publishers such as Springer. Recognized for his outstanding contributions to research, he was ranked among the world's top 2% of scientists by Stanford University and Elsevier (2023–2024). He also serves as a reviewer and chief guest editor for several reputable international journals. Additionally, Dr. Almulhim is a Lead Author for the Intergovernmental Panel on Climate Change (IPCC) Special Report on Climate Change and Cities (2027), commissioned by the United Nations Environment Programme (UNEP).

Contributors

Ismaila Rimi Abubakar Department of Urban and Regional Planning, Ahmadu Bello University, Zaria, Nigeria

Muhtasim Ahmed Remote Sensing Division, Center for Environmental and Geographic Information Services (CEGIS), Dhaka, Bangladesh

Abdulaziz I. Almulhim Department of Urban and Regional Planning, College of Architecture and Planning, Imam Abdulrahman Bin Faisal University, Dammam, Saudi Arabia

Mohammad Rakibul Islam Anik Department of Architecture, Rajshahi University of Engineering & Technology, Rajshahi, Bangladesh

Ernestine Atangana Centre for Environmental Management, University of the Free State, Bloemfontein, South Africa

P. Balakrishnan Department of Mechanical Engineering, SRM Institute of Science and Technology, Kattankulathur, India

Parama Bannerji Department of Geography, Nababarrackpur Prafulla Chandra Mahavidyalaya, Kolkata, India

Debasish Batabyal Amity Institute of Travel and Tourism, Amity University, Kolkata, India

Heba Bedair Department of Botany, Faculty of Science, Tanta University, Tanta, Egypt

Sujayita Bhattacharjee S.I.W.S N.R. Swamy College of Commerce & Economics and Smt. Thirumalai College of Science (Autonomous), Mumbai, India

Gouri Sankar Bhunia Independent Researcher, Kharagpur, West Bengal, India

Soumen Bramha Department of Geography, Subarnarekha Mahavidyalaya, Jhargram, West Bengal, India

André Brown School of Architecture, Victoria University of Wellington, Wellington, New Zealand

Derrick Frimpong Buabeng Department of Land and Spatial Sciences, Namibia University of Science and Technology, Windhoek, Namibia

Uday Chatterjee Department of Geography, Bhatter College Dantan (Autonomous), Kharagpur, West Bengal, India

Maheea Emtiaz Chowdhury Department of Urban and Regional Planning, Chittagong University of Engineering & Technology (CUET), Chittagong, Bangladesh

Lineker Max Goulart Coelho Department of Engineering Technology and Didactics, Technical University of Denmark, Ballerup, Denmark

José Rafael Núñez Collado School of Architecture, Victoria University of Wellington, Wellington, New Zealand

Shataparna Das Department of Architecture, Shahjalal University of Science and Technology, Sylhet, Bangladesh

Subrata Das Department of Architecture, Shahjalal University of Science and Technology, Sylhet, Bangladesh

Subhankar Dutta Amity Research Center, Amity University, Kolkata, India

Soumya Ghosh Natural and Medical Sciences Research Center, University of Nizwa, Nizwa, Oman

S. R. Kamalesh Principal Govt. Bilasa Girls Autonomous P.G. College, Bilaspur, Chhattisgarh, India

Ashkan Khalife Mathematical and Physical Sciences, College of Arts and Sciences, University of Nizwa, Nizwa, Oman

Maxwell Fobi Kontor School of Architecture, Victoria University of Wellington, Wellington, New Zealand

S. Kumar Ericsson India Global Private Limited, Noida, India

K. Madhivadhani Department of Planning, School of Architecture and Planning, Anna University, Chennai, India

Sharmila Majumdar Amity Research Center, Amity University, Kolkata, India

Philip Mensah Department of Land and Spatial Sciences, Namibia University of Science and Technology, Windhoek, Namibia

Md. Shah Jalal Misuk Department of Urban and Regional Planning, Chittagong University of Engineering & Technology (CUET), Chittagong, Bangladesh

Suchitra Mohanty Amity Research Center, Amity University, Kolkata, India

Farha Moon Department of Architecture, Shahjalal University of Science and Technology, Sylhet, Bangladesh

Md. Habibun Nabi Department of Urban and Regional Planning, Chittagong University of Engineering & Technology (CUET), Chittagong, Bangladesh

Sumanta Nayek Department of Environmental Science, Amity Institute of Environmental Sciences, Amity University, Kolkata, India

Paul J. Oberholster Centre for Environmental Management, University of the Free State, Bloemfontein, South Africa

D. S. Parihar Department of Geography, Kumaun University, Nainital, India

Susanta Patra Department of Geography, Subarnarekha Mahavidyalaya, Jhargram, West Bengal, India

Kazi Jihadur Rashid Department of Geography, Florida State University (FSU), Tallahassee, FL, USA

Saroj Senapati Annapurna Memorial College of Education, Kakdwip, India

Madhuri Sharma Department of Geography & Sustainability, University of Tennessee, Knoxville, TN, USA

P. Sudharsanamurthy Department of Planning, School of Architecture and Planning, Anna University, Chennai, India

Jobyda Sultana Department of Urban and Regional Planning, Chittagong University of Engineering & Technology (CUET), Chittagong, Bangladesh

Rowshon Ara Toma Remote Sensing Division, Center for Environmental and Geographic Information Services (CEGIS), Dhaka, Bangladesh

Kemo Usto Department of Engineering Technology and Didactics, Technical University of Denmark, Ballerup, Denmark

Eric Yankson Department of Architecture, Planning and Construction, Namibia University of Science and Technology, Windhsoek, Namibia

Part I

Urban Water Challenges and Governance for Sustainability

1 Governing Urban Water Systems in an Era of Climate Stress and Rapid Urbanization

Abdulaziz I. Almulhim
and Ismaila Rimi Abubakar

Abstract

Urban water systems are increasingly under strain due to the dual pressures of rapid urbanization and climate change, particularly in the Global South with weak institutional capacities and increasing water demand. Traditional centralized governance models are proving inadequate in the face of emerging environmental, technological, and socio-economic complexities. This chapter explores the evolving role of decentralized governance in urban water management, highlighting the shift from authoritative providers to adaptive coordination. Drawing on case studies from five cities: Cape Town, Riyadh, Medellín, Chennai, and Jakarta; this chapter investigates diverse governance approaches and opportunities to build more climate-resilient, equitable, and sustainable urban water systems, highlighting innovations such as demand-side management, smart infrastructure, civic-led restoration, and participatory accountability. The analysis reveals that effective water governance in the Global South must evolve beyond service provision to enable governments to act as facilitators, regulators, and coordinators. The findings emphasize the urgency for institutional reform, cross-sectoral collaboration, and inclusive decision-making in transforming water governance to meet twenty-first-century challenges. The chapter proposes a conceptual framework for water governance centered on multi-level governance, participatory planning, climate-risk integration, technological innovation, and equity. This transformative framework advocates for a bottom-up, integrated governance paradigm that bridges national aspirations with local realities to ensure water security in an era of global transformation. It can support the achievement of Sustainable Development Goals (SDG) 6 and SDG 11 by fostering resilient, inclusive, and sustainable urban water systems.

Keywords

Urban water management · Climate change · Urbanization · Governance framework · Global south · Equity

A. I. Almulhim (✉)
Department of Urban and Regional Planning, College of Architecture and Planning, Imam Abdulrahman Bin Faisal University, Dammam, Saudi Arabia
e-mail: aialmulhim@iau.edu.sa

I. R. Abubakar
Department of Urban and Regional Planning, Ahmadu Bello University, Zaria, Nigeria
e-mail: raismailarimi@abu.edu.ng

U. Chatterjee, A. I. Almulhim (eds.), *Urban Waterscapes*, Sustainable Development Goals Series,
https://doi.org/10.1007/978-3-032-20649-7_1

1.1 Introduction

Rapid urbanization and the accelerating impacts of climate change are placing unprecedented pressure on urban water systems, especially in the Global South. In many of these regions, urban growth is occurring at unsustainable rates, often outpacing the capacity of infrastructure to support expanding populations (Abubakar & Aina, 2018; Almulhim & Cobbinah, 2023). Arid and coastal regions are particularly vulnerable, as they face already-depleted water resources compounded by erratic rainfall, prolonged droughts, sea-level rise, flooding, saline intrusion, and deteriorating water quality (Almulhim & Abubakar, 2023; World Bank, 2021). Traditional, centralized governance models increasingly struggle to adapt to these dynamic socio-environmental challenges (Pahl-Wostl, 2017). As a result, urban water management requires innovative approaches and significant resource investments.

Rapid urbanization in the Global South further exacerbates pressure on water systems as the demand for drinking water rises, wastewater production increases, and land alterations intensify surface runoff and flooding (Abubakar & Mu'azu, 2022: McDonald et al., 2014). Simultaneously, climate change disrupts hydrological cycles, degrading the availability and quality of water resources, severely and disproportionately impacting vulnerable communities, particularly those in informal settlements lacking the infrastructure and resources to withstand environmental shocks (Lourenço et al., 2020). The convergence of climate stress, governance failures, and deepening inequalities has pushed cities like Cape Town, Chennai, Jakarta, and São Paulo to the frontlines of global water crises (Rodina, 2019). In this context, adaptable governance frameworks that respond to both climate and urbanization pressures are more urgently needed than ever.

Historically, central governments have been responsible for the provision and oversight of urban water infrastructure and services. However, the interconnected nature of modern cities and challenges posed by rapid urbanization and climate change have rendered these conventional models inadequate (Gleick, 2003; Bakker, 2010). Similarly, technological advancement, ecosystem vulnerabilities, and socio-ecological complexities demand more flexible and context-sensitive governance. It is now widely recognized that water governance involves more than managing the relationship between populations and natural resources (Almulhim & Abubakar, 2025) to reframing their roles as facilitators of participatory governance and coordinators within more adaptive, decentralized frameworks (Akther & Evans, 2024; Wen et al., 2015). This shift signals a growing responsibility for developing emerging governance structures to confront these challenges, particularly in rapidly urbanizing Southern cities.

Persistent gaps in governance frameworks further complicate efforts to address these challenges. Many cities maintain hierarchical and fragmented institutional arrangements, often lacking cross-sectoral collaboration in areas such as spatial planning, climate integration, and water policy (Harshadeep & Young, 2020). In much of the global South, inadequate funding, limited technical capacity, and fragile infrastructure hinder the ability to implement effective solutions. Although physical water scarcity is often cited as the root cause of crises, it is increasingly recognized that governance deficiencies, what Hering et al. (2013) termed "governance droughts," are at the heart of the problem. These institutional shortcomings, coupled with resource scarcity, represent a fundamental barrier to addressing water management challenges in a changing urban climate.

Despite the shift in their roles, governments remain pivotal actors in shaping urban water outcomes. Their influence spans policymaking, resource allocation, regulatory development, stakeholder engagement, and ensuring equity and sustainability (Ferdowsi et al., 2024). There is a growing consensus that a more integrated and holistic approach is required—one that aligns urban planning, disaster risk reduction, public health, and climate adaptation strategies (Mume et al., 2024). This transformation is both technological and institutional, calling for inclusive governance practices that prioritize marginalized groups and incorporate diverse stakeholders in decision-making. Promising models are emerging, including smart water systems, nature-based solutions (NBSs), and participatory planning (UNEP, 2023). Yet, there remains a lack of global understanding of how cities—particularly in the Global South—are reforming governance in response to mounting water challenges (Yu et al., 2012).

Considering these dynamics, there is a critical need to embed climate resilience, equity, and innovative governance into urban water systems. Resilient infrastructure is essential to withstand future climate shocks. Ensuring equitable access is particularly urgent for marginalized and vulnerable populations. Finally, forward-thinking governance—emphasizing digital tools, community participation, and flexible institutions—can improve the effectiveness and responsiveness of urban water management. Addressing these factors through comprehensive policy frameworks will help cities develop dynamic and inclusive systems capable of managing the dual challenges of climate change and rapid urban growth (Crosson et al., 2020).

This chapter investigates the evolving role of government in managing urban water systems under the twin pressures of climate change and urbanization. It examines how traditional governance models are being reshaped, identifies emerging strategies for coping with increasingly complex and multi-scalar water issues, and explores barriers and opportunities for achieving sustainable, inclusive, and climate-resilient urban water management, with a particular focus on the Global South. It is guided by the following research questions:

1. What are the key governance challenges facing urban water systems in the context of rapid urbanization and escalating climate risks?
2. How are governments adapting their roles—from planners and providers to enablers and coordinators—in promoting sustainable urban water management?
3. What innovative policy responses and governance strategies are emerging in cities confronting water-related crises?
4. How can urban water governance frameworks be strengthened to enhance resilience, inclusivity, and long-term sustainability?

1.2 Materials and Methods

This chapter employs a qualitative comparative case study approach to investigate the evolving role of governance in managing urban water systems in the context of climate change and rapid urbanization. A qualitative methodology is suitable for examining complex governance dynamics, institutional interactions, and the contextual variations of urban areas (Creswell & Poth, 2018). Given the interdisciplinary nature of urban water issues—spanning environmental, political, social, and technological domains—this approach provides the necessary depth and flexibility to explore adaptation and innovation within governance systems. It is especially effective for understanding governance phenomena that are context-dependent and not easily quantified (Flyvbjerg, 2006).

Through a multi-case international comparison, this research conducts within-case analysis to identify shared governance patterns while also recognizing unique, localized responses. It draws on a wide array of secondary sources on the nexus between urbanization, water policy, governance, and climate change, including peer-reviewed journal articles, published between 2010 and 2024. Policy documents from international organizations such as the World Bank, United Nations, and UNEP, as well as national urban planning agencies, are also included. Additional materials comprise municipal water management policies, reports on climate resilience, integrated urban water management strategies, and publications from non-governmental organizations and think tanks.

Data collection was conducted using databases such as Google Scholar, Scopus, and Web of Science, with filters applied for geographic

Table 1.1 Water governance types and major supply problems across the case study cities

City	Country (region)	Governance system	Water supply problems
Cape Town	South Africa (Sub-Saharan Africa)	Crisis-driven decentralization	Drought, demand management
Riyadh	Saudi Arabia (Middle East)	Centralized smart systems	Water scarcity in arid zones
Medellín	Colombia (Latin America)	Public-private collaboration	Equity, infrastructure, risk management
Chennai	India (South Asia)	Fragmented governance	Flooding, water quality, urban poverty
Jakarta	Indonesia (Southeast Asia)	Centralized top-down planning	Land subsidence, floods, poor access

region and publication year. A thematic analysis was employed to interpret the data, identify recurring patterns, and classify governance-related themes. Three overarching thematic categories were developed in alignment with the chapter's three thematic areas: urban water challenges, urbanization and climate change nexus, and urban water sustainability and governance responses. These themes frame the analysis presented in the results section.

The selection of five case study cities was guided by the following criteria intended to capture diverse governance systems and varying levels of climate and urbanization pressure (Table 1.1).

- The presence of integrated water-related impacts from both climate change and urban growth
- Variation in governance structures (e.g., centralized, decentralized, mixed, and participatory)
- Regional diversity across Africa, Asia, Latin America, and the Middle East
- Socio-economic and infrastructural diversity, encompassing informal settlements and rapidly modernizing cities.

1.3 Results

This section presents the findings, thematically organized around the following three core focus areas: urban water challenges, urbanization and climate change nexus, and urban water sustainability and governance responses.

1.3.1 Urban Water Challenges

This subsection highlights urban water challenges across the five selected cities—Cape Town, Riyadh, Medellín, Chennai, and Jakarta—including water scarcity, flooding, aging infrastructure, and the prevalence of informal settlements (Fig. 1.1). These cities represent diverse urban environments facing complex and interlinked crises, ranging from acute hydrological shocks to persistent governance inefficiencies. Collectively, they exemplify the growing fragility of urban water systems under intensifying climatic and demographic pressures.

Cape Town and Riyadh are prominent examples where water scarcity poses a critical threat. In 2018, Cape Town nearly exhausted its water supply, facing the infamous "Day Zero" scenario. This crisis, driven by dependence on a single water source, population growth, and prolonged droughts, prompted the city to implement aggressive demand-reduction policies. These included increased water tariffs, widespread public education campaigns, and behavioral nudging—successfully reducing water demand by 50% in just one year (Enqvist & Ziervogel, 2019). Meanwhile, Riyadh relies on desalination for over 60% of its water supply, which is costly, energy-intensive that exacerbates environmental pressures (Baig et al., 2020). With only 89.5 m^3 per capita annual renewable water availability, Saudi Arabia falls far below the international water scarcity threshold of 500 m^3 (Fazel et al., 2022).

In Chennai and Jakarta, seasonal flooding is a persistent problem, intensified by unregulated urban expansion, shrinking wetlands, and

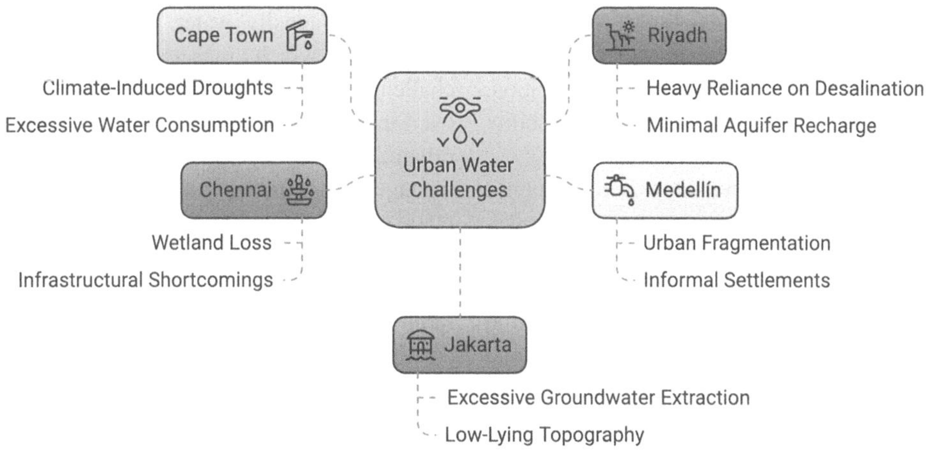

Fig. 1.1 Major water availability challenges in case-study cities

inadequate stormwater infrastructure. Jakarta is currently one of the world's fastest-sinking cities, subsiding at an average rate of 25 cm per year (Abidin et al., 2015). In Chennai, the devastating 2015 floods displaced over 1.8 million people and caused damages exceeding \$3 billion, largely due to poor drainage systems and a failure to protect ecological buffers such as wetlands.

Medellín stands out for tackling both climate risk and social equity in its water governance. With approximately 30% of its population living in informal settlements, Medellín's publicly owned utility, Empresas Públicas de Medellín (EPM), undertook a comprehensive modernization of water infrastructure. These efforts—built on strong public-private partnerships and participatory governance—increased access to drinking water to 98.4% and sanitation services to 97.2% of the population (Piedrahíta et al., 2002; International Water Association, 2022). The city also implemented NBSs, such as river restoration and green infrastructure, to reduce flooding risks and support ecosystem resilience (Sepúlveda Murillo et al., 2019). Medellín thus offers a strong example of how Global South cities can integrate equity, climate adaptation, and water resilience in response to rapid urbanization (Dajer Barguil, 2017).

Both Jakarta and Chennai reveal stark infrastructural deficiencies. In Jakarta, over 30% of the population resides in kampungs—informal settlements that are both flood-prone and poorly serviced. These neighborhoods are often located along rivers and canals, exposing residents to pollution and waterborne diseases (Technology Executive Committee, 2023). Similarly, Chennai's informal settlements lack proper drainage and sanitation, leading to chronic waterlogging during monsoon seasons. Health crises such as cholera and dengue are common. For example, studies in Indonesia indicate that diarrhea incidence is more than three times higher in informal areas compared to planned neighborhoods.

Governance failures, including institutional fragmentation, top-down planning, and a lack of community engagement—are central obstacles to efficient urban water management. In Chennai, governance is fragmented across multiple agencies: the Chennai Metropolitan Water Supply and Sewerage Board, the Greater Chennai Corporation, and the Public Works Department (Mishra et al., 2020). This division results in inefficiencies and duplicated efforts, particularly during crisis response (Harishankar & Vedamuthu, 2019).

Jakarta exemplifies centralized, top-down governance. The controversial National Capital Integrated Coastal Development (NCICD) seawall project, designed to combat coastal flooding, has been criticized for ignoring ecological, cultural, and social concerns (Permanasari, 2019). Such technocratic, one-size-fits-all solutions often fail to engage with local realities. Riyadh, by contrast, has implemented centralized smart water technologies to monitor and optimize water usage. While efficient at the macro scale, these systems lack local adaptability and public participation, raising concerns about long-term sustainability and public trust (Anguelovski et al., 2018).

However, Cape Town's Day Zero response showcased a participatory governance model, where public transparency, behavioral nudging, and widespread engagement helped avoid catastrophe. By displaying household consumption data publicly, the city fostered accountability and civic responsibility (Enqvist & Ziervogel, 2019). Medellín similarly combines bottom-up civic participation with public-private partnerships, using mechanisms such as participatory budgeting to expand water services to marginalized areas (Dajer Barguil, 2017).

Trust in governance is also a recurring theme. In Jakarta and Chennai, privatization has triggered public backlash due to perceived inequality and lack of transparency (Ray & Shaw, 2018). In Jakarta, a public interest lawsuit led to the cancellation of a 25-year water privatization contract after it was revealed that private companies failed to serve poorer areas (Bakker, 2010). In Riyadh, public skepticism toward desalination persists, stemming from its environmental impacts and lack of community consultation (Ferdowsi et al., 2024). Meanwhile, Medellín's focus on social inclusion and accountability has fostered greater trust and more equitable water access (Botero-Mesa & Roca-Servat, 2019). Urban water challenges stem not only from environmental degradation or resource scarcity but also from fragmented governance structures and exclusionary policies. Without coordinated, participatory, and context-sensitive approaches, governance failures are likely to exacerbate existing vulnerabilities rather than resolve them.

1.3.2 Urbanization and Climate Change Nexus

This section explores the complex nexus between urbanization and climate change, emphasizing their combined impact on urban water systems. It highlights how rapid urban growth and climate variability intensify water stress and examines the extent to which cities are incorporating these challenges into their water planning efforts. The overlapping impacts of climate change and urbanization have made urban water resource management immensely complex. Changing climatic conditions and increasingly erratic weather patterns—such as irregular rainfall and prolonged droughts—pose serious risks. Simultaneously, rapid urban growth increases the complexity, density, and size of cities, placing further strain on already limited water supplies. As urban areas expand, natural ecosystems that support hydrological cycles are often degraded or destroyed, exacerbating water scarcity.

Chennai's economy has been among the fastest growing in India (Sobha, n.d.). However, the city faces recurring floods and water shortages due to unrestrained road construction and booming real estate development (Natarajan & Kalloikar, 2017). More than 60% of the city's wetlands—crucial to stormwater management—have been lost. A similar pattern exists in Cape Town, where peri-urban sprawl has increased suburban water consumption, further stressing a limited water supply (Rodina, 2019). In Jakarta, the over-extraction of groundwater, combined with river reliance, contributes to land subsidence at rates of 10 to 25 cm per year (Abidin et al., 2015), heightening flood vulnerability. Riyadh also faces growing pressure on its water resources due to rapid urbanization.

According to Baig et al. (2020), Riyadh's water supply relies heavily on desalination, an expensive and energy-intensive process. This dependence, compounded by urban expansion, has driven significant investments in water infrastructure and technology. Yet, the flooding caused by seasonal rainstorms is worsened by a lack of vegetation, with urban development disrupting the natural water cycle—especially in informal settlements without proper sanitation systems. In Medellín, steep informal settlements are critically vulnerable to both flooding and landslides. Rapid urban growth, coupled with rainfall, exacerbates these risks. Community-based water governance and participatory budgeting have helped build resilience in these water-stressed areas

(Dajer Barguil, 2017), demonstrating a model of locally sensitive and inclusive planning. Climate change introduces a new layer of complexity to urban water management, with droughts, irregular rainfall, and rising temperatures becoming more common. For instance, Riyadh experiences virtually no rainfall during its prolonged and intense summer months (Hasanean & Almazroui, 2015). Meanwhile, desalination—its primary water source—further strains environmental resources (Dasari et al., 2021). In Jakarta, heavy rainfall and rising sea levels are leading to increased pluvial flooding. Without urgent interventions, it is estimated that a third of Jakarta could be submerged by 2050 (World Bank). These threats underscore the need for "innovation-centered frameworks" that respond to both climate and urbanization pressures.

Medellín's poor neighborhoods, where infrastructure is minimal, face especially acute water and sanitation challenges (Dajer Barguil, 2017). City planners have since incorporated climate preparedness into water policies to improve resilience through community engagement. However, the city still struggles to integrate climate adaptation into water management. Administrative silos often lead to fragmented approaches, with water and climate issues addressed separately (Anguelovski, Irazábal-Zurita, & Connolly, 2018). This fragmented governance is particularly evident in cities like Jakarta and Chennai. Jakarta has prioritized large-scale infrastructure such as sea walls, while often neglecting upstream watershed management and community inclusion (Romano & Akhmouch, 2019). Likewise, Chennai's metropolitan development plan lacks water-sensitive urban design elements, exacerbating flood risks. Cape Town, in contrast, has made significant progress by incorporating climate risks into water planning. Following the Day Zero crisis, the city diversified its water sources, launched aquifer recharge programs, and adopted climate modeling to guide decisions (Rodina, 2019). This highlights the importance of integrating climate adaptation strategies into governance to build long-term resilience.

Governments have taken varying approaches to the urbanization–climate–water nexus. Cape Town's decentralized governance allowed for effective citizen engagement during the Day Zero crisis (Enqvist & Ziervogel, 2019). Riyadh, by contrast, employs a centralized, technology-driven system that lacks community integration and long-term sustainability planning (Baig et al., 2020). Medellín's participatory budgeting empowers residents, particularly in informal settlements, to influence water infrastructure decisions (Dajer Barguil, 2017). Meanwhile, Jakarta's reliance on large-scale infrastructure has sidelined grassroots involvement and failed to address root causes of flooding (Victorianus et al., 2020). Ultimately, the intertwined forces of urbanization and climate change exert profound stress on urban water ecosystems. Cape Town, Medellín, and Riyadh provide distinct case studies of governance approaches attempting to address these challenges. Persistent gaps in cross-sectoral coordination, however, hinder efforts to create integrative and sustainable solutions. Strengthening urban water systems in the face of rapid growth and climate change requires community-driven adaptive governance, integration of climate risk into hydrologic planning, balance between engineered and ecological approaches, and a renewed focus on ecosystem preservation (Table 1.2).

Table 1.2 Urbanization impacts and water system challenges in case study cities

City	Urbanization impact	Water system challenges
Cape Town	Expanding suburbs, tourism demand	Overreliance on dams, increased per capita use
Riyadh	High-rise, energy-intensive growth	Heavy desalination dependence
Medellín	Hillside expansion, informal growth	Limited infrastructure in informal settlements
Chennai	Wetland encroachment	Urban flooding, loss of flood buffers
Jakarta	Vertical expansion, land subsidence	Sea-level rise, groundwater depletion

1.3.3 Urban Water Sustainability and Governance Responses

This section examines how governments are adapting to growing urban water challenges, with a focus on decentralization, stakeholder engagement, and the adoption of smart technologies. It highlights a global shift from traditional centralized approaches toward participatory planning, nature-based solutions, digital water systems, and co-managed governance models, as demonstrated across five selected case studies. The accelerating effects of climate change and urbanization are forcing a re-evaluation of urban water systems, particularly in perennially water-stressed cities in the Global South. With increasing populations, these cities face mounting water scarcity, prompting governments to move away from centralized control toward more participatory and decentralized systems integrated with modern technology. The goal is to achieve water sustainability, resilience, and equity in urban contexts.

Governments are increasingly decentralizing water governance to address local priorities and demands more effectively. This enables greater flexibility, innovation, and responsiveness. A prominent example is Cape Town, which narrowly avoided catastrophic supply exhaustion during the 2018 "Day Zero" water crisis. During this period, the city devolved a substantial portion of its water management functions to local authorities and stakeholders. A Water Resilience Task Team was formed, comprising civil society members, municipal employees, and technical experts (Taing et al., 2019). The city also implemented innovative demand-management strategies, including behavioral nudges that informed residents about their water usage relative to their peers. This approach reduced consumption by over 50% in just 1 year (Warner & Meissner, 2021). Smart metering and tariff restructuring were pivotal to managing the crisis, demonstrating the effectiveness of devolved governance models.

Similarly, Medellín, Colombia, has decentralized its public utility, Empresas Públicas de Medellín (EPM), granting it considerable autonomy. The EPM collaborates with community stakeholders in service design and partners with environmental groups to protect watersheds and manage stormwater (Dajer Barguil, 2017). This governance framework not only alleviates water stress in the short term but also builds long-term resilience through community participation. Stakeholder engagement and participatory planning also enhance the equity of water governance. In Chennai, India, civil society organizations collaborate with municipal authorities in restoring urban lakes and tanks, often working with the Environmentalist Foundation of India. Since 2016, over 60 water bodies have been restored using a combination of traditional and modern conservation techniques, fostering community ownership and enhancing flood mitigation. In São Paulo, Brazil, low-income residents participate in participatory budgeting to allocate funding for drainage and sanitation improvements. This has improved municipal-citizen trust and addressed water infrastructure needs in vulnerable communities (Touchton & Wampler, 2020). These examples illustrate how inclusive governance helps ensure that the water needs of marginalized populations are addressed effectively.

The adoption of smart technologies is another key development in urban water governance. Digital tools such as smart meters, leak detection systems, and data-driven monitoring enhance the performance, transparency, and resilience of water systems. Riyadh has developed one of the Middle East's most sophisticated smart water grids, including 30 new water and sanitation facilities with a combined capacity of 117,000 m^3/day (National Water Company, 2025). The city has also introduced digital customer engagement platforms that allow residents to monitor water usage, report leaks, and receive conservation advice. These efforts exemplify how smart water systems can transform management in water-scarce regions. Cape Town's post-crisis reforms included installing water management devices in high-use households that restricted supply after preset daily limits were exceeded, enabling more effective demand control (Rodina, 2019). The NBSs represent one of

Table 1.3 Comparison of urban water governance strategies among the case study cities

City	Strategy focus	Key practices and innovations	Outcomes
Cape Town	Crisis governance, decentralization	Water resilience task team, smart meters, public campaigns	50% reduction in water use, diversified sources
Riyadh	Smart infrastructure	Digital metering, leak detection, and desalination expansion	Reduced losses, enhanced customer engagement
Medellín	Participatory governance	Community-inclusive utility model, public-private watershed management	Improved access, reduced risks in informal settlements
Chennai	Nature-based, community-driven	Lake restoration, bio-swales, civil society co-management	Flood mitigation, groundwater recharge
São Paulo	Equity-oriented planning	Participatory budgeting, "Green and Blue" infrastructure	Better access in underserved areas, resilience building
Jakarta	Top-down flood control	Sea walls, drainage megaprojects, minimal community input	Persistent flood risks, socio-political tensions

the most innovative strategies in urban water governance that enhance water management and promote biodiversity and climate resilience by restoring urban forests, wetlands, and permeable surfaces (Ahmed et al., 2022). Chennai, for instance, has integrated such solutions by reviving urban lakes and wetlands to support groundwater recharge and mitigate flooding. Similarly, bioswales, rain gardens, and permeable pavements are used to reduce runoff and increase infiltration (Rosado et al., 2024). São Paulo's "Green and Blue Network" connects parks, rivers, and recreational areas to improve stormwater drainage and reduce urban heat, with a focus on socioeconomically vulnerable neighborhoods (Mehta et al., 2021). Medellín has implemented green corridors and river restoration projects to combat flooding and urban heat islands, improving both climate resilience and water management (UNEP, 2019).

Despite these advancements, many cities still struggle with legacy top-down approaches. Jakarta's NCICD project, which relies on large-scale infrastructure like sea walls to control flooding, illustrates the risks of single-solution strategies that lack local knowledge or community involvement. Although heavy investment is directed toward infrastructure, social and ecological considerations are often neglected. As a result, many of Jakarta's poorest citizens remain highly vulnerable, and public dissatisfaction continues to grow (Technology Executive Committee, 2023). In contrast, cities like Cape Town and Medellín have embraced decentralization and participatory models. During Cape Town's Day Zero crisis, authorities empowered citizens to take responsibility for water conservation, encouraging grassroots engagement and fostering a culture of cooperation (Mavundla et al., 2025). Medellín similarly adopted community-based, cross-sectoral governance models that align infrastructure planning with local needs, providing a more flexible and responsive approach (UNEP, 2019). Case studies from Cape Town, Medellín, and Riyadh demonstrate the benefits of integrating participatory planning, smart technologies, and decentralized governance to address urban water challenges. Jakarta, by contrast, illustrates the shortcomings of rigid, expert-led systems that lack adaptability and inclusiveness. For cities facing the twin pressures of rapid urbanization and climate change, sustainable and equitable water management requires a shift from centralized control to more participatory, technologically enabled, and ecologically informed governance (Romano & Akhmouch, 2019) (Table 1.3).

1.4 Discussion

Analyzing the urban water governance frameworks of Cape Town, Riyadh, Medellín, Chennai, and Jakarta reveals significant variation in how cities respond to the dual pressures of climate change and urbanization. While each city faces unique challenges, several recurring gaps emerge, particularly regarding underdeveloped governance structures. These shortcomings underscore the need for systemic reforms that extend beyond technical fixes to address institutional and political complexities. Rapid population growth, compounded by climate change, places tremendous strain on urban water systems, many of which are already approaching a breaking point. From a policy perspective, urban water systems require not only technical management but also high levels of organizational capacity and political will. As urban populations grow and climate risks intensify, the necessity for adaptive, responsive, and participatory governance becomes more pressing. This chapter affirms the growing consensus that traditional centralized and hierarchical water governance structures are increasingly inadequate. Instead, cities must move toward more flexible, devolved, and inclusive models to meet emerging urban water challenges (Frey et al., 2021).

A central insight from the case studies is the pervasive neglect of governance. Despite differing contexts, all five cities share common systemic issues that hinder effective water management. A key challenge is institutional fragmentation, which leads to poor coordination among agencies. In Chennai, for instance, separate departments manage water supply, sanitation, stormwater, and drainage, resulting in inefficiencies and resource misallocation (Tajuddin & Dąbrowski, 2021). In Jakarta, top-down governance excludes communities from decision-making processes, leading to a significant gap between policy intentions and on-the-ground realities—exemplified by the NCID project's failure to account for local and ecological contexts (Technology Executive Committee, 2023). Over-centralization remains another critical issue in cities like Riyadh and Jakarta, where national-level control dominates urban water governance. Riyadh's system, heavily reliant on desalination and technological solutions, lacks meaningful input from municipal governments or local communities (Baig et al., 2020). While such approaches may be efficient in the short term, they often fall short in terms of long-term sustainability and local adaptability. Weak community participation is also a significant gap across several cities. In Riyadh and Jakarta, citizen engagement is limited to information dissemination and advocacy rather than genuine involvement in policy formulation or implementation. This exclusion breeds mistrust and weakens public accountability (Bensús, 2021). In contrast, Cape Town and Medellín demonstrate how community involvement, through participatory budgeting and grassroots initiatives, can enhance both governance outcomes and public trust (Rodina, 2019; Dajer Barguil, 2017).

Another crucial shortfall is the limited integration of climate risk into urban governance. Many cities, including Chennai, lack comprehensive climate adaptation strategies within their planning frameworks, leaving them vulnerable to natural disasters like flooding (Tajuddin & Dąbrowski, 2021). While most cities do not yet employ climate-informed planning models, Cape Town stands out for incorporating climate projections and diversifying its water sources as part of a long-term adaptation strategy (Rodina, 2019). To address these issues, several fundamental reforms are necessary. First, governance must be strengthened at multiple levels. Empowering local governments and improving coordination between municipal and national authorities can streamline decision-making and enhance responsiveness (Romano & Akhmouch, 2019). Medellín offers a strong example, where the integration of utility services into municipal functions has created a more coherent and effective governance structure (Dajer Barguil, 2017). Second, institutionalizing participatory planning is essential. Community involvement ensures that policies are grounded in local realities, increasing their effectiveness and legitimacy. Chennai's lake restoration efforts and São Paulo's participatory budgeting initiatives both illustrate how citizen engagement can drive sustainable and inclusive outcomes (Tajuddin & Dąbrowski, 2021; UN-Habitat, n.d.).

Integrating climate risk into water governance must also be prioritized. Cities need to adopt risk modeling, scenario-based planning, early warning systems, and long-term adaptation strategies to mitigate climate impacts. Cape Town's use of climate models and aquifer recharge initiatives exemplify effective climate-resilient planning (Rodina, 2019). The adoption of digital water tools and improved transparency is another vital reform. Smart meters, leak detection systems, and real-time monitoring can greatly enhance efficiency and accountability. Riyadh's smart grid and digital customer platforms demonstrate how technology can improve both operational performance and user engagement (National Water Company, 2025). Promoting water equity in informal settlements is also critical. These communities often lack access to basic services, exacerbating their vulnerability to climate and health risks. Flexible tenure arrangements and community-driven infrastructure projects, such as those in Medellín, are essential for inclusive water governance (Dajer Barguil, 2017). Additionally, cities should scale and adapt best practices from successful case studies. Cape Town's decentralized crisis management and demand reduction strategies, as well as Medellín's community-based governance and nature-based solutions, offer transferable models for other water-stressed urban areas (Rodina, 2019; Dajer Barguil, 2017).

The findings of this study hold important implications for policy frameworks aligned with the SDGs, particularly SDG6 (Clean Water and Sanitation) and SDG11 (Sustainable Cities and Communities). Achieving these goals requires a radical transformation in urban water governance—toward systems that are participatory, resilient, and equitable. This entails developing cross-sectoral governance structures, integrating climate adaptation into planning, and fostering co-governance that incorporates local knowledge and priorities (Romano & Akhmouch, 2019; Yang & Zhu, 2017). Building climate-resilient infrastructure—both green and grey—must go hand in hand with improving access to information and enhancing public trust. Inclusive and accountable governance is not only essential for managing current urban water demands but also for preparing cities to face future climate uncertainties.

1.5 Proposed Framework for Urban Water Governance

This chapter proposes a new urban water governance framework in response to the escalating challenges posed by rapid urbanization and climate change (Fig. 1.2). The framework integrates multi-level governance, participatory planning, climate risk adaptation, digital innovation, and equity-focused strategies to build resilient and sustainable urban water systems. It advocates for a fundamental shift in responsibility—empowering communities as active participants, while redefining the role of governments from direct service providers to facilitators and regulators. This conceptual model reframes the constraints introduced by urbanization and climate change, aiming to ensure that future water systems are not only technically sound but also socially inclusive and environmentally sustainable. The analysis underscores a critical reality: governments can no longer function solely as service providers. In the face of increasing complexity, they must shift toward facilitating, regulating, and coordinating adaptive, participatory, automated, and data-driven systems. Multi-Level Governance (MLG) is central to bridging national priorities with local realities. The framework is grounded in the following core principles:

- Decentralization enables context-sensitive and accountable service delivery.
- Climate integration strengthens planning and infrastructure resilience.
- Equity and inclusion are essential for any meaningful governance transformation.
- Technological innovation, when coupled with transparent regulatory frameworks, builds public trust and improves service efficiency.

Addressing these pillars supports progress toward SDGs—specifically SDG 6 (Clean Water and Sanitation) and SDG 11 (Sustainable Cities and Communities)—by improving water access and enhancing urban resilience. This research affirms the urgent need to adapt urban water governance to address the interconnected pressures of urban growth and climate impacts. The case studies—

Fig. 1.2 Proposed framework for climate-resilient urban water governance

Cape Town, Riyadh, Medellín, Chennai, and Jakarta—highlight that water insecurity stems not only from environmental stressors but also from institutional fragmentation, governance silos, and inadequate policy frameworks. Each city contributed valuable lessons:

- Cape Town demonstrated pioneering demand-side management and decentralized drought resilience.
- Medellín showcased equity-driven participatory governance through accountable utilities and civic engagement.
- Chennai illustrated co-managed, nature-based solutions led by civil society in restoring community water tanks.
- Riyadh implemented transparent smart infrastructure and data-enabled efficiency measures.
- In contrast, Jakarta revealed the limitations of overly centralized, technocratic governance that ignored local ecological knowledge and community participation.

These diverse cases point to a shifting governance paradigm. The traditional command-and-control model centered on state authority is no longer viable. Governments must instead serve as enablers—facilitating collaboration among civil society, the private sector, and local communities to co-create inclusive, effective water solutions. This transition is especially urgent in the Global South, where cities face rapid population growth and limited institutional capacity. In this context, multi-level governance is indispensable. Horizontal and vertical integration across national, municipal, and neighborhood levels ensures urban water planning is both top-down and bottom-up (Foster, 2020). Similarly, climate adaptation and social equity must be considered at all planning levels. From early warning systems to water-sensitive urban design, proactive planning is needed to protect critical infrastructure and vulnerable populations. Data-informed decision-making is also essential. Cities like Riyadh and Cape Town are using smart meters, digital dashboards, and predictive analytics to enhance water management, reduce losses, and improve transparency (Aldegheishem, 2023; Bwalya, 2019). However, the successful implementation of such technologies depends on robust legal and regulatory frameworks, especially in informal or underserved areas where trust in governance remains low. Finally, this framework aligns with international development objectives, particularly SDG 6 and SDG 11, which emphasize the need for access to water and sanitation, and the development of inclusive, resilient urban environments. Achieving these goals requires integrated, participatory, and forward-looking governance structures.

1.6 Conclusion

Urban water governance systems are critically lagging in addressing the mounting challenges of rapid urbanization and climate change in many developing countries. Despite these challenges, many cities are increasingly adopting insightful and innovative governance approaches. This chapter shows that cities such as Cape Town, Riyadh, Medellín, Chennai, and Jakarta exemplify how water insecurity stems not only from environmental factors but also from inefficient infrastructure, fragmented governance, and weak institutions. Cape Town's demand management and behavioral drought resilience during the Day Zero crisis, and Medellín's participatory governance model, coupled with a strong sense of managerial accountability, demonstrate significant progress. Chennai highlights the role of civil society in nature-based solutions, particularly through the restoration of community water tanks. Riyadh showcases the potential of transparent, smart infrastructure and the transformative capacity of technology-driven governance. Conversely, Jakarta underscores the shortcomings of over-centralized, technocratic governance models that overlook community involvement and local ecological knowledge.

These case studies collectively illustrate a critical shift in urban water governance from centralized, state-centric models to more integrated, inclusive, and participatory frameworks. Governments should transition from direct service providers to facilitators, regulators, and coordinators who empower civil society, the private sector, and local communities to co-create sustainable solutions. This transformation is especially urgent in the Global South, where cities experience the fastest population growth and often have the weakest governance abilities. A multi-level governance approach is essential for addressing the complexity of water challenges in urban areas. Horizontal and vertical collaboration between national, municipal, and neighborhood actors enables more comprehensive and context-specific water governance. In parallel, addressing social equity and climate resilience requires multidimensional planning—ranging from early warning systems to water-sensitive urban design—to safeguard both critical infrastructure and vulnerable populations. Data-informed decision-making is equally vital. Cities like Riyadh and Cape Town are reducing water losses and increasing governance transparency using smart meters, digital dashboards, and predictive analytics. However, these technological advancements must be supported by robust legal frameworks to ensure equitable access, particularly for marginalized populations in informal settlements.

While this chapter provides valuable insights, it does not capture the full diversity of global urban contexts. Notably, regions such as Southeast Asia, Sub-Saharan Africa, and the Caribbean remain underrepresented. Further research is needed to broaden the geographic scope and deepen the comparative analysis of urban water governance practices. However, this chapter contributes to achieving the SDGs—specifically clean water and sanitation (SDG6) and sustainable cities and communities (SDG11) by proposing a framework for implementing responsive, multi-sectoral, and participatory governance systems that prioritize both social equity and climate resilience. It underscores the urgent need to reform conventional, top-down governance models in favor of progressive, bottom-up frameworks. Doing so will allow cities to build sustainable and inclusive urban water systems that serve all residents, especially those who are most underserved and vulnerable.

Funding The research was not supported by any funding.

Conflicts of Interest None.

Informed Consent Not applicable.

Authors Contribution Both authors have contributed equally to this work.

Data Availability Statement All utilized data are included in the chapter.

AI Disclosure During the preparation of this work, the author(s) used Grammarly to check the grammar, punctuation, and clarity. After using this tool, the author reviewed and edited the content as needed and take full responsibility for the content of the publication.

References

Abidin, H. Z., Andreas, H., Gumilar, I., & Brinkman, J. J. (2015). Study on the risk and impacts of land subsidence in Jakarta. *Proceedings of the International Association of Hydrological Sciences, 372*, 115–120. https://doi.org/10.5194/piahs-372-115-2015

Abubakar, I. R., & Aina, Y. A. (2018). Achieving sustainable cities in Saudi Arabia: Juggling the competing urbanization challenges. In *E-Planning and collaborations: Concepts, tools, and applications* (Vol. 1–3, pp. 234–255).

Abubakar, I. R., & Mu'azu, N. D. (2022). Household attitudes toward wastewater recycling in Saudi Arabia. *Utilities Policy, 76*, 101372.

Ahmed, F., Sharma, S., Loc, H. H., & Fai, C. M. (2022). *Nature based solutions for sustainable urban storm water management in Global South: A short review*. Earth ArXiv (California Digital Library). California Digital Library. https://doi.org/10.31223/x5405j

Akther, S., & Evans, J. (2024). Emerging attributes of adaptive governance in the global south. *Frontiers in Environmental Science, 12*, 1372157. https://doi.org/10.3389/fenvs.2024.1372157

Aldegheishem, A. (2023). Assessing the Progress of smart cities in Saudi Arabia. *Smart Cities, 6*(4), 1958. https://doi.org/10.3390/smartcities6040091

Almulhim, A. I., & Abubakar, I. R. (2023). Developing a sustainable water conservation strategy for Saudi Arabian cities. *Groundwater for Sustainable Development, 23*, 101040.

Almulhim, A. I., & Abubakar, I. R. (2025). Exploring household water conservation practices: A structural equation Modeling approach. *Journal of Open Innovation: Technology, Market, and Complexity, 11*(1), 100486.

Almulhim, A. I., & Cobbinah, P. B. (2023). Urbanization-environment conundrum: An invitation to sustainable development in Saudi Arabian cities. *International Journal of Sustainable Development & World Ecology, 30*(4), 359–373.

Anguelovski, I., Irazábal-Zurita, C., & Connolly, J. J. T. (2018). Grabbed urban landscapes: Socio-spatial tensions in green infrastructure planning in Medellín. *International Journal of Urban and Regional Research, 43*(1), 133–156. https://doi.org/10.1111/1468-2427.12725

Baig, M. B., Alotibi, Y., Straquadine, G. S., & Alataway, A. (2020). Water resources in the Kingdom of Saudi Arabia: Challenges and strategies for improvement. In A. Al-Zubari (Ed.), *Water policies in MENA countries* (pp. 135–160). Springer. https://doi.org/10.1007/978-3-030-29274-4_7

Bakker, K. (2010). *Privatizing water: Governance failure and the world's urban water crises*. Cornell University Press.

Bensús, V. (2021). Improving local governance with citizen engagement? *City, 25*, 88. https://doi.org/10.1080/13604813.2021.1885913

Botero-Mesa, M., & Roca-Servat, D. (2019). Water rights and everyday Ch'ixi practices in the barrio El Faro in Medellín, Colombia. *Water, 11*(10), 2062. https://doi.org/10.3390/w11102062

Bwalya, K. J. (2019). The smart city of Johannesburg, South Africa. In *Elsevier eBooks* (p. 407). Elsevier BV. https://doi.org/10.1016/b978-0-12-816169-2.00020-1

Creswell, J. W., & Poth, C. N. (2018). *Qualitative inquiry and research design: Choosing among five approaches* (4th ed.). Sage Publications.

Crosson, C., Achilli, A., Zúñiga-Terán, A. A., Mack, E. A., et al. (2020). Net zero urban water from concept to applications: Integrating natural, built, and social Systems for Responsive and Adaptive Solutions. *ACS ES&T Water, 1*(3), 518. https://doi.org/10.1021/acsestwater.0c00180

Dajer Barguil, D. (2017). Charting uncharted waters: Participatory budgeting as a novel transitional justice guarantee of non-recurrence of the conflict in Colombia? *GIGAPP Estudios Working Papers, 66*, 197–230. https://www.oidp.net/docs/repo/doc207.pdf

Dasari, H. P., Desamsetti, S., Langodan, S., Viswanadhapalli, Y., & Hoteit, I. (2021). Analysis of outdoor thermal discomfort over the Kingdom of Saudi Arabia. *GeoHealth, 5*(6), e2020GH000370. https://doi.org/10.1029/2020gh000370

Enqvist, J. P., & Ziervogel, G. (2019). Water governance and justice in Cape Town: An overview. *WIREs Water, 6*(4), e1354. https://doi.org/10.1002/wat2.1354

Fazel, H. K., Abdo, S. M., Althaqafi, A., Eldosari, S. H., Zhu, B.-K., & Safaa, H. M. (2022). View of Saudi Arabia strategy for water resources Management at Bishah, Aseer Southern Region Water Assessment. *Sustainability, 14*(7), 4198. https://doi.org/10.3390/su14074198

Ferdowsi, A., Piadeh, F., Behzadian, K., Mousavi, S.-F., & Ehteram, M. (2024). Urban water infrastructure: A critical review on climate change impacts and adaptation strategies. *Urban Climate, 58*, 102132. https://doi.org/10.1016/j.uclim.2024.102132

Flyvbjerg, B. (2006). Five misunderstandings about case-study research. *Qualitative Inquiry, 12*(2), 219–245. https://doi.org/10.1177/1077800405284363

Foster, S. (2020). Global policy overview of groundwater in urban development—A tale of 10 cities! *Water, 12*(2), 456. https://doi.org/10.3390/w12020456

Frey, K., Ramos, R. F., dos Anjos, L. A. P., Milz, B., & Jacobi, P. R. (2021). Polycentric water governance in the urban global south. In *World sustainability series* (p. 47). Springer International Publishing. https://doi.org/10.1007/978-3-030-76624-5_4

Gleick, P. H. (2003). The human right to water. *Water Nepal, 10*(1), 1–12. https://doi.org/10.3126/wn.v10i1.97

Harishankar, R. R., & Vedamuthu, R. (2019). Evaluating the functioning mechanisms of 'TANK systems' in Peri-urban areas of Chennai, India—Land use change as the determinant. *Water, 11*(6), 1219. https://doi.org/10.3390/w11061219

Harshadeep, N. R., & Young, W. J. (2020). Disruptive technologies for improving water security in large river basins. *Water, 12*(10), 2783. https://doi.org/10.3390/w12102783

Hasanean, H., & Almazroui, M. (2015). Rainfall: Features and variations over Saudi Arabia, a review. *Climate, 3*(3), 578–626. https://doi.org/10.3390/cli3030578

Hering, J. G., Waite, T. D., Luthy, R. G., Drewes, J. E., & Sedlak, D. L. (2013). A changing framework for urban water systems. *Environmental Science & Technology, 47*(19), 10721. https://doi.org/10.1021/es4007096

International Water Association (IWA). (2022). *Water and sanitation for the poorest in Colombia*. https://iwa-network.org/wp-content/uploads/2022/08/1-Water-and-sanitation-for-the-poorest-in-Colombia.pdf

IPCC. (2022). Climate change 2022: Impacts, adaptation, and vulnerability. Contribution of Working Group II to the Sixth Assessment Report of the Intergovernmental Panel on Climate Change. https://www.ipcc.ch/report/sixth-assessment-report-working-group-ii/

Lourenço, I. B., de Oliveira, A. K. B., Marques, L. S., Barbosa, A. A. Q., & Miguez, M. G. (2020). A framework to support flood prevention and mitigation in the landscape and urban planning process regarding water dynamics. *Journal of Cleaner Production, 277*, 122983. https://doi.org/10.1016/j.jclepro.2020.122983

Mavundla, K. P., Okedi, J., Kalumba, D., & Armitage, N. (2025). Estimation of infiltration parameters for groundwater augmentation in Cape Town, South Africa. *Hydrology, 12*(4), 87. https://doi.org/10.3390/hydrology12040087

McDonald, R., Weber, K., Padowski, J., & Flörke, M. (2014). Water on an urban planet: Urbanization and the reach of urban water infrastructure. *Global Environmental Change, 27*, 96–105. https://doi.org/10.1016/j.gloenvcha.2014.04.022

Mishra, B. K., Chakraborty, S., & Kumar, P. (2020). Urban water governance: Concept and pathway. In Kumar et al. (Eds.), *Sustainable solutions for urban water security* (pp. 161–174). Springer. https://doi.org/10.1007/978-3-030-53110-2_8

Mume, A. A., Turyasingura, B., Abdi, E., Umer, Y., Amanzi, L. N., Uwimbabazi, A., et al. (2024). Impact of climate change on the environment: A synthesis study. *Asian Journal of Research in Agriculture and Forestry, 10*(2), 86. https://doi.org/10.9734/ajraf/2024/v10i2288

Natarajan, P., & Kalloikar, S. (2017). Urban resilient integrated water management pathways, to achieve sustainable water resources development in Chennai metropolitan city, Tamil Nadu. *India. Water Practice & Technology, 12*(3), 564. https://doi.org/10.2166/wpt.2017.057

National Water Company. (2025, April 8). NWC starts implementation of 30 water and sanitation projects in Riyadh Region at nearly SR 2b. https://www.nwc.com.sa/En/MediaCenter/News/Pages/NWC-starts-implementation-of-30-water-and-sanitation-projects-in-Riyadh-Region-at-nearly-SR-2b.aspx

Pahl-Wostl, C. (2017). Governance of the water-energy-food security nexus: A multi-level coordination challenge. *Environmental Science & Policy, 92*, 356–367. https://doi.org/10.1016/j.envsci.2017.07.017

Permanasari, E. (2019). Reading political insinuation in urban forms: Saving the sinking Jakarta through Giant Sea wall project. *Geographia Technica, 14*, 56. https://doi.org/10.21163/gt_2019.141.19

Piedrahíta, F. J., Zuleta, F. A., & Manjarrés, J. F. (2002). Initiatives to meet the challenge of providing water services to the most underprivileged communities: Tools and opportunities—The case of Empresas Públicas de Medellín (EEPPM). *Water Science and Technology, 46*(6–7), 355–362. https://doi.org/10.2166/wst.2002.0700

Ray, B., & Shaw, R. (2018). Implications of water insecurity and future prospects in Asian cities. In *Disaster risk reduction* (p. 413). https://doi.org/10.1007/978-981-10-8947-3_24

Rodina, L. (2019). Water resilience lessons from Cape Town's water crisis. *Wiley Interdisciplinary Reviews: Water, 6*(4), e1376. https://doi.org/10.1002/wat2.1376

Romano, O., & Akhmouch, A. (2019). Water governance in cities: Current trends and future challenges. *Water, 11*(3), 500. https://doi.org/10.3390/w11030500

Rosado, D., Fárez-Román, V., Müller, F., Nambi, I., & Fohrer, N. (2024). Rethinking urban water management through drivers-pressures-states-impacts-responses framework application in Chennai, India. *Environmental Management, 74*(5), 970–988. https://doi.org/10.1007/s00267-024-02022-z

Sepúlveda Murillo, F. H., Chica Olmo, J., & Soto Builes, N. M. (2019). Spatial variability analysis of quality of life and its determinants: A case study of Medellín, Colombia. *Social Indicators Research, 144*(3), 1233–1256. https://doi.org/10.1007/s11205-019-02088-x

Sobha. (n.d.). *Fastest growing cities in India*. Sobha. Retrieved April 19, 2025, from https://www.sobha.com/blog/fastest-growing-cities-in-

india/#:~:text=4.,contributing%20to%20their%20population%20growth

Taing, L., Chang, C., Pan, S., & Armitage, N. (2019). Towards a water secure future: Reflections on Cape Town's day zero crisis. *Urban Water Journal, 16*(7), 530. https://doi.org/10.1080/1573062x.2019.1669190

Tajuddin, N., & Dąbrowski, M. (2021). Enabling socio-ecological resilience in the global south: Insights from Chennai, India. *Sustainability, 13*(19), 10522. https://doi.org/10.3390/su131910522

Technology Executive Committee. (2023). Urban flood management in Jakarta: Case study. United Nations Framework Convention on Climate Change. https://unfccc.int/ttclear/misc_/StaticFiles/gnwoerk_static/TEC_NSI/b37e396ecf0b42879e913c50c75883e9/46cd95e108ae48cea7196244d2721393.pdf

Touchton, M., & Wampler, B. (2020). Public engagement for public health: Participatory budgeting, targeted social programmes, and infant mortality in Brazil. *Development in Practice, 30*(5), 681. https://doi.org/10.1080/09614524.2020.1742662

UNEP. (2023). *Water governance for sustainability: Nature-based solutions and community participation in water management.* United Nations Environment Programme.

UN-Habitat. (n.d.). *Urban governance.* United Nations Human Settlements Programme. Retrieved April 13, 2025, from https://unhabitat.org/topic/urban-governance

United Nations Environment Programme. (2019). Medellín shows how nature-based solutions can keep people and planet cool. https://www.unep.org/news--and-stories/story/medellin-shows-how-nature-based-solutions-can-keep-people-and-planet-cool

Victorianus, D., Widayati, N., & Supriyatna, E. (2020). Floating neighbourhood to anticipate Jakarta flooding. In *IOP Conference Series Materials Science and Engineering* (Vol. 1007, No. 1, p. 12018). IOP Publishing. https://doi.org/10.1088/1757-899x/1007/1/012018

Warner, J. F., & Meissner, R. (2021). Cape Town's "day zero" water crisis: A manufactured media event? *International Journal of Disaster Risk Reduction, 64*, 102481. https://doi.org/10.1016/j.ijdrr.2021.102481

Wen, B., van der Zouwen, M., Horlings, E., van der Meulen, B., & van Vierssen, W. (2015). Transitions in urban water management and patterns of international, interdisciplinary and intersectoral collaboration in urban water science. *Environmental Innovation and Societal Transitions, 15*, 123–139. https://doi.org/10.1016/j.eist.2014.03.002

World Bank. (2021). *Managing urban water systems in the face of climate change: Global perspectives.* World Bank Group.

Yang, J., & Zhu, X. (2017). Adapting urban water utilities to climate uncertainties: A case study of Wuhan, PRC. *Procedia Engineering, 198*, 496. https://doi.org/10.1016/j.proeng.2017.07.104

Yu, C., Brown, R. R., & Morison, P. (2012). Co-governing decentralised water systems: An analytical framework. *Water Science & Technology, 66*(12), 2731. https://doi.org/10.2166/wst.2012.489

2 Water Crisis Situation in Darjeeling Hill Town, and Strategies for Sustaining Water Resources

Suchitra Mohanty, Sharmila Majumdar, Parama Bannerji, Debasish Batabyal, Subhankar Dutta, and Sumanta Nayek

Abstract

Darjeeling, 'The Queen of the Hills', is one of the most popular hill towns in the state of West Bengal, India. The town receives one of the highest rainfalls in the country (average precipitation of 3092 mm). Despite this, Darjeeling has been facing severe water crises over the years, especially during the peak tourist seasons. The town's population is projected to increase from 1.62 lakh in 2023 to 2 lakh in 2031. The town is also experiencing an increase in the tourist inflow (~3.55 lakh/year). These factors are driving the town's water demand. Rainfall-dependent natural springs are the primary source of water supply in the town. However, due to human encroachment, unlawful and illegal activities, only 14 out of 26 springs are presently active. The town's drinking water is primarily supplied from the Twin Senchal lakes (South and North Lakes). According to the estimates of the Darjeeling Water Works Department, the demand-supply gap of water in the town is 67% per day (including 25% due to leakages). To counter the water crisis, the Government launched the Atal Mission for Rejuvenation and Urban Transformation scheme and the Balason River Project. Despite taking up those initiatives, the water crisis in Darjeeling town persists. The present study helps in identifying the existing gaps and provides valuable insights into understanding the urban water footprint and managing water sustainability in other hill cities in India and around the world. It also offers a profound understanding of the urban water footprint, which is essential and aligns with SDGs (6.4) for efficient and sustainable management of fresh water resources to ensure water security.

Keywords

Urban development · Population growth · Water resources · Demand-supply gap · Sustainable management

S. Mohanty · S. Majumdar · S. Dutta
Amity Research Center, Amity University, Kolkata, India

P. Bannerji
Department of Geography, Nababarrackpur Prafulla Chandra Mahavidyalaya, Kolkata, India

D. Batabyal
Amity Institute of Travel and Tourism, Amity University, Kolkata, India

S. Nayek (✉)
Department of Environmental Science, Amity Institute of Environmental Sciences, Amity University, Kolkata, India

U. Chatterjee, A. I. Almulhim (eds.), *Urban Waterscapes*, Sustainable Development Goals Series,
https://doi.org/10.1007/978-3-032-20649-7_2

2.1 Introduction

Climate change and anthropogenic changes are significantly impacting mountain hydrology across the globe. Particularly in the Himalayan region, towns and cities are expanding with unprecedented rates of population and development pressure, which is made worse by the effects of climate change. The nature and degree of this impact depend upon local human behaviour as well as the climatic pressure that a particular mountain region must withstand. It also depends on how that region is handling these crises not only in the present-day context but also how it incorporates these insights in its long-term planning.

Historically mountains have attracted large numbers of tourists and to exploit that opportunity to the fullest, many mountain regions have chosen the path of quick urbanisation, building tourist resorts, and secondary homes, attracting other formal and informal businesses, and so on. With increased business opportunities, these regions are also experiencing increasing rural-to-urban migration, which further puts pressure on existing infrastructure. Urban development relies on resources through water supply, energy requirements and is influenced by population density, socioeconomic conditions, and interactions between human and natural systems (Liu & Yang, 2012, McDonald et al., 2014). Thus, the proliferation of businesses and unplanned urbanization are exposing these regions to immense ecological threats. Water is one of the most vulnerable resources in most mountainous areas due to seasonal variations in its quantity and quality. With changes in the land-use pattern, road construction, infrastructural development, and the growth of tourism, water demand is rising. All of these contribute to the depletion and scarcity of water supplies (Jong, 2015; Mitra et al., 2020).

Darjeeling, 'The Queen of the Hills,' is one of the most popular tourist destinations in India. However, water sustainability pertaining to urban water management often appears at odds. The local municipality authority, which oversees Darjeeling town's water supply, has been struggling with the issues for a long time. Lack of an effective water management system may increase competition for water, increase water price, limit people's access to water, lead to political disputes and social unrest, and damage the environment (Yang et al., 2016). The town's water supply is primarily dependent on few springs and Senchal lakes. The town still relies on the early nineteenth-century water infrastructure which was developed by the then British ruler (Koner & Samanta, 2021a). The water infrastructure of the state has remained largely inadequate to meet the growing water demand in the town. The town has a painful history of suffering due to water scarcity and also due to poor quality of water. The problem is still lingering. The crisis of drinking water escalates, particularly during the summer season (Planning Commission of India, 2008), and growing tourism is accentuating the situation further. Here, crisis of water primarily reflects on the city's limited capacity to serve the limited number of people. The age-old water infrastructure of the city cannot cope with the increasing demand.

Darjeeling's water crisis is an instance of a complex interplay of numerous factors that have rendered the city's water supply unsustainable (Samanta & Koner, 2016). In light of SDG 6.4, which addresses water shortages by requiring a substantial increase in water-use efficiency across all sectors and the guarantee of sustainable freshwater withdrawals and supplies, analyzing urban water dynamics becomes imperative to address water scarcity. Therefore, it is essential to consider all major relevant factors under a conceptual framework (Fig. 2.1) that integrates geographical, environmental, and socioeconomic conditions in and outside urban areas. The most promising way to address the issues of the region's escalating water scarcity is by practicing sustainable water harvesting and resource management. In order to accomplish this, it is crucial to adopt recent technology, the application of modern tools like GIS and information from satellite imagery, community involvement, appropriate policy intervention, the use of traditional knowledge, and water management techniques (Bomjon, 2002).

Taking into consideration the city's geographic location and socioeconomic conditions, the present research study intends to analyze the

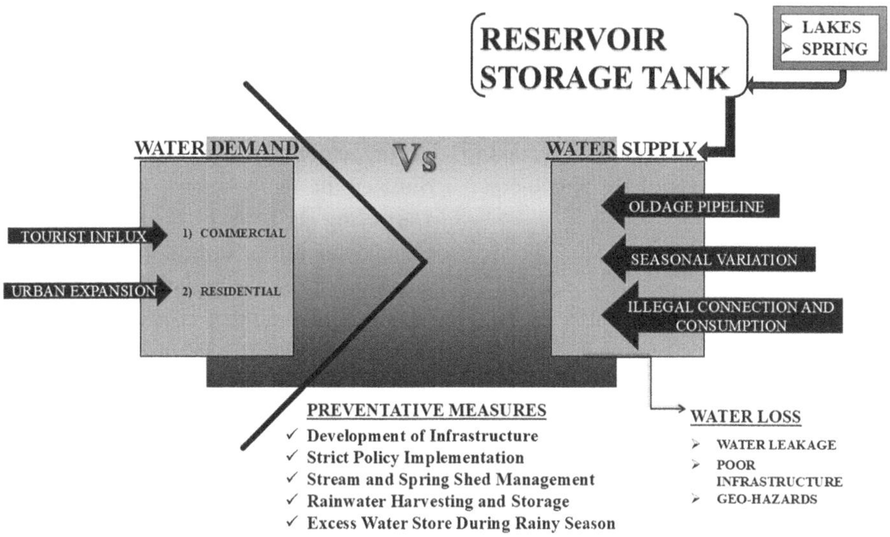

Fig. 2.1 Conceptual framework of water supply system in Darjeeling

existing state of the municipal water supply system by identifying the gaps and measuring distant effects and flows of urban water usage. For this purpose, both qualitative and quantitative methodologies are used to gain insights into the ground reality. The specific objectives are to (1) analyze the town's water availability and management; (2) examine the district's water demand-supply situation in relation to Darjeeling Hill Town; and (3) recommend appropriate policy measures for improved water management in Darjeeling Town.

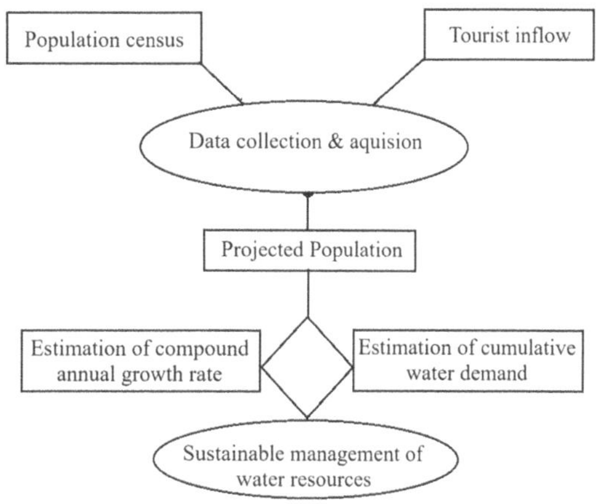

Fig. 2.2 Flow chart of methodology

2.2 Methodology

The sequential steps adopted for this study is displayed in Fig. 2.2.

2.2.1 Data Collection and Aquations

The information regarding the total population was compiled from Census 2011 report (Population Census, 2011a, 2011b). Data on tourist inflow for the Darjeeling district are collected from The India Watch. The data on per capita availability of water is collected from published sources.

2.2.2 Estimation of Projected Population

The projected population was taken from Census 2011 for the period during 2011 and 2021–2030 (Population Census, 2011a, 2011b). Based on this available data (2021–2030) on population, the Compound Annual Growth Rate (CAGR) was estimated (from 2031 to 2036) by using the ordinary least square method. The exponential function was found best and applied to compute annual compound growth (Dandekar, 1980).

$$Yt = Y_0(1+r) \tag{2.1}$$

Where Yt, is the population in the year t; Y_0 is the population in the base year; and r is the compound growth rate to be estimated. Now, taking the natural log on both sides of the above equation, we get,

$$\text{In}\, Yt = \text{In}\, Y_0 + t\,\text{In}(1+r) \tag{2.2}$$

Putting In Y_t = Y, In Y_0 = a, In (1 + r) = b in Eq. (2.2)

$$\begin{aligned} &\text{we get}, Y = a + bt \text{ and } 1 + r = \exp^b \\ &\text{Therefore}, r = \left(\exp^b - 1\right) \end{aligned} \tag{2.3}$$

So the growth rate is calculated by multiplying r by 100.

2.2.3 Estimation of Demand for Water

The data on per capita water demand were obtained from Ministry of Jal Shakti, (2020) and Darjeeling Waterworks Department (2012). It was observed that the per capita water requirement was varying from 135 L/da to 70 L/day. Accordingly, the projected demand for water was calculated by using both the requirements (i.e., 135 and 70 L/day). Per capita water demand for the projected period was estimated by multiplying the estimated population base by the per capita water requirement for both Darjeeling district as well as Darjeeling town. Further, Darjeeling is a very popular destination for tourists, attracting both national and international crowds. However, there is very little historical data available on the inflow of tourists in Darjeeling. That makes prediction even more difficult. As a result, the demand for water driven by the tourists was calculated separately as per the availability of industry data during 2017–2022. So, it is rationally expected that the demand for water in Darjeeling town will be over and above the estimated demand for water based on the projected population.

2.3 Geographic Details of the Darjeeling Town

Darjeeling district is situated in the northern part of the eastern Indian state, of West Bengal. Situated in the southern part of that district, Darjeeling hill town is surrounded by high Himalayan hills. Several perennial and semi--perennial springs supply water to its valley areas. The area of the town is just 10.75 sq. km. It is located between 26° 31″ to 27° 31″ North latitude and 87° 59″ and 88° 53″ East Longitude. The average elevation of this place is 6982 ft. During summer, the temperature of this place generally lies between 16 ° C–17 ° C; it drops down 5 ° C – 6 ° C during winter (Bhutia, 2017; Drew & Rai, 2018; Mitra et al., 2020). The place receives one of the highest rainfalls in the country, which is approximately 309.2 cm. Despite that, the place faces one of the most severe water crises in the state (Ghatani & Khawas, 2020). To understand the reasons behind that, it is important to evaluate the existing water demand-supply condition of this place. That will shed some light on the resiliency of the existing water infrastructure of the town.

2.3.1 Availability of Water Resources in the Darjeeling Town

Natural springs, both perennial and non-perennial springs are the primary source of water in Darjeeling town (Tamang, et al., 2020). The discharged water from 26 springs from the catchment area of Senchal Forest and wildlife sanctuary is stored in an arrestor tank. From there, using a masonry conduit line, water is brought to twin Senchal Lakes. While rainfall is the main source of water, for the collection and distribution of water, the topography of the place plays a crucial role as well. For that, it is extremely important to preserve the catchment area and save it from the challenges of deforestation and illegal construction.

The local municipal authority oversees the public water supply of Darjeeling town from the Senchal Lakes. For the actual distribution of water to the town, it relies on the centralized water infrastructure, that was developed by the British in the early nineteenth century. Even the twin Senchal Lakes were constructed way back in 1910 (South Lake) and 1932 (North Lake) (Ghatani & Khawas, 2020). The total capacity of these lakes is approximately 33 million gallons, which is meant to support a maximum of 15,000 people. Over time, sporadic initiatives like the installation of water supply stations, like Khangkhola Station, Rambi water line, Sindhap Lake, Bangla Khola, and Bokshi Jhora are carried out to complement the existing water infrastructure. However, these initiatives are far less than what is required to meet the growing water demand of the town. The existing water networks in Darjeeling town comprise approximately 35 km of transmission main and 83 km of distribution main pipelines. The water flow through these pipelines is regulated using several valves. But there had been very little maintenance or restoration work. Apart from ad-hoc repair works, there is no regular or statutory routine to check for the reconstruction, realignment, or replacement work of the old pipelines, valves, etc. In the absence of any collective measures to solve the issue locally, for example, through the installation of overhead water storage tanks, etc., individuals are relying on buying water from the tankers that are illegally transporting water to the town from the nearby water sources. This arrangement is not only creating equity-related problems, but it is also encouraging water mafias to take control of the local water governance (Kovácsa et al., 2019).

2.3.2 Rainfall Pattern

Rainfall helps in recharging the springs in the Darjeeling town. The availability of water largely depends on the amount of rainfall this region receives especially during the summer months (Ghatani & Khawas, 2020). In the Darjeeling district, the rainy season starts in June and lasts until December. The month of July has the highest recorded rainfall in the districts, followed by August and September. The average annual rainfall in the region is 2459 mm (2014–2021) (Govt. of India, 2022). The Government of West Bengal, however, mentioned the average annual rainfall of the town as 3092 mm (Government of West Bengal, 2023). Darjeeling town is not in the vicinity of any big river, but the place receives very high rainfall. Even with that, the acute problem of water supply starkly highlights the limitations of its existing infrastructure capacity to support the growing water demand (Kovácsa et al., 2019).

2.4 Water Demand in the Darjeeling Town

Since the water supply infrastructure was created and constructed early nineteenth century, Darjeeling's current water supply system is unable to meet the demands of the city's growing population. Because pipelines and storage tanks are not routinely maintained, the feeder and supply pipes are rusty, twisted, and insufficiently strong to store water. The water sources' catchments are neither protected or conserved. Consistent population growth and sustained year-on-year growth in tourists are putting pressure on various resources. Since the water infrastructure of Darjeeling town has historically remained inadequate, these factors are further aggravating the water crises for the locals as well as for the transient crowds, which in this case are the tourists.

2.4.1 Population Growth

Darjeeling is one of the most populous hill destinations in India. Tea, Timber, and Tourism are considered the three key driving factors for the growth as well as the development of the Darjeeling town (Koner & Samanta, 2021a). During the past few decades, Darjeeling has witnessed an increase in population growth, which has led to rapid urbanisation. This trend is expected to continue even in the future. Population in Darjeeling district and the Darjeeling town is projected as 2,070,000 and 154,000 in 2021 (Population Census Darjeeling District, 2011a; Population Census Darjeeling Town, 2011b, Population Census Darjeeling District, 2011a). It was estimated to touch 2,263,497 and 228,187 in 2036 for the Darjeeling district and the Darjeeling town, respectively.

2.4.2 Tourist Inflow

Like any other hill town, Darjeeling has immense potential to ensure community benefits through the sustainable development of tourism. But for that, it was essential to consider the demand side as well as supply side components of tourism which influence the ecology of this place. (Batabyal & Das, 2022). Sustainable approaches to tourism have become essential to maintain a desirable tourism environment as it contributes more to the tourist experience (Bhutia, 2015; Basak et al., 2021). During the colonial period, this hill town was designed to support a population of only 15,000. According to recent studies, this town received 3.5 lakh domestic tourists and 50,000 foreign tourists in 2015 (Bhutia, 2015). Further, the Indian Himalayan hill station, Darjeeling, has been experiencing immense urban transformation due to explosive population growth and increased tourism activities. These are not only damaging town's architectural design and heritage, but also accentuating water demand. (Dasgupta & Garg, 2021). Many studies have highlighted that during the peak season; Darjeeling Municipality has to cater to the water demand of an additional 10,000 tourists per day. Taking the consumption norms of World Health Organization (WHO) as a benchmark, this suggests, that for these additional tourists the water demand augments by 155,000 gallons every day (Mondal & Roychowdhury, 2019). According to an industry report (The India Watch, 2022), tourist footfalls (national and international) in the Darjeeling district are estimated at around 3,60,000 in 2022 (Table 2.1). However, the Hindustan Times reported that Darjeeling hilly town in particular, receives around 2,00,000 tourists every year (The Hindustan Times, 2023).

Table 2.1 Tourist inflow and calculated water demand in Darjeeling town (2017–2022)

Year	Tourist inflow in Darjeeling Town[a]	Water demand in Darjeeling Town (ML/Day)[b]	Water demand in Darjeeling Town (ML/Day)[c]
2017	550,000	74	39
2018	600,000	81	42
2019	350,000	47	25
2020	180,000	24	13
2021	170,000	23	12
2022	360,000	49	25

[a]The India Watch: https://theindiawatch.com/service-industry/a-usd-100-million-tourism-market--darjeeling-will-continue-to-attract-the-hospitality--business

[b]Calculated as 135 L/Day, (Ministry of Jal Shakti, 2020)

[c]Calculated as 70 L/day, (Darjeeling Waterworks Department, 2012)

2.4.3 Water Demand vs Water Supply

For Darjeeling town, water availability per day is estimated at around 3.8 ML/day in 2021. Out of which, around 25% of total water gets wasted due to various leakages. In addition to this, hospital, the army, and St Pauls require a fixed supply of around 0.5 ML/day. Thus, the net availability of water (after deducting fixed supply and leakages) is estimated at around 2.3 ML/day (Table 2.2). Therefore, the water deficit per day is calculated as 9.2 ML/day for Darjeeling town with a per capita water requirement of 70 L/day (Darjeeling Waterworks Department, 2012) without accounting for tourist footfalls. The demand-supply gap of water requirement in the town would be widened with the inclusion of tourist footfalls. As per the Ministry of Jal Shakti, Govt. of India (2020), the per capita water demand in Darjeeling town was registered as 135 (L/Day) during 2011–2036. On the other hand, as per the Darjeeling Water Supply Department, the per capita water demand in the town was recorded as 70 (L/Day) during the same time (Koner & Samanta, 2021b). Hence, it was evident from the Jal Shakti data, that the water demand-supply gap in Darjeeling town was more severe compared to other research findings and also as per the recommendations of the Darjeeling Water Supply Department. While attempting to quantify the severity of the crises, various studies have reported that the average water demand of the town is more than15 lakh gallons. However, the town only receives 5 Lakh gallons of municipal water supply (Bhutia, 2017). Based on the population as well as tourist inflows in the Darjeeling district, an estimated demand for water has been projected. Following that, the per capita requirement of daily water-use recommended by the Ministry of Jal Shakti (2020) is 135 L/day. The water demand for Darjeeling district is calculated as 288 Million L/day (ML/day) in 2024, which would be around 306 ML/day in 2036 (Fig. 2.3). Whereas water demand for Darjeeling town (accounts for around 10% population of Darjeeling district) is calculated as 22 ML/day in 2024, which would be around 31 ML/day in 2036 (Fig. 2.4).

Table 2.2 An estimation of gap in water demand and supply in Darjeeling town (2021)

Water demand and supply (ML/day)	Darjeeling town
Water demand (town population)	11.0
Water demand (Hospital, Army, St. Pauls)	0.5
Total demand	11.5
Total supply (ML/day)	3.8
Leakages (25%)	1.0
Fixed supply (Hospital, Army, St. Pauls)	0.5
Total water available	2.3
Water deficit (demand-supply gap)	9.2

However, according to the waterworks department, around 70 L per capita per day (Koner & Samanta, 2021a, 2021b) water should be considered as the minimum requirement of water by the people in Darjeeling town. In this study, the water demand data of 70 (L/Day) has been considered for analysis keeping in mind the socio-physical nature of Darjeeling town. Based on that, it is estimated that the Darjeeling district and the town require 158 ML/day and 16 ML/day respectively. Therefore, it is estimated that the total demand for water (in 2036) in the Darjeeling district would be between 158 to 306 ML/day, whereas in Darjeeling town it would be between 16 to 31 ML/day. In addition, total water demand by the tourists is estimated to grow from 49 ML/day and

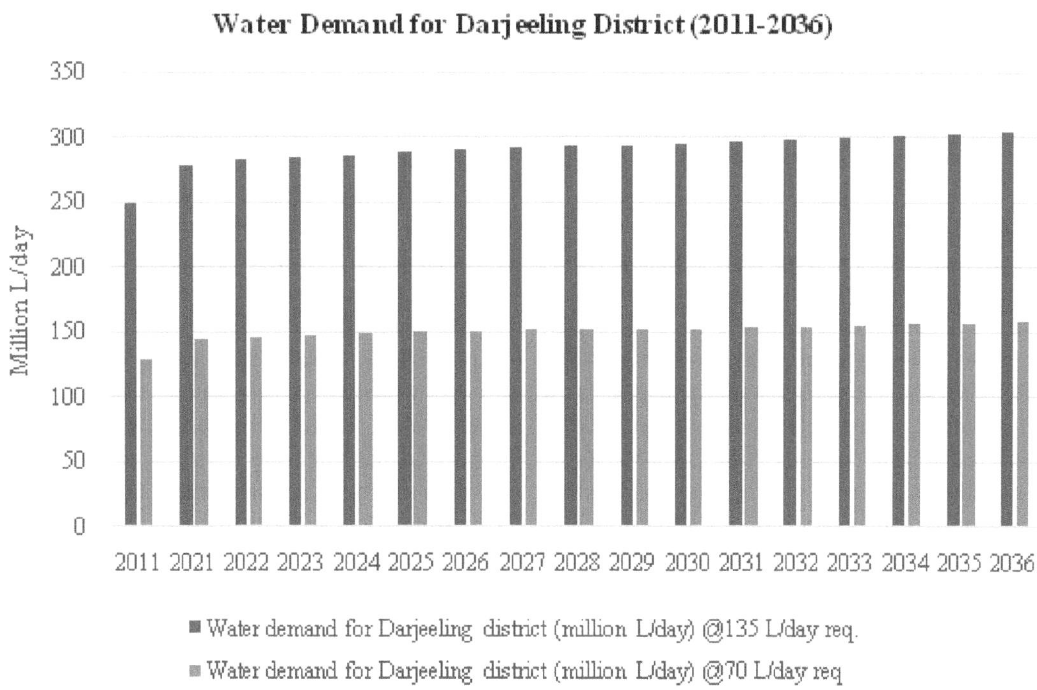

Fig. 2.3 Estimated water demand for Darjeeling district during 2011–2036

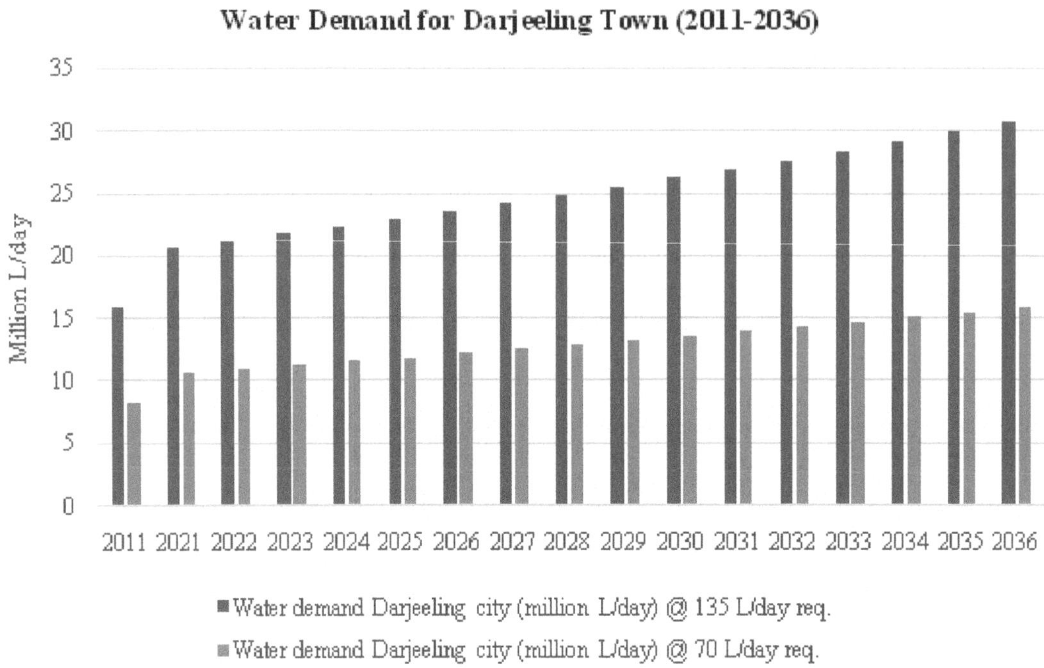

Fig. 2.4 Estimated water demand for Darjeeling town during 2011–2036

25 ML/day with recommended per capita of 135 L and 70 L per day as presented in Table 2.1.

2.5 Urban Development and Planning for Water Management in Towns of Fragile Ecosystems

An adequate supply of drinking water is a prerequisite for any region to grow and expand as envisioned. Urban water management includes a number of components ranging from urban drainage systems, hydropower plants, waste water treatment plant etc. For proper water management in urban areas, keeping in mind the SDG 6 (Sustainable Development Goals), the challenge of compromised access to safe, and affordable water sanitation, leading to decreased water quality and ecosystem issues, must be addressed. Hence, management of water resources should be considered within the framework of a state-led, all-regional development. Watersheds and river basins require regional planning, which might be integrated into a long-term national plan (Khawas, 2005).

Considering the case of Darjeeling town, it is located in fragile mountainous ecosystem where water supply takes mostly of the form of natural sources. However, it has a seasonal floating tourist migration which adds to its pressure to supply water. The water management should be augmented in such a way that it keeps in mind both the environmental and economic concerns. The town needs a water infrastructure management which is holistic without pressurising the tourism sector due to an imbalance between demand and supply of water. Therefore, it should of utmost priority to maintain tourist visits along with decrease in the leakages and increase in the reuse of water in specific sectors to reduce the imbalance. Small scale and less expensive planning are required to maintain water supply system such as restoring the storage capacity through renovation of existing lakes and reservoir, and by building subsidiary tanks to serve the entire city. Regular monitoring of the long-standing pipelines to reduce water loss through leakages, and storage of excess water during the rainy season. From environmental perspective, illegal constructions in and around spring sources and lake catchment areas must be stopped in order to sustain the springs' flow and inflow of lake water, followed by a well-planned spring- shed and water-shed management system. The administration should take prequestionnaire measures regarding loss of vegetation in the vicinity of urban settlement area (Koner & Samanta, 2021a).

To mitigate the water crisis, the government launched the Atal Mission for Rejuvenation and Urban Transformation scheme and the Balason River Project. However, both projects could not take off due to various issues (Bhutia, 2017; Koner & Samanta, 2021b). Therefore, the water supply bottlenecks continue to haunt the Darjeeling town. While looking for a viable solution, it is important to carry out a comparative study of different water supply systems and try to weigh their advantages and disadvantages by considering a significant volume of floating population. For attending to water sufficiency and for sustainable town development, while estimating the water demand of the town it is also important to examine the types of tourism and volume of tourists. This is because, factors like duration of the trip, purpose of visit, and per-capita spending all these variables are important drivers of water demand. Another plausible solution can be promoting nearby and alternative tourist destinations. This will divert tourist traffic and thereby reduce the pressure that is being created due to the floating population (Mondal & Roychowdhury, 2019).

2.6 Suggestive Measures

Water security is one of the main global issues triggered by development activities and is also essential to reaching the Sustainable Development Goals (SDGs). In order to empower and sustain urban improvement, it is essential to use water responsibly and protect it. Water resource management regulations must be strengthened, and universal access to water must be guaranteed to confront the issues associated with human development and livable cities.

To address the water crisis in the Darjeeling town area, it is important to examine the existing water supply infrastructure; and identify the possible pinch points. That will ensure the effectiveness of necessary policy interventions; and reduce the chances of failure or delay. In the hilly territory of the Darjeeling area, the natural water supply is particularly sourced from spring water (Tamang et al., 2020). While considering the water crisis mainly in the summer season, it is essential to carry out rainwater and spring water conservation initiatives at proper places. Accordingly, different types of water reservoirs with appropriate treatment facilities are to be developed across Darjeeling town. The town also needs a uniform water supply system by renovating the old and complex water distribution system. The repair work of leaked water reservoirs in the Rockville Tank and its key distribution must be carried out with high importance. Accordingly, it can save water and meet approximately 15%–20% of overall water demand in the Darjeeling town. Wherever, the new pipelines are to be laid, it should possibly be laid with adequate soil cover from top to evade any type of illegal tapping. Besides, in order to ensure supplying of pure drinking water, all the old and worn-out water pipes of the private lines should be disconnected from the drain and Jhora. Also, the local authority would need to work with the local NGOs for addressing water crisis in the town (Bhutia, 2017).

2.7 Conclusions

Water administration in Darjeeling Town exemplifies complex socioecological processes and policy problems, particularly for the area with greater environmental vulnerabilities. The policy initiatives failed to address the escalating concerns. This comprehensive approach integrates existing key components (both local and distant) and improve qualitative understanding of the relationships between wide range of parameters/factors that affect water dynamics from a holistic perspective, which is crucial to effective and sustainable management.

Based on this study, some remedial steps will be useful to reduce the water demand-supply gap in Darjeeling town. For instance, a suitable hill-based water policy guideline such as the National Mission for Sustaining the Himalayan Ecosystem (NMSHE) or the Integrated Watershed Management Programme (IWMP) needs to be developed and implemented. The objective behind such type of policy guideline will be to provide apt methods about how to conserve the existing water supply resources in Darjeeling town. It will also shed some light on what type of new water projects can be taken up to improve future water supply. Besides that, to restrict the water loss through pipe leakage, and careless wastage (unnecessary tap water running), and ensure a smooth supply of water across the town, pipeline maintenance and laying of new pipes are required on very urgent basis. Moreover, during the monsoon season, the excess rainwater must be stored for rainwater harvesting. Subsequently, during summer or peak tourist season, the stored rainwater can be purified through a water treatment plant, and accordingly, this water can be used to meet the surging demand. Apart from that, special attention needs to be given to conserving natural spring water and Senchal Lake. For that, pollution management, especially to protect natural water bodies that supply substantial amount of water to Darjeeling town, is of utmost importance. In addition, appropriate governance related to accomplishing sustainable water management practices, community participation and awareness program about the importance of water conservation, and strict punishment (financial penalty, in some cases imprisonment) associated with water pollution and careless water wastage may seem beneficial to address the issue in a holistic way.

Considering the widespread concern about the portability of drinking water, this study offers valuable insights into various aspects of the hydrogeological setup and water supply in the hill towns. An in-depth assessment combining spring flow rate, microbial richness, and catchment area ecology must be undertaken in order to develop multidimensional strategies for the water conservation and sustainable management of urban water resources.

Acknowledgement The authors wish to thank Amity University Kolkata, for providing infrastructure and facilities to conduct this research work.

Funding The research study was not supported by any funding.

Potential Conflicts of Interest On behalf of all authors, the corresponding author states that there is no conflict of interest.

Informed Consent Informed consent was obtained from all individual participants included in the study.

Authors Contribution All authors have made fair contribution to this work.

Data Availability Statement The datasets used for the current study are available from the corresponding author and can be obtained on reasonable request.

References

Basak, D., Bose, A., Roy, S., Chowdhury, I. R., & Sarkar, B. C. (2021). Understanding sustainable homestay tourism as a driving factor of tourist's satisfaction through structural equation Modeling: A case of Darjeeling Himalayan region, India. *Current Research in Environmental Sustainability, 3*, 100098. https://www.sciencedirect.com/science/article/pii/S2666049021000748

Batabyal, D., & Das, D. K. (2022). Socio-economic development through tourism: An investigative study for the Himalayan state Sikkim, India. In N. C. Jana, A. Singh, & R. B. Singh (Eds.), *Livelihood enhancement through agriculture, tourism and health. Advances in geographical and environmental sciences* (pp. 237–246). Springer. https://doi.org/10.1007/978-981-16-7310-8_12

Bhutia, S. (2015). Sustainable tourism development in Darjeeling Hills of West Bengal, India: Issues and challenges. *Global Journal of Human-Social Science, 15*(3), 1–10. https://globaljournals.org/GJHSS_Volume15/1-Sustainable-Tourism-Development.pdf

Bhutia, S. (2017). A situational analysis of water resources in Darjeeling municipal town: Issues and challenges. *International Journal of Research in Geography, 3*(4), 52–60. https://www.arcjournals.org/pdfs/ijrg/v3-i4/7.pdf

Bomjon, S. (2002). Water resources management in the Eastern Himalayan urban ecosystem, Abstract #2665. EGS XXVII General Assembly, Nice, 21–26 April 2002.

Census. (2011a). Darjiling (Darjeeling) District – Population 2011–2024, Population Census Data.

https://www.census2011.co.in/census/district/1-darjiling.html
Census. (2011b). Darjiling Population 2024, Population Census Data. https://www.census2011.co.in/census/city/193-darjiling.html
Dandekar, V. M. (1980). Introduction to Lonavala seminar on data base methodology for the study of growth rates in agriculture. *Indian Journal Of Agricultural Economics, 35*(2), 1–12.
Darjeeling Waterworks Department. (2012). A report on water supply system of Darjeeling municipal area. Darjeeling, India.
Dasgupta, S., & Garg, P. (2021). The urban development and heritage conundrum: The challenges of heritage conservation in the Hill-Station of Darjeeling. *Journal of Urban Design, 26*(2), 184–208. https://doi.org/10.1080/13574809.2020.1770584
Drew, G., & Rai, R. P. (2018). Connection amidst disconnection: Water struggle, social structures, and geographies of exclusion in Darjeeling. In T. Middleton & S. Shneiderman (Eds.), *Darjeeling reconsidered* (pp. 219–239). Oxford University Press.
Ghatani, S., & Khawas, V. (2020). Understanding issues and challenges of urban water management in the Himalayan towns: A study of Darjeeling town. *Journal of Xi'an University of Architecture and Technology, XII*(III), 4778.
Government of India. (2022). *Report on central ground water board Eastern region, National Aquifer Mapping and Management Plan, Darjeeling District, West Bengal.* Ministry of Jal Shakti Department of Water Resources, River Development and Ganga Rejuvenation. https://cgwb.gov.in/cgwbpnm/public/uploads/documents/17026226165961340933file.pdf
Government of West Bengal. (2023). Darjeeling Demography. https://darjeeling.gov.in/demography/
Jong, C. (2015). Challenges for mountain hydrology in the third millennium. *Frontiers in Environmental Science, Sec. Interdisciplinary Climate Studies, 3*, 38. https://doi.org/10.3389/fenvs.2015.00038
Khawas, V. (2005). Conservation and management of water resource in Sikkim Himalaya: Some suggestions. *Spatio-Economic Development Record, 12*(4), 27–31.
Koner, K., & Samanta, G. (2021a). Urban environment and sustainable water supply: A comprehensive analysis of Darjeeling City, India. *Environment, Development and Sustainability, 23*, 17459–17482. https://doi.org/10.1007/s10668-021-01396-y
Koner, K., & Samanta, G. (2021b). 'Where does water go'? A critical analysis of nature of water crisis in Darjeeling City, India. *Applied Water Science, 11*(12), 184. https://doi.org/10.1007/s13201-021-01523-0
Kovács, E. K., Hemant, O., Kaustuv, R. N., Thomas, N., Chetan, A., Devendra, C., Ngamindra, D., Kamal, D., Vijay, G., Tikeshwari, J., Natasha, K. M., Anvita, P., Nuvodita, S., Vishal, S., Rajesh, T., & Bhaskar, V. (2019). A political ecology of water and small-town urbanisation across the lower Himalayas. *Geoforum, Elsevier, 107*, 88–98. https://doi.org/10.1016/j.geoforum.2019.10.008
Liu, J., & Yang, W. (2012). Water sustainability for China and beyond. *Science, 337*, 649–650. https://doi.org/10.1126/science.1219471
McDonald, R. I., Weber, K., Padowski, J., Flörke, M., Schneider, C., Green, P. A., Gleeson, T., Eckman, S., Lehner, B., Balk, D., Boucher, T., Grill, G., & Montgomery, M. (2014). Water on an urban planet: Urbanization and the reach of urban water infrastructure. *Global Environmental Change, 27*, 96–105. https://doi.org/10.1016/j.gloenvcha.2014.04.022
Mitra, S., Roy, A. K., & Tamang, L. (2020). Assessing the status of changing channel regimes of Balason and Mahananda River in the sub-Himalayan West Bengal, India. *Earth Systems and Environment, 4*, 409–425. https://doi.org/10.1007/s41748-020-00160-y
Mondal, T. K., & Roychowdhury, P. (2019). Water scarcity in Himalayan Hill town: A study of Darjeeling municipality, India. In B. Ray & R. Shaw (Eds.), *Urban drought. Emerging water challenges in Asia* (pp. 363–383). Disaster Risk Reduction. Springer. https://doi.org/10.1007/978-981-10-8947-3_21
Per Capita Availability of Water, Ministry of Jal Shakti. (2020). Posted on: 02 Mar 2020 5: 40PM, by press information bureau, Govt. of India, New Delhi. https://pib.gov.in/PressReleasePage.aspx?PRID=1604871
Planning Commission of India. (2008). Problems of hilly habitations in areas covered by Hill Areas Development Program (HADP)/ Western Ghats Development Program (WGDP). http://planningcommission.nic.in/aboutus/committee/wrkgrp11/tg11_hillarea.Pdf
Samanta, G., & Koner, K. (2016). Urban political ecology of water in Darjeeling, India. *South Asian Water Studies, 5*(3), 42–57.
Tamang, L., Chhetri, A., & Chhetri, A. (2020). Sustaining local water sources: The need for sustainable water management in the Hill Towns of the Eastern Himalayas. In S. Bandyopadhyay, H. Magsi, S. Sen, & T. Ponce Dentinho (Eds.), *Water management in South Asia. Contemporary South Asian studies.* Springer. https://doi.org/10.1007/978-3-030-35237-0_8
The Hindustan Times. (2023). Darjeeling to reimpose tourist tax after more than a decade. https://www.hindustantimes.com/cities/kolkata-news/darjeeling-to-reimpose-tourist-tax-after-more-than-a-decade-101701159025324.html
The India Watch. (2022). Darjeeling a USD 100 million tourism market. https://theindiawatch.com/service-industry/a-usd-100-million-tourism-market--darjeeling-will-continue-to-attract-the-hospitality--business
Yang, W., Hyndman, D. W., Winkler, J. A., Viña, A., Deines, J., Lupi, F., Luo, L., Li, Y., Basso, B., Zheng, C., Ma, D., Li, S., Liu, X., Zheng, H., Cao, G., Meng, Q., Ouyang, Z., & Liu, J. (2016). Urban water sustainability: Framework and application. *Ecology and Society, 21*(4), 4. https://doi.org/10.5751/ES-08685-210404

3 Urban Growth, Climate Change, and Extreme Floods: Envisioning Mumbai's Predicament in the Context of Sustainability

Sujayita Bhattacharjee and Madhuri Sharma

Abstract

Climate change is a harsh reality confronting the twenty-first-century world. It has been intensifying and leading to increased frequencies of extreme weather events day by day. As per the United Nations Environment Program (UNEP), extended periods of global climate change have led to extreme weather patterns, contributing to an increased likelihood of floods. Global warming poses a range of climate hazards, increasing the flooding vulnerabilities for coastal cities like Mumbai. Further, with the intensification of urban growth activities and the strong desire to live by the water bodies, the geological eventualities and flooding frequencies have been increasing in the cities of the global south. This study explores the vulnerabilities of extreme floods in the city of Mumbai due to climate change, and based on these analyses, we try to envision the city's predicament in the context of urban sustainability. We use the PESTEL (i.e., Political, Economic, Social, Technological, Environmental, and Legal) framework to systematically evaluate the elements that influence flood resilience in Mumbai. Findings show that flood risks have increased more than in the past due to a combination of more extreme and unpredictable patterns of monsoon rainfall, ongoing sea-level rise, and human intervention. Residents of the informal settlements bear most of the impact of flooding risks due to their location in flood-prone areas and lack of resilient infrastructure. We recommend developing an integrated flood management policy and a multi-dimensional governance framework, aligning with the United Nations Sustainable Development Goals to resolve the issue.

Keywords

Climate change · Floods · Hydrological events · Sustainability · Vulnerabilities · Urbanization

3.1 Introduction

Climate change is the greatest challenge confronting our planet at the present time. One of its key manifestations is reflected in the increased average temperatures around the Earth, which trigger several adverse effects. Elevated global temperatures lead to a rise in sea levels and a rise

S. Bhattacharjee
S.I.W.S N.R. Swamy College of Commerce & Economics and Smt. Thirumalai College of Science (Autonomous), Mumbai, India

M. Sharma (✉)
Department of Geography & Sustainability, University of Tennessee, Knoxville, TN, USA
e-mail: msharma3@utk.edu

U. Chatterjee, A. I. Almulhim (eds.), *Urban Waterscapes*, Sustainable Development Goals Series,
https://doi.org/10.1007/978-3-032-20649-7_3

in the frequency of extreme weather phenomena, including storms, droughts, floods, and a variety of tropical illnesses (UNEP, 2017; WHO, 2023; Furtak & Wolińska, 2023). Cities are one of the biggest contributors to climate change for being the major sources of greenhouse gas emissions (GHG) from urban industries and lifestyle activities, even though these have detrimental impacts on the infrastructure, housing, basic services, human livelihoods, and health of their citizens (UNEP, 2017). There are two main ways in which urban expansion affects the local, regional, and global climate. First, changes in urban characteristics, such as rising urban heat, affect local temperatures, atmospheric flows, and the frequency and intensity of lightning strikes (Esfandeh et al., 2022); and second, changes in the transport and dispersion of air pollutants affect the regional and global climate without any regard for geographic boundaries (Baklanov et al., 2016). These interconnected aspects point out the fact that cities are both contributors to climate change as well as victims of the phenomenon by themselves.

Despite comprising only 2% of the world's surface area, cities are responsible for over 70% of GHG and more than two-thirds of the world's energy use, thereby standing at the forefront of the worldwide climate crisis (World Bank, 2023). Floods and other extreme weather phenomena are increasingly frequent due to climate change and fast urbanization all over the world (Wang et al., 2022; Rentschler et al., 2023). Unplanned urban growth and uncontrolled rural-to-urban migration are exacerbating the problems by increasing slum size and numbers, overcrowding, pollution, shortages in affordable housing, inadequate access to waste management and sanitation, and increased susceptibility to natural disasters, which are made worse by climate change (Sharma, 2017; Bhattacharjee & Sattar, 2022; Sharma & Abhay, 2022; World Bank, 2023). Urbanization and climate change together contribute to a majority of urban flooding incidents (Alexander et al., 2019; Pour et al., 2020), altering natural hydrological cycles by disrupting surface water drainage, reducing groundwater recharge, and increasing runoff due to impervious surfaces. Flooding is a serious natural hazard that is becoming increasingly common and has major impacts on human safety, the economy, and society (Zhou et al., 2012; Yin et al., 2016). Furthermore, flooding impacts people differently regarding their everyday lives, livelihoods, and economy, even though urbanization and climate change may affect the frequency and intensity of floods (Kundzewicz et al., 2014; Dube et al., 2022). The threat of flooding is especially increasing across the cities of the developing world, which contain concentrated urban populations within limited and well-planned urban infrastructure. This occurs because uncontrolled urban growth raises the risks of flooding in cities due to dense populations and unauthorized urban housing, often located in flood zones, along with localized changes in hydrological and hydrometeorological conditions such as frequent heavy rainfalls (Hemmati et al., 2020; Sun et al., 2021).

Urban growth also exacerbates the loss of vegetation and green coverage, baring the lands, increasing the impervious surfaces, and hence increasing the urban heat index (Ajjur & Al-Ghamdi, 2022). Impervious surfaces such as buildings, asphalt, etc., disrupt the natural hydrological cycle by reducing the capacity of the land to absorb water, leading to water run-off at high speeds, frequently overflowing drainage systems, and causing floods (Ajjur & Al-Ghamdi, 2022). Besides urban growth, precipitation extremes associated with climate change expose people and infrastructure to urban flooding (Ahmed et al., 2018). The frequency and severity of hazards are aggravated by climate change, but cities also contribute to these changes through increased consumption of energy, burning of fossil fuels, and economic activities (Roy, 2009). Changing land use patterns and rapid urbanization, coupled with the adverse effects of climate change, disrupt the delicate balance of urban hydrology, affecting surface water flow, groundwater levels, and floodplain dynamics. In recent years, urban flooding has been recognized as a natural disaster that poses a severe threat to human life and property, along with the potential to negatively affect the environment of urban areas worldwide (Bibi & Kara, 2023). In recent decades, there has been an increase in both the frequency and intensity of urban floods (Wang et al., 2022).

Over the last few decades, a significantly larger share of global urbanization has been occurring in developing countries, with rivers and coastlines forming the primary locations of many such large cities (Grimm et al., 2008). The combined effects of urban expansion and climate change put many cities at risk, especially those that are in low-lying coastal areas, making them highly vulnerable to flooding. The fast-urbanizing cities of the developing world also have several characteristics that unequally affect their large population by having an unequal distribution of wealth and opportunities, such as a significant share being forced to live in the slums and even as homeless, amidst a lack of basic amenities (Bhattacharjee & Sattar, 2021b; Bhattacharjee & Sattar, 2021a; Bhattacharjee & Sattar, 2022; Chakraborty et al., 2023). The exposure and sensitivity to extreme climatic events under overstretched or unwise urban management increase due to the rapid growth of the urban population and marginalization of the urban poor, forcing them into risk-prone low-lying areas and the development of dense informal settlements by degradation of flood-regulating ecosystems, along with sea level rise due to global warming (Sterzel et al., 2020). Hurricanes and big storm waves coincide with extremely high tides, making low-lying shoreline areas more susceptible to floods (Griggs & Reguero, 2021). The combination of expanding coastal hazards along with coastal urbanization and the growing density of population in coastal locations is what makes adaptation so urgent (Griggs & Reguero, 2021).

3.1.1 Mumbai—Urban Growth, Climate Change, and Extreme Floods

Mumbai, among the world's most populous cities, is increasingly becoming vulnerable to extreme flooding because of its rapid urbanization and climate change (Tripathy et al., 2024). Bolstered by a complex network of rivers, creeks, and stormwater drains, the city is built on low-lying islands that are not capable of handling the increasing rainfall intensity that is expected under climate change. Aggressive land reclamation practiced for decades has erased Mumbai's coastal and estuarine landscape (Anand, 2023), eliminating most functions for natural flood-buffering. Rapid, often unplanned urbanization has resulted in the convergence of natural drainage and floodplains, increasing vulnerability to flooding. Every year, the monsoons produce widespread waterlogging, disrupting transport networks, damaging infrastructure, and disproportionately affecting vulnerable populations inhabiting informal settlements (Singh et al., 2023). In addition to the flooding risk, sea level rise, high tides, and storm surge are growing risks of coastal inundation, impacting the city's sustainability and potential disaster preparedness.

Urban development in Mumbai has mostly progressed chaotically, as economic activity and rural-urban migration developed the region randomly without satisfactory planning (Randolph, 2024). The reclamation of land from the sea, initiated during the British colonial period, fundamentally changed the drainage processes and runoff from surfaces in Mumbai, modifying the natural flood defense in the city (Anand, 2023). The increasing prevalence of changed characteristics and impermeable surfaces has led to an impeded recharge of groundwater, with monsoon flooding becoming increasingly intense. Informal settlements that comprise almost half of Mumbai's population continue to experience the highest impacts from urban flooding due to insufficient drainage and unplanned infrastructure (Doshi & Garschagen, 2024). Climate change will perpetuate the risk of floods in Mumbai, where coastal soil erosion and varied rainfall patterns will likely exacerbate flooding. Studies have increasingly shown the growing frequency of intense precipitation with an upward trajectory over Mumbai (Mann et al., 2023). Whilst recent records have reported monsoon precipitation with increasing intensity. Furthermore, there is ongoing sea-level rise with climate change, with thermal expansion of ocean waters and melting glacial ice contributing to global sea-level rise, which is likely to threaten the coastal zones surrounding the city of Mumbai. High tides and storm surge activities created by cyclones that form over the Arabian Sea threaten low-lying areas like Dharavi and the eastern suburbs (Thomas & Lekshmy, 2022).

Despite being an economic hub, the city's urban planning deficiencies—such as inadequate infrastructure, poor land-use regulations, and ineffective environmental policies—have significantly contributed to increased climate vulnerabilities, particularly flooding, heat stress, and air pollution. Additionally, weak urban policies and insufficient climate adaptation strategies hinder the city's resilience. Flood risk management is fragmented in Mumbai, with multiple organizations, including the Brihanmumbai Municipal Corporation (BMC), the Maharashtra state government, and the National Disaster Management Authority, working in silos. The drainage capacity, which was largely constructed during the British era, is old, dilapidated, and inadequate for present levels of precipitation (Awasthi, 2023). Projects such as the Brihanmumbai Storm Water Disposal System (BRIMSTOWAD), which is intended to improve drainage capacity, have met with challenges such as bureaucracy and financial limitations, which have slowed progress (Gupta, 2007). Even after many extreme weather events, Mumbai's urban planning policies are still reactive instead of proactive. For example, the Development Plan 2034 does not engage fully with climate resilience planning explicitly referenced in zoning and land-use policies, nor has there been any enforcement of Environmental Impact Assessments (EIAs) on infrastructure, which may accelerate ecological fragility.

A crucial component of current environmental and urban planning issues is the link between climate change, urban floods, and sustainability. Creating policies that support sustainable urban growth while reducing and adjusting to the effects of flooding and climate change requires an understanding of this relationship for the integration of hydrological resilience into urban planning. Based on this rationale, we explore the nexus of climate change, urban growth, and extreme floods in Mumbai, and based on the understanding gained, we attempt to envision the city's predicament from a sustainability perspective.

This study represents a novel exploration of the literature on the interrelated effects of climate change, urban development, and extreme flooding in Mumbai. It provides a new vantage point by placing the city of Mumbai into a global development narrative about climate change, urbanization, and exposure to extreme weather risks. The literature demonstrates that urban flooding is on the rise globally through increased temperatures, sea level rise, and unplanned urbanization. This research uniquely analyzes Mumbai as a coastal megacity with rapid urbanization, situated in the Global South. This research brings together global climate necessities with local urban realities through an interdisciplinary lens. Furthermore, by including hydrological resilience as a part of urban planning, it establishes a sustainability-minded approach to understanding urban flood vulnerabilities and assessing adaptation and mitigation approaches for more effective, contextualized climate-resilient policies in urban planning. By emphasizing the protection and restoration of water-related ecosystems, this study directly aligns with SDG 6.6, highlighting the role of sustainable urban water management in safeguarding rivers, wetlands, and coastal areas, and ensuring resilient urban environments that benefit both people and ecosystems (Fig. 3.1).

3.2 Methodology

3.2.1 Study Area

This study focuses on Mumbai, a coastal city with a mean elevation of 11 meters above sea level, located in the western part of India in the state of Maharashtra. Mumbai is a reclaimed cluster of seven islands (Colaba, Little Colaba, Isle of Bombay, Mazgaon, Parel, Worli, and Mahim) (Tare, 2019), that has attained its present shape following a series of land reclamation schemes (Bhattacharjee & Sharma, 2023). It is essentially formed out of the sea, characterized by its unwavering need for open space and con-

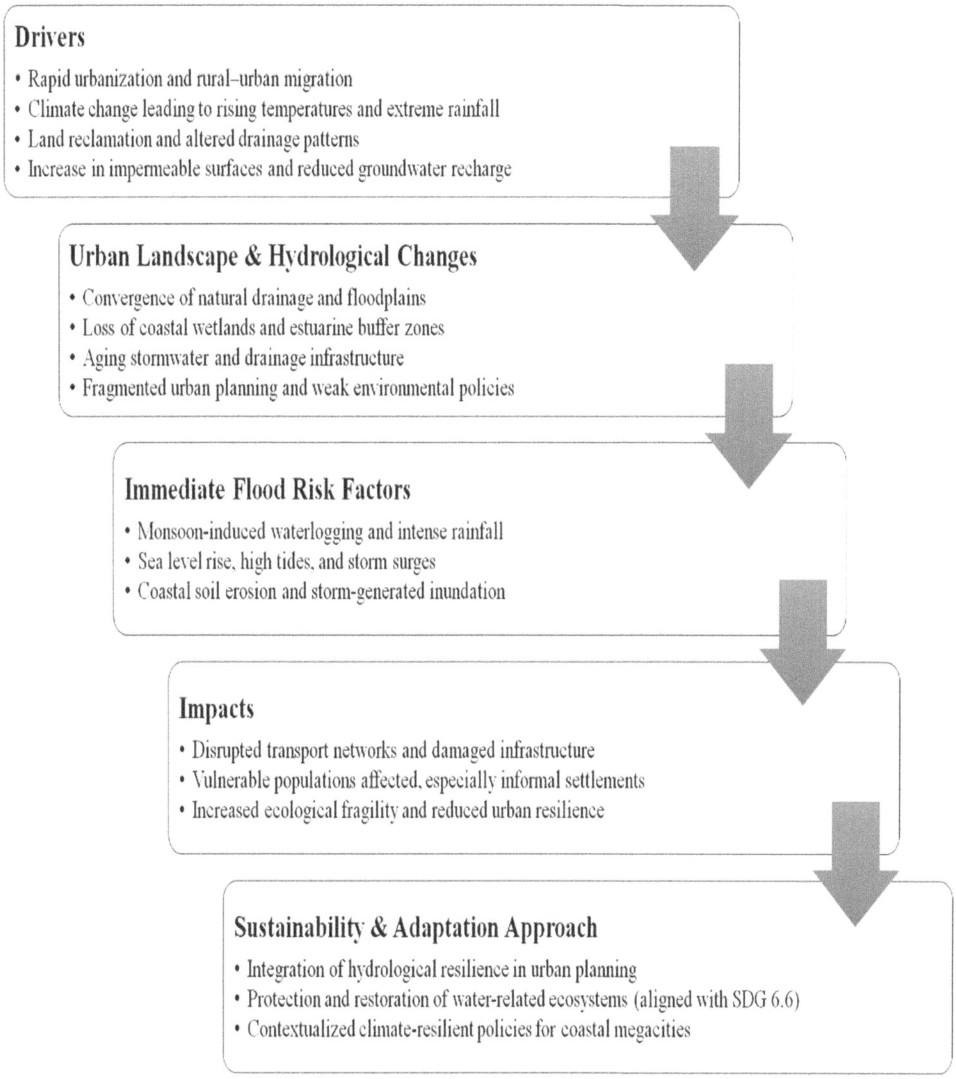

Fig. 3.1 Flow chart of climate change, urbanization, and flooding in Mumbai (Source: Prepared by the authors)

nectivity (Patel, 2021). While it is broadly situated in the physiographic division of coastal plains and islands of India, within this broad division, it lies under the micro-division of Maharashtra Littoral (Census of India, 2011a; Census of India, 2011b). Administratively, Mumbai is divided into two districts—the Mumbai City District and the Mumbai Suburban District, both governed by the Municipal Corporation of Greater Mumbai (MCGM) (Fig. 3.2). Although the average annual rainfall received in Greater Mumbai is 2208.82 mm, it is increasing annually at the rate of 5.18 mm (Mann et al., 2023). The city has a hot-humid climate with an annual mean temperature of 27.2 °C, summer highs reaching 40 °C and over 75% humidity, while winters remain mild and moderate (Banerjee et al., 2022). According to the latest census data, the population of Greater Mumbai is 12.44 million (Census of India, 2011a; Census of India, 2011b). It is noteworthy that Mumbai is positioned as geographically vulnerable (Bhattacharjee & Sattar, 2025), which, when coupled with the intensification of monsoons and outdated drainage systems, has led to repeated crises during monsoon seasons that result in

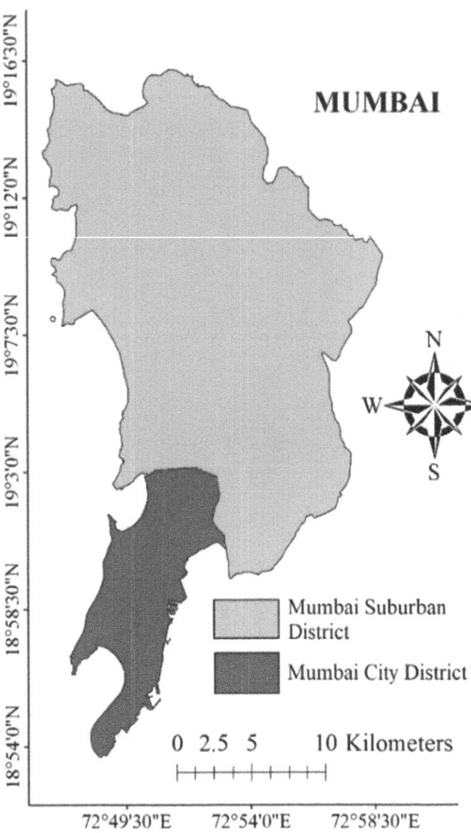

Fig. 3.2 The study area—Mumbai (Source: Prepared by the authors)

emergency modes of disruption to infrastructure, lives, and public safety.

3.2.2 Research Design

The researchers adopt a descriptive qualitative mixed-methods approach, driven by a mix of textual analysis, policy analysis, and spatial analysis within a wider critical urban studies context. The authors operate on the understanding that flood vulnerability is far more than a hydrological or engineering problem. Urban flooding manifests from uneven development, allowing developments on flood-prone, vulnerable lands, poor urban planning, and sub-optimal institutional arrangements that intersect in a socio-environmental problem. This analytical approach deepens our understanding of the reflective dynamics between climate change, urban development, and extreme urban flooding patterns in Mumbai. The methodology for this study is premised on the interpretive tradition of urban geography and sustainability studies, where understanding the socio-political and environmental dimensions of urban change, and their impacts on different parts of the city, is central to this research. Here, we have used a variety of secondary materials, which ranged from peer-reviewed journal articles and book chapters to news articles, planning documents, reports from government and NGO organizations, and other credible online sources. The rationale for using these pieces of evidence is to suitably account for the historical, spatial, and institutional elements of flooding and the ongoing urban change in Mumbai. Using the available literature, the researchers produced a thematic retelling of Mumbai's urban flooding issues and constructed reasonable and contextualized conjectures of the pathways forward towards a sustainable urban resilience ecology.

The use of PESTEL provides an overall framework for examining the political, economic, social, technological, environmental, and legal aspects contributing to flood vulnerability in Mumbai. The PESTEL analysis assesses aspects of public policy failure, governance shortcomings, economic factors, and social inequality/race that impact flooding urban settings, as well as technology-related failures in improved urban flood risk management, an overall environmental decline attributed to rapid urbanization, and poorer legal and institutional/capacity in enforcing land use and human climate adaptation measures. The PESTEL framework connects these aspects and provides an in-depth understanding of the conditions that led to flooding in Mumbai, fills in gaps in the current literature, and hence offers a valuable indication of sustainability considerations for urban planning and policy revision in the future. In addition, we have used satellite imagery and maps to support the analysis presented in this study. The software used for mapping in this study is Google Earth Pro 7.1 and ArcGIS 10.8. A methodological flow chart of the study is presented in Fig. 3.3.

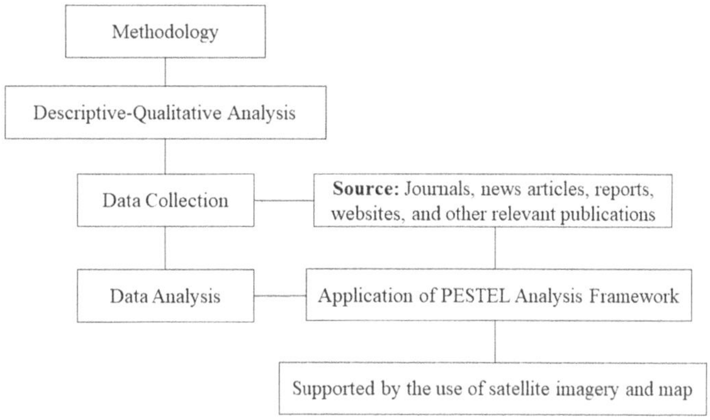

Fig. 3.3 Methodological flow chart of the study (Source: Prepared by the authors)

This study aims to accomplish the following objectives:

- Delineate the underlying sources of flooding in Mumbai, through a regionally, temporally, and spatially intersectional reading of flooding within a PESTEL framework;
- By drawing from multi-disciplinary knowledge bases, reduce disciplinary divides across environmental science, social policy, and urban planning;
- Advancement of scholarly and policy dialogue towards integrated, inclusive, adaptive systems of governance in rapidly urbanizing coastal megacities.

The overall purpose of this research is to draw attention to the more widespread climate vulnerability discourse in urban India while promoting resilient and sustainable urban development pathways. It adds to globalized conversations on urban adaptation and disaster risk reduction by featuring a place-characteristic, social, cultural, political, and economic context, a rich understanding of flooding in Mumbai, which may act as either a model or a cautionary tale for other coastal megacity contexts addressing these problems.

3.3 Results and Discussion

This chapter contextualizes Mumbai's floods within the larger picture of urban water management in a sustainable way. It reflects on past successes and failures, examines the pressures of climate and urbanization, and aims to imagine a better, resilient future. The chapter develops an example-specific and potentially globally relevant story to support this book's argument that resilient waterscapes are the aspect of urban sustainability within climate uncertainty.

3.3.1 Recent Flood Events in Mumbai

The topographic characteristics of Mumbai make it prone to heavy rainfall (Rana et al., 2014). Strong westerly and south-westerly monsoons flowing across the Arabian Sea have an orographic impact that causes Mumbai, situated on the windward side of India's Western Ghats, to receive significant torrential rainfall (Tawde, 2013; Rana et al., 2014). However, despite the heavy rainfall, waterlogging and floods in the city are comparatively new. Urban flooding incidents have intensified in Mumbai since the early 2000s, despite the historically high rainfall levels. The worst flooding nightmare witnessed by Mumbai was on July 26th, 2005, which brought the city to a grinding halt, leading to the death of more than 500 people, injuring many others, and causing severe damage to both public and private property (The Indian Express, 2017). This reveals systemic weaknesses in drainage, land use, and disaster preparedness.

The city has seen multiple instances of intense rainfall and flooding throughout the monsoon season since the Mumbai flooding in 2005. When examining the recent occurrences of floods and waterlogging during the rainy season in Mumbai from the year 2014 to 2023 (Table 3.1), we realize that the chaos and crises have become the defining characteristics of this metropolis. Mumbai has several issues each year because of heavy rains, flooding, and waterlogging, including traffic delays, building collapses, disruptions to rail and aviation services, and loss of life and property.

The data from 2014 to 2024 indicates consistent breakdowns in core life-support systems such as transport, drainage, and emergency services in all monsoons over the last 4 years, including multiple instances of casualties, damage to property, and disruption of civic life. On the heels of this flood-inflicted disruption in civic life, we learn that myriads of people in Mumbai share an acute and chronic hydrometeorological vulnerability, made worse by unsustainable urban growth and poorly-resourced stormwater networks.

3.3.2 Role of Climate Change and Urban Growth on Flooding in Mumbai

Floods in Mumbai result from the varied effects of climate change and urban growth experienced by the region. An increase in the frequency and severity of extreme weather events, such as storms and torrential rains, is associated with climate change. Previous studies carried out in the context of South

Table 3.1 Major impacts of rainfall, floods, and waterlogging in Mumbai (2014 to 2024)

Year	Month	Nature of impact due to rainfall, floods, and waterlogging
2014	July	Traffic jams
2015	June	Roads were blocked, rail services were interrupted, flights were delayed, and power-generating plants had to be shut down
2016	June	The movement of local trains was affected
2017	August	Traffic jams, trains, and flight services were affected, a building collapsed in the Bhendi Bazaar area, and drainage systems were clogged
2018	June	Waterlogging, delayed trains, and flights
	July	Streets and railway tracks were flooded, local train services were delayed, and *Dabbawala* services were disrupted
2019	July	Overflowing of the Mithi river, pre-emptive evacuation of 1000 people in Kurla, transport sector severely affected, railway tracts were under water, and some suburban and regional trains were suspended, flights were interrupted due to the closure of runway, the collapse of a wall in Malad, and the death of several people in slums
2020	August	Uprooting trees, damaging cranes, and tearing down hoardings and awnings
2021	June	Waterlogging in several places, flooding of the railway tracks, and a brief suspension of suburban rail services
	July	Collapse of a wall, landslides in multiple locations, death of several people, and inundation of a water purification plant
2022	May	The transport sector was severely affected, with severe issues with road traffic, and a tree-felling incident
	July	Traffic diversions
2023	June	Traffic snarls, heavy flooding, building collapses, death and injury, rainwater pumping machinery blocked due to garbage, several tree-felling incidents
2024	July	Severe rainfall (300 mm in about 6 h), bus and train service disrupted, and traffic jams
	September	Water logging, flights, and local train services were delayed, and schools and colleges were closed

Source: Compiled by the authors from various sources (The Economic Times, 2014; Davies, 2015; BBC, 2015; The Economic Times, 2016; Davis, 2017; Hindustan Times, 2017; India Today, 2018; The Times of India, 2018; Hindustan Times, 2018; Flood List, 2019a; Flood List, 2019b; India Today, 2019; Tare, 2019; Tare, 2020; Bhalerao & Singh, 2020; The New Indian Express, 2021; Ellis-Petersen, 2021; Bhalerao, 2022; The Times of India, 2022; India Today, 2023; The Indian Express, 2023)

Asia (Kumar et al., 2006; May, 2011; Meehl & Arblaster, 2003) as well as studies on Mumbai (Rana et al., 2014), have indicated the intensification of monsoon rains in this area, increasing the risk of flooding and extended periods of heavy rain. Besides, the distribution and timing of rainfall in Mumbai are impacted by monsoon patterns, which are influenced by the changing climatic patterns (Shastri et al., 2015; Shastri et al., 2019). Long dry spells punctuated by intense rainfall bursts have become increasingly common in recent years, and these periods are more prone to overflow infrastructure and flood sewers in the densely populated metropolis (Ellis-Petersen, 2021).

The recent instances of heavy rainfall in Mumbai have also overwhelmed its 150-year-old drainage infrastructure, which was originally designed to withstand 2.5 cm of rain per hour during no high tides (The Times of India, 2018). The unprecedented urban growth and the changing patterns of land use have altered Mumbai's natural drainage patterns. The number of impermeable surfaces, such as concrete and asphalt, has increased significantly, thereby decreasing the land's capacity to absorb water as well as increasing thermal temperatures. This has led to increased runoff during periods of intense precipitation, exacerbating flooding. Earlier, the Mithi River used to play a key role as a stormwater drainage for Mumbai (Naik & Kanekar, 2022). However, over time, the haphazard urban growth of Mumbai and the growing pressure of its population have turned the river into a convenient site for dumping waste, sewage, and other forms of garbage that clog the drainage system (Shelar, 2019; Bhattacharjee & Sharma, 2025). The devastation and fatalities brought about by the severe flooding in Mumbai in July 2005 were mostly attributed to the river's poor condition (Flood List, 2019a). While in the Global North, cities like New York, Vancouver, and Amsterdam are increasingly addressing their flood risks through adaptive infrastructure (Axelsson et al., 2021), cities like Mumbai in the Global South continue to struggle with outdated drainage systems and a lack of coordinated urban planning.

Sea levels are also rising globally because of climate change, increasing the risk of coastal flooding, particularly during storm surges (Mimura, 2013; Lindsey, 2023). Mumbai has been identified as a city susceptible to the effects of sea level rise. Slums have come to be concentrated in low-lying locations close to the sinking coasts, rendering them susceptible to flooding (The Times of India, 2018). Several slums have emerged along the low-lying coastal areas of Mumbai, such as the slum settlement along the Versova Beach of Mumbai (Fig. 3.4). In this figure, it can be seen that the slum settlements along the beach are extremely close to the coastline, posing a serious threat to the lives of the inhabitants. However, this pattern is not unique to Mumbai. In Dhaka, Bangladesh too, unplanned slums on flood-prone river banks threaten the lives of millions (Pu et al., 2024). In Lagos, coastal settlements like Makoko, which comprise the city's lowest-income communities, are also at risk from rising sea levels and inadequate defense measures (Ndimele et al., 2024).

In many instances, poor urban planning and overstressed infrastructure exacerbate the effects of climate change. The effects of extreme weather events can be made worse by inadequate flood defenses, poorly planned drainage systems, and inadequate stormwater management (Bibi & Kara, 2023). The devastating flood that occurred in late July 2005 was an eye-opener, exposing Mumbai's vulnerability to exceptionally heavy rainfall as well as the inadequacies of the drainage system to handle such a deluge. It showed the problems associated with the fast-urbanizing cities of the Global South, with infrastructural drawbacks and insufficient disaster preparedness increasing climate-related vulnerabilities. In the wake of this, Mumbai reevaluated its disaster management plans and infrastructure, following which improvements were made to transportation infrastructure, drainage systems, and preparedness for harsh weather events. With programs such as BRIMSTOWAD, microscale pumping stations for floods, and Mithi River restoration, the city has adopted and implemented best practices for resilient solutions. However, encroachments, waste dumping, and poor planning and implementation have limited the city's adaptation to flooding events, causing stagnation in the over-

Fig. 3.4 A slum area along the Versova Beach of Mumbai (Plotted on Google Earth Pro 7.1). Source: Prepared by the authors

all improvement process (Bhattacharjee & Sharma, 2025). While the coastal reclamation projects have worsened waterlogging, mangrove conservation remains crucial but is under threat. Thus, even today, Mumbai is struggling with the problems of floods and waterlogging during the rainy season. Moreover, not unlike the catastrophic flood in Mumbai in July 2005, recent floods in Jakarta (Indonesia) (Ogie et al., 2018), Xi'an (China) (Liu et al., 2023), are testimony to the fact that when heavy rainfall takes place, it reveals the weaknesses in infrastructure, failures in urban planning, and how better socio-economic means can prevent and reduce the impact of extreme weather anywhere.

3.3.3 Envisioning a Sustainable Predicament for Mumbai

We envision a sustainable predicament for Mumbai amidst its flood challenges by applying the PESTEL analysis, which is a framework used for assessing the political, economic, social, technological, environmental, and legal factors. By using it, we suggest a multidimensional decision-making model to deal with the problem at hand. The PESTEL analysis is discussed below:

(a) ***Political Factors***

Politically, laws that give priority to flood control and sustainable urban planning can be promoted and put into practice. Political goals can be aligned with the sustainability of the environment over the long run. In 2010, Mumbai unveiled an enhanced and more extensive plan for disaster relief and emergency operations, building on the Disaster Risk Management Plan (DRM) of 2005 (United Nations, n.d.). Based on the recent floods experienced by the city, more emphasis is necessary on the aspect of flood management in Mumbai. However, the aspect of indifference by policymakers to the city's needs, couched in terms of 'global standards of living,' has not escaped attention (The Hindu, 2018). In addition, cooperation across local, regional, national, and government levels can be strengthened to guarantee concerted efforts in putting into practice sustainable flood prevention strategies.

(b) ***Economic Factors***

Finance and investments for environmentally friendly flood infrastructure projects can be obtained by means of international funding, public-private partnerships, and private-sector incentives. For example, (a) the Green Climate Fund (GCF), which was created in accordance with the United Nations Framework Convention on Climate Change (UNFCCC), offers financing for climate change-related initiatives, such as ecologically appropriate flood infrastructure (GCF, 2023); and (b) the Asian Development Bank provides funding for a range of eco-friendly infrastructure projects across Asia, including programs aimed at reducing flood risk (Asian Development Bank, 2023). Additionally, performing a cost-benefit analysis to assess the long-term advantages and economic viability of sustainable flood management measures in relation to possible damages can help avert economic damages in such initiatives.

(c) ***Social Factors***

By including local knowledge, building community-based early warning systems, and integrating the local communities of Mumbai in the decision-making process, we can promote social resilience. The native communities had a deeper understanding of the water systems and the intricate interdependence between the mangroves and these systems for their survival; however, as time passed, the settlements experienced significant urbanization with the reclamation of land (Bhadgaonkar & Bhadgaonkar, 2022). The native local communities of Mumbai, such as the '*Koli* and *Agri* communities, have indigenous knowledge related to the land.' (Choudhari, n.d.). Involving them in the decision-making processes can fetch great benefits. It is imperative to guarantee that sustainable flood management plans consider social equity considerations, averting disproportionate effects on susceptible communities like the slum dwellers of Mumbai, and encourage comprehensive adaptation tactics.

(d) ***Technological Factors***

Mumbai has implemented smart flood management strategies such as artificial intelligence forecasting, geographic information systems (GIS) flood mapping, and automated water pumps to support early warning and risk assessment practices. However, the efficiency of these technologies is limited by their lack of integration into urban planning, poor drainage maintenance, and the absence of real-time monitoring across the entire city. Nature-based solutions are less prioritized over more concrete-heavy projects. To make a long-term impact, Mumbai must strengthen governance and invest in policy-driven integration of Artificial Intelligence (AI), Geographic Information System (GIS), Internet of Things (IoT), and flood wall infrastructure that is sustainable and technology-driven. For example, (a) Digital Elevation Models (DEMs) can be used to estimate the flow of water under various conditions and generate maps of floodplains (Tanaka et al., 2019; AL-Areeq et al., 2023); (b) to reduce the risk of flooding, GIS can help with the planning and design of infrastructure, such as stormwater drainage systems (Samuel, 2019); and (c) by incorporating current weather conditions, GIS may include real-time meteorological data, such as rainfall readings, into flood models to produce more accurate predictions (Piadeh et al., 2022). Building resilience against floods requires investing in the upkeep and upgrading of technology infrastructure, such as water pumping stations, drainage systems, and smart city solutions.

(e) ***Environmental Factors***

Ecological considerations can be incorporated into urban development by protecting wetlands, green spaces, and natural ecosystems that naturally reduce the impact of flooding. During the past 27 years in Mumbai, the Coastal Regulation Zone notification has been altered 34 times, which has allowed urban development rather than saving its coasts (Kukreti, 2019). Such counterproductive actions hinder ecological considerations and create urban hazards. Instead, it is important to take ecological considerations into account. For example, the saltpan lands of Mumbai must be protected and never opened for development. However, it is noteworthy that the recent removal of their wetland status has rendered the saltpans of Mumbai vulnerable to urban development (Khergamker, 2019; Venkatraman, 2019). This policy shift opened saltpans for consideration in housing development projects. Against this backdrop and in consonance with the acute space shortage in Mumbai, the Greater Mumbai Development Plan (DP) 2034 identified 2100 hectares of saltpan lands to be allocated for the construction of 1 million affordable housing units (Thakur, 2018). This has given rise to a worrisome situation. Furthermore, climate adaptation techniques can be integrated into long-term flood control plans to mitigate the effects of climate change in the area. Green infrastructure can be used to facilitate natural stormwater management to alleviate flooding; in addition, preserving and restoring mangrove ecosystems along the shore can help guard against coastal erosion and storm surges by acting as natural barriers.

(f) ***Legal Factors***

The creation of new legislative frameworks that support sustainability and ensure that flood management activities comply with current environmental norms and standards can be advocated. Establishing and upholding stringent land-use laws that deter building in regions at high risk of flooding can encourage resilient and sustainable urban development. The recent discussions on opening the salt-pan lands of Mumbai for developmental purposes must be restricted by stringent environmental laws and regulations to protect the city from further inundation and flooding (Malik, 2019).

In this way, Mumbai can strategically plan and implement sustainable flood management strategies by taking these Political, Economic, Social, Technological, Environmental, and Legal elements into account. This all-encompassing strategy considers the wider background and several factors that affect the city's ability to withstand flooding.

3.4 Recommendations

A multidimensional decision-making model is proposed in this chapter through the PESTEL analysis, which aims to address Mumbai's flood challenges comprehensively by considering diverse factors and fostering sustainability across political, economic, social, technological, environmental, and legal dimensions. Accordingly, the following recommendations are made:

(a) There is a need for integrated flood management policies in Mumbai that would entail cooperation between the municipal corporation, state, and central governments. This can be put into practice by reinforcing the Mumbai Climate Action Plan (MCAP) with flood management in urban planning, creating interagency flood task forces comprising municipal corporations, the Maharashtra government, and national agencies to streamline communication, and aligning with the National Disaster Management Authority (NDMA) guidelines for urban flood management.

(b) Community engagement is vital to disaster preparedness. Actionable steps can include community-based early warning systems in flood-prone areas, as developed in Kerala, where using SMS, mobile alerts, and training local disaster response teams, flood preparedness is developed (Shaji, 2025). Similarly, systems can be developed in slums and vulnerable areas of Mumbai by collaborating with members of Mumbai's native fishing and coastal communities to exchange knowledge on tidal patterns and drainage susceptibility.

(c) Flood management strategies should focus on vulnerable groups in Mumbai, particularly slum dwellers. Strategies should comprise slum rehabilitation with disaster-resistant housing designs to facilitate or allow for safe relocation rather than forced eviction, collaborative governance approaches that have resident representatives from informal settlements involved in planning decisions in the city's political framework, and implementing small-scale drainage solutions (such as porous roads or community rainwater storage tanks) in informal settlements.

(d) The MCGM can leverage technology such as Artificial Intelligence (AI), Geographic Information System (GIS), and remote sensing technology for improved flood prediction and response. Initiatives could include—continued investment in AI-derived flood forecasting models like those used in Chennai and Bangalore, expanding the use of GIS-derived flood risk mapping throughout the city to determine high-risk flood areas to inform, the modification and land use/review planning, and adopting actual in-time meteorological data into the MCGM's Disaster Management Cell system.

(e) Mumbai must also ensure flooding is managed by enhancing and protecting the natural floodwater buffers. These interventions would seek to do so by simultaneously enhancing the protection of mangroves, wetlands, and green space-generating zoning laws; instituting nature-based solutions such as bioswales, green roofs, and urban forests; and implementing Environmental Impact Assessments (EIA) on all new developments that affect natural drainage.

(f) The governance in Mumbai is needed to provide a clear legal framework. These regulations would look to further preclude the destruction or alteration of wetlands by formally enforcing compliance with the Coastal Regulation Zone (CRZ) laws and instituting stronger legal repercussions for illegal encroachments in flood-prone areas.

(g) Flood management must prioritize Mumbai's vulnerable populations, particularly slum dwellers. Steps must include—slum rehabilitation with flood-resilient housing designs, ensuring safe relocation rather than forced eviction; participatory governance models, where informal settlement residents are part of urban planning decisions, and implementation of localized drainage solutions (e.g., permeable roads, community rainwater storage) in informal settlements.

(h) To finance the flood management projects, funds can be used from municipal budgets, government grants, public-private partnerships (PPPs), climate finance, and contributions from developers. Corporate Social Responsibility (CSR) programs, and environmental and/or ecological compensation and mitigation fees can also combine to help fund various resilience measures. Besides these, the creation of urban flood resilience bonds can be considered for enhancing urban flood resilience measures and creating benefits for the communities involved.

3.5 Conclusion

In this chapter, by using Mumbai as a case study, we examine how cities are at the same time both complicit in and vulnerable to climate change. In many ways, the overall argument is full of paradoxes. While urban development is both a driver of economic and demographic growth, it also creates socio-ecological fragility through unregulated urbanization, environmental degradation, and a lack of sufficient infrastructure. Here, we present an interdisciplinary and integrative approach to diagnosing urban flood susceptibility, one that goes beyond technocratic solutions and looks critically at socio-political, economic, and legal structures embedded in the potpourri of climate-related urban risk. Our findings illustrate the necessity of developing policies that marry environmental sustainability with social equity and reflect our broader global aspirations to mitigate inequities in exposure to climate hazards. This chapter is relevant in terms of strategic policy as it offers a variety of practical recommendations for local governments, urban planners, and international development agencies. It highlights the importance of upgrading existing drainage and stormwater management systems, preserving and restoring natural systems, implementing reforms to improve institutional capacity for governance, and developing and implementing urban planning that is inclusive of vulnerable groups in our society. This research is very closely aligned with the United Nations Sustainable Development

Goals, in particular SDG # 6, SDG # 11, and SDG # 13.

Although this study is focused on Mumbai, the findings and recommendations on urbanization, climate change, and social vulnerabilities have broader implications for fast-growing urban centers across the globe that face the social complexities of climate change, urbanization, and socio-economic vulnerability. This research contributes toward efforts to operationalize sustainability frameworks in cities and establishes a basis for future comparative studies. It creates opportunities to build on other research in areas such as the impact of Urban Heat Islands (UHI) on changing flood patterns, using nature-based approaches to climate-proof urban infrastructure, and socio-economic vulnerability mapping as a preparedness technique, and integration of AI-based early warning systems and smart technologies to support urban flood prediction and mitigation, etc.

This chapter serves as a clarion call to researchers, public policy-makers, and international stakeholders to reflect on urban development through a climate-resilient lens. It adds to the global understanding of how sustainable urbanization and just climate adaptation must be the foundation of any long-term strategy to respond to urban disasters in the Anthropocene.

AI Disclosure Statement We declare that we have not used any AI in this study.

References

Ahmed, F., Moors, E., Khan, M. S. A., Warner, J., & Terwisscha van Scheltinga, C. (2018). Tipping points in adaptation to urban flooding under climate change and urban growth: The case of the Dhaka megacity. *Land Use Policy, 79*, 496–506. https://doi.org/10.1016/j.landusepol.2018.05.051

Ajjur, S. B., & Al-Ghamdi, S. G. (2022). Exploring urban growth–climate change–flood risk nexus in fast growing cities. *Scientific Reports, 12*(1), 12265. https://doi.org/10.1038/s41598-022-16475-x

AL-Areeq, A. M., Sharif, H. O., Abba, S. I., Chowdhury, S., Al-Suwaiyan, M., Benaafi, M., Yassin, M. A., & Aljundi, I. H. (2023). Digital elevation model for flood hazards analysis in complex terrain: Case study from Jeddah, Saudi Arabia. *International Journal of Applied Earth Observation and Geoinformation, 119*, 103330. https://doi.org/10.1016/j.jag.2023.103330

Alexander, K., Hettiarachchi, S., Ou, Y., & Sharma, A. (2019). Can integrated green spaces and storage facilities absorb the increased risk of flooding due to climate change in developed urban environments? *Journal of Hydrology, 579*, 124201. https://doi.org/10.1016/j.jhydrol.2019.124201

Anand, N. (2023). Anthroposea: Planning future ecologies in Mumbai's wetscapes. *Environment and Planning D: Society and Space, 41*(4), 683–706. https://doi.org/10.1177/02637758231183439

Asian Development Bank. (2023). ADB's work on climate change and disaster risk management. Accessed January 11, 2024, from https://www.adb.org/what-we-do/topics/climate-change/overview

Awasthi, M. (2023). The story repeats itself: Why Mumbai gets choked every monsoon. In: The Sunday Guardian Live. Accessed March 25, 2025, from https://sundayguardianlive.com/news/the-story-repeats-itself-why-mumbai-gets-choked-every-monsoon

Axelsson, C., Soriani, S., Culligan, P., & Marcotullio, P. (2021). Urban policy adaptation toward managing increasing pluvial flooding events under climate change. *Journal of Environmental Planning and Management, 64*(8), 1408–1427. https://doi.org/10.1080/09640568.2020.1823346

Baklanov, A., Molina, L. T., & Gauss, M. (2016). Megacities, air quality and climate. *Atmospheric Environment, 126*, 235–249. https://doi.org/10.1016/j.atmosenv.2015.11.059

Banerjee, S., Middel, A., & Chattopadhyay, S. (2022). A regression-based three-phase approach to assess outdoor thermal comfort in informal micro-entrepreneurial settings in tropical Mumbai. *International Journal of Biometeorology, 66*(2), 313–329. https://doi.org/10.1007/s00484-021-02136-7

BBC. (2015). India monsoon rains bring busy Mumbai to grinding halt. BBC News.

Bhadgaonkar, K., & Bhadgaonkar, J. (2022). The 'forgotten' water ecosystems of Mumbai. Accessed December 6, 2023, from https://www.downtoearth.org.in/blog/environment/the-forgotten-water-ecosystems-of-mumbai-84732

Bhalerao, S. (2022). Heavy overnight rainfall throws parts of Mumbai out of gear; flooding, traffic diversions follow. The Indian Express.

Bhalerao, S., & Singh, L. (2020). Massive flooding and destruction after heavy rain, strong winds batter Mumbai. The Indian Express.

Bhattacharjee, S., & Sattar, S. (2021a). Pandemic and urban sustainability: Analyzing the Covid-19 scenario in Mumbai, India. *Geography, Environment, Sustainability, 14*(1), 17–24.

Bhattacharjee, S., & Sattar, S. (2021b). COVID-19 pandemic and the poor in the urban spaces of India with special reference to Mumbai. *Equality, Diversity and Inclusion: An International Journal, 41*(1), 64–79. https://doi.org/10.1108/EDI-07-2020-0196

Bhattacharjee, S., & Sattar, S. (2022). Life in the slums of Mumbai: Struggle for survival amidst the unsanitary conditions. In A. L. Singh & S. Jamal (Eds.), *Sanitation for urban poor in Indian cities: Issues and challenges* (pp. 31–47). B. R Publishing Corporation.

Bhattacharjee, S., & Sattar, S. (2025). Sustainability at crossroads: The conflicting needs of affordable housing development versus saltpan conservation in Mumbai, India. In *Proceedings of the international conference of contemporary affairs in architecture and urbanism-ICCAUA* (Vol. 8, No. 1, pp. 556–562). https://doi.org/10.38027/ICCAUA2025EN0283

Bhattacharjee, S., & Sharma, M. (2023). Polycentric urbanism and the growth of new economic hubs in Mumbai, India. In U. Chatterjee, N. Bandyopadhyay, M. D. Setiawati, & S. Sarkar (Eds.), *Urban commons, future smart cities and sustainability* (1st ed., pp. 169–186). Springer International Publishing.

Bhattacharjee, S., & Sharma, M. (2025). Ramifications of anthropogenic activities on urban riverine ecosystem: A study of Mithi River in Mumbai. In S. C. Pal, U. Chatterjee, M. D. Setiawati, & D. Ruidas (Eds.), *Remotely sensed Rivers in the age of Anthropocene* (pp. 209–228). Springer Nature Switzerland.

Bibi, T. S., & Kara, K. G. (2023). Evaluation of climate change, urbanization, and low-impact development practices on urban flooding. *Heliyon, 9*(1), e12955. https://doi.org/10.1016/j.heliyon.2023.e12955

Census of India. (2011a). *District census handbook: Mumbai Suburban District*. Directorate of Census Operations Maharashtra. Govt. of India Press.

Census of India. (2011b). *District census handbook: Mumbai District*. Directorate of Census Operations Maharashtra. Govt. of India Press.

Chakraborty, A., Sharma, M., & Abhay, R. K. (2023). Colonial imprints in contemporary urban livability: An inter-ward analysis of Kolkata. *Geo Journal, 88*(1), 543–559. https://doi.org/10.1007/s10708-022-10606-7

Choudhari, G. (n.d.). *The Kolis – An introduction*. Jio Institute. Accessed September 2, 2024, from https://exhibits.jioinstitute.edu.in/spotlight/the-kolis-an-introduction

Davies, R. (2015). Heavy rain and high tide combine to flood Mumbai—300 mm of rain in 24 Hours—FloodList. Accessed December 18, 2023, from https://floodlist.com/asia/heavy-rain-high-tide-floods-mumbai-june-2015

Davis, R. (2017). India—Deadly floods in Maharashtra, Mumbai streets under water (updated)—FloodList. Accessed December 18, 2023, from https://floodlist.com/asia/india-floods-maharashtra-august-2017

Doshi, D., & Garschagen, M. (2024). Actor-specific adaptation objectives shape perceived roles and responsibilities: Lessons from Mumbai's flood risk reduction and general considerations. *Regional Environmental Change, 24*(4), 164. https://doi.org/10.1007/s10113-024-02315-3

Dube, K., Nhamo, G., & Chikodzi, D. (2022). Flooding trends and their impacts on coastal communities of Western Cape Province, South Africa. *Geo Journal, 87*(Suppl 4), 453–468. https://doi.org/10.1007/s10708-021-10460-z

Ellis-Petersen, H. (2021). *Dozens dead in Mumbai after 'monstrous' monsoon rains cause landslide*. The Guardian.

Esfandeh, S., Danehkar, A., Salmanmahiny, A., Sadeghi, S. M. M., & Marcu, M. V. (2022). Climate change risk of urban growth and land use/land cover conversion: An in-depth review of the recent research in Iran. *Sustainability, 14*(1), 338. https://doi.org/10.3390/su14010338

Flood List. (2019a). India—Heavy rain and floods cause major disruption in Mumbai. Accessed December 15, 2023, from https://floodlist.com/asia/india-floods-mumbai-september-2019

Flood List. (2019b). India—Dozens dead after torrential rainfall in Mumbai and Maharashtra—Flood List. In: Flood List. Accessed December 18, 2023, from https://floodlist.com/asia/india-floods-mumbai-maharashtra-july-2019

Furtak, K., & Wolińska, A. (2023). The impact of extreme weather events as a consequence of climate change on the soil moisture and on the quality of the soil environment and agriculture—A review. *Catena, 231*, 107378. https://doi.org/10.1016/j.catena.2023.107378

GCF. (2023). About GCF. In: Green Climate Fund. Accessed January 11, 2024, from https://www.greenclimate.fund/about

Griggs, G., & Reguero, B. G. (2021). Coastal adaptation to climate change and sea-level rise. *Water, 13*(16), 2151. https://doi.org/10.3390/w13162151

Grimm, N. B., Faeth, S. H., Golubiewski, N. E., Redman, C. L., Wu, J., Bai, X., & Briggs, J. M. (2008). Global change and the ecology of cities. *Science, 319*(5864), 756–760. https://doi.org/10.1126/science.1150195

Gupta, K. (2007). Urban flood resilience planning and management and lessons for the future: A case study of Mumbai, India. *Urban Water Journal, 4*(3), 183–194. https://doi.org/10.1080/15730620701464141

Hemmati, M., Ellingwood, B. R., & Mahmoud, H. N. (2020). The role of urban growth in resilience of communities under flood risk. *Earths Future, 8*(3), e2019EF001382. https://doi.org/10.1029/2019EF001382

Hindustan Times. (2017). Mumbai floods: Why India's cities are struggling with extreme rainfall. Hindustan Times.

Hindustan Times. (2018). Pre-monsoon showers flood Mumbai, navy to step in if required. Hindustan Times.

India Today. (2018). Maximum city submerged: Mumbai rains in 20 images. In: India Today. Accessed December 18, 2023, from https://www.indiatoday.in/india/photo/maximum-city-submerged-mumbai-rains-in-20-images-1281559-2018-07-10

India Today. (2019) Mumbai rains: Downpour brings Mumbai on its knees again, 35 killed across Maharashtra. India Today.

India Today. (2023). Heavy rain in Mumbai, CM Shinde reviews flood prevention system. India Today.

Khergamker, G. (2019). Centre, Maharashtra play tag to 'develop' salt pans. In: Down to Earth. Accessed January 17, 2024, from https://www.downtoearth.org.in/news/environment/centre-maharashtra-play-tag-to-develop-salt-pans-63206

Kukreti, I. (2019). Coastal regulation zone notification: What development are we clearing our coasts for. In: Down to Earth. Accessed December 25, 2023, from https://www.downtoearth.org.in/coverage/governance/coastal-regulation-zone-notification-what-development-are-we-clearing-our-coasts-for-63061

Kumar, K. R., Sahai, A. K., Kumar, K. K., Patwardhan, S. K., Mishra, P. K., Revadekar, J. V., Kamala, K., & Pant, G. B. (2006). High-resolution climate change scenarios for India for the 21st century. *Current Science, 90*(3), 334–345.

Kundzewicz, Z. W., Kanae, S., Seneviratne, S. I., Handmer, J., Nicholls, N., Peduzzi, P., Mechler, R., Bouwer, L. M., Arnell, N., Mach, K., Muir-Wood, R., Brakenridge, G. R., Kron, W., Benito, G., Honda, Y., Takahashi, K., & Sherstyukov, B. (2014). Flood risk and climate change: Global and regional perspectives. *Hydrological Sciences Journal, 59*(1), 1–28. https://doi.org/10.1080/02626667.2013.857411

Lindsey, R. (2023). Climate change: Global Sea level | NOAA Climate.gov. Accessed September 2, 2024, from http://www.climate.gov/news-features/understanding-climate/climate-change-global-sea-level

Liu, J., Yang, X., & Ren, S. (2023). Research on the impact of heavy rainfall flooding on urban traffic network based on road topology: A case study of Xi'an City, China. *Land, 12*(7), 1355. https://doi.org/10.3390/land12071355

Malik, F. (2019). Mumbai will drown if development is allowed on salt pans. Hindustan Times.

Mann, R., Gupta, A., Dhorde, A., & Sharma, S. (2023). Observed trends and coherent changes in daily rainfall extremes over Greater Mumbai, 1985–2020. *Theoretical and Applied Climatology, 151*(3), 1889–1910. https://doi.org/10.1007/s00704-022-04354-4

May, W. (2011). The sensitivity of the Indian summer monsoon to a global warming of 2°C with respect to pre-industrial times. *Climate Dynamics, 37*(9), 1843–1868. https://doi.org/10.1007/s00382-010-0942-8

Meehl, G. A., & Arblaster, J. M. (2003). Mechanisms for projected future changes in south Asian monsoon precipitation. *Climate Dynamics, 21*(7), 659–675. https://doi.org/10.1007/s00382-003-0343-3

Mimura, N. (2013). Sea-level rise caused by climate change and its implications for society. *Proceedings of the Japan Academy Series B, Physical and Biological Sciences, 89*(7), 281. https://doi.org/10.2183/pjab.89.281

Naik, K., & Kanekar, S. (2022). Walking along the Mithi River: Exploring a 17 KM long trained river and urban settlements on its "Banks." In: WRI INDIA. Accessed December 6, 2023, from https://wri-india.org/blog/walking-along-mithi-river-exploring-17-km-long-trained-river-and-urban-settlements-its-banks

Ndimele, P. E., Ojewole, A. E., Mekuleyi, G. O., Badmos, L. A., Agosu, C. M., Olatunbosun, E. S., Lawal, O. O., Shittu, J. A., Joseph, O. O., Ositimehin, K. M., Ndimele, F. C., Ojewole, C. O., Abdulganiy, I. O., & Ayodele, O. T. (2024). Vulnerability, resilience and adaptation of Lagos coastal communities to flooding. *Earth Science, Systems and Society, 4*(1), 10087. https://doi.org/10.3389/esss.2024.10087

Ogie, R. I., Holderness, T., Dunn, S., & Turpin, E. (2018). Assessing the vulnerability of hydrological infrastructure to flood damage in coastal cities of developing nations. *Computers, Environment and Urban Systems, 68*, 97–109. https://doi.org/10.1016/j.compenvurbsys.2017.11.004

Patel, A. (2021). Mumbai: The city that was born of the sea may eventually be consumed by it. In: Scroll.in. Accessed December 12, 2023, from https://scroll.in/magazine/1004151/mumbai-the-city-that-was-born-of-the-sea-may-eventually-be-consumed-by-it

Piadeh, F., Behzadian, K., & Alani, A. M. (2022). A critical review of real-time modelling of flood forecasting in urban drainage systems. *Journal of Hydrology, 607*, 127476. https://doi.org/10.1016/j.jhydrol.2022.127476

Pour, S. H., Wahab, A. K. A., Shahid, S., Asaduzzaman, M., & Dewan, A. (2020). Low impact development techniques to mitigate the impacts of climate-change-induced urban floods: Current trends, issues and challenges. *Sustainable Cities and Society, 62*, 102373. https://doi.org/10.1016/j.scs.2020.102373

Pu, C., Yang, F., & Wang, X. (2024). Flood risk assessment of slums in Dhaka city. *Geocarto International, 39*(1), 2341802. https://doi.org/10.1080/10106049.2024.2341802

Rana, A., Foster, K., Bosshard, T., Olsson, J., & Bengtsson, L. (2014). Impact of climate change on rainfall over Mumbai using distribution-based scaling of global climate model projections. *Journal of Hydrology: Regional Studies, 1*, 107–128. https://doi.org/10.1016/j.ejrh.2014.06.005

Randolph, G. F. (2024). Does urbanization depend on in-migration? Demography, mobility, and India's urban transition. *Environment & Planning A, 56*(1), 117–135. https://doi.org/10.1177/0308518X231180609

Rentschler, J., Avner, P., & Hallegatte, S. (2023). Tracking urban flood exposure: Global trends since 1985. Accessed December 18, 2023, from https://blogs.worldbank.org/sustainablecities/tracking-urban-flood-exposure-global-trends-1985

Roy, M. (2009). Planning for sustainable urbanisation in fast growing cities: Mitigation and adaptation issues addressed in Dhaka, Bangladesh. *Habitat International, 33*(3), 276–286. https://doi.org/10.1016/j.habitatint.2008.10.022

Samuel, A. (2019). Using GIS for effective stormwater runoff management – Powered by Nobel Systems. Accessed January 11, 2024, from https://nobelsystemsblog.com/2019/12/21/using-gis-for-effective-stormwater-runoff-management/

Shaji, K. A. (2025). Kerala sets another example of disaster preparedness with a coordinated early warning system. In: Down To Earth. Accessed March 31, 2025, from https://www.downtoearth.org.in/natural-disasters/kerala-sets-another-example-of-disaster-preparedness-with-a-coordinated-early-warning-system

Sharma, M. (2017). Quality of life of labour engaged in the informal economy in the National Capital Territory of Delhi, India. *Khoj: The International Peer Reviewed Journal of Geography, 4*(1), 14. https://doi.org/10.5958/2455-6963.2017.00002.9

Sharma, M., & Abhay, R. K. (2022). Urban growth and quality of life: Inter-district and intra-district analysis of housing in NCT-Delhi, 2001–2011–2020. *GeoJournal, 87*(4), 797–819. https://doi.org/10.1007/s10708-021-10570-8

Shastri, H., Ghosh, S., Paul, S., Shafizadeh-Moghadam, H., Helbich, M., & Karmakar, S. (2019). Future urban rainfall projections considering the impacts of climate change and urbanization with statistical–dynamical integrated approach. *Climate Dynamics, 52*(9), 6033–6051. https://doi.org/10.1007/s00382-018-4493-8

Shastri, H., Paul, S., Ghosh, S., & Karmakar, S. (2015). Impacts of urbanization on Indian summer monsoon rainfall extremes. *Journal of Geophysical Research: Atmospheres, 120*(2), 496–516. https://doi.org/10.1002/2014JD022061

Shelar, J. (2019). *How the Mithi was destroyed*. The Hindu.

Singh, H., Nielsen, M., & Greatrex, H. (2023). Causes, impacts, and mitigation strategies of urban pluvial floods in India: A systematic review. *International Journal of Disaster Risk Reduction, 93*, 103751. https://doi.org/10.1016/j.ijdrr.2023.103751

Sterzel, T., Lüdeke, M. K. B., Walther, C., Kok, M. T., Sietz, D., & Lucas, P. L. (2020). Typology of coastal urban vulnerability under rapid urbanization. *PLoS One, 15*(1), e0220936. https://doi.org/10.1371/journal.pone.0220936

Sun, X., Li, R., Shan, X., Xu, H., & Wang, J. (2021). Assessment of climate change impacts and urban flood management schemes in Central Shanghai. *International Journal of Disaster Risk Reduction, 65*, 102563. https://doi.org/10.1016/j.ijdrr.2021.102563

Tanaka, K., Fujihara, Y., Hoshikawa, K., & Fujii, H. (2019). Development of a flood water level estimation method using satellite images and a digital elevation model for the Mekong floodplain. *Hydrological Sciences Journal, 64*(2), 241–253. https://doi.org/10.1080/02626667.2019.1578463

Tare, K. (2019). Mumbai floods: How BMC and Co. sank the city again. India Today.

Tare, K. (2020). Why South Mumbai got flooded this monsoon. India Today.

Tawde, S. A. (2013). *Investigation of orographically induced rainfall over Western Ghats and its association with other monsoon parameters*. Master thesis, Indian Institute of Remote Sensing, ISRO.

Thakur, P. (2018). Allowing developers to construct on Mumbai's salt pan lands may do more harm than good. In: Housing News. Accessed January 17, 2024, from https://housing.com/news/allowing-developers-construct-mumbais-salt-pan-lands-may-harm-good/

The Economic Times. (2014). Pics: Floods across India cause devastation – Heavy rain causes water logging in Mumbai. The Economic Times.

The Economic Times. (2016). Mumbai rain: Heavy showers lash city. The Economic Times.

The Hindu. (2018). Flood of despair: Mumbai's flooding woes. The Hindu.

The Indian Express. (2017). Mumbai rains: A grim reminder of July 26, 2005, the day the city came to a grinding halt. In: The Indian Express. Accessed December 17, 2024, from https://indianexpress.com/article/india/mumbai-rains-july-26-2005-floods-maharashtra-water-logging-4819676/

The Indian Express. (2023). From building collapse to flooding & traffic snarls: Key points as monsoon hits Mumbai. The Indian Express.

The New Indian Express. (2021). Blame game on after 60 mm rain in an hour floods Mumbai. The New Indian Express.

The Times of India. (2018). Monsoon reaches Mumbai, expect flood to follow. The Times of India.

The Times of India. (2022). Heavy rain lashes Mumbai; flooding in low-lying areas, road traffic hit. The Times of India.

Thomas, S., & Lekshmy, P. R. (2022). Recent trends in tropical cyclones over the Arabian Sea and the vulnerability of India's west coast. *Arabian Journal of Geosciences, 15*(23), 1713. https://doi.org/10.1007/s12517-022-10992-9

Tripathy, S. S., Chaudhuri, S., Murtugudde, R., Mhatre, V., Parmar, D., Pinto, M., Zope, P. E., Dixit, V., Karmakar, S., & Ghosh, S. (2024). Analysis of Mumbai floods in recent years with crowdsourced data. *Urban Climate, 53*, 101815. https://doi.org/10.1016/j.uclim.2024.101815

UNEP. (2017). Cities and climate change. In: UNEP – UN Environment Programme. Accessed December 18, 2024, from http://www.unep.org/explore-topics/resource-efficiency/what-we-do/cities/cities-and-climate-change

United Nations. (n.d.). Mumbai disaster risk management master plan. In: United Nations | Department of Economic and Social Affairs Sustainable Development. Accessed January 11, 2024, from https://sdgs.un.org/partnerships/mumbai-disaster-risk-management-master-plan

Venkatraman, T. (2019). How much of Mumbai's salt pans can be developed? Master plan in the works. Hindustan Times.

Wang, J., Wah, Y. C., & Cao, S.-J. (2022). Urban development in the context of extreme flooding events. *Indoor and Built Environment, 31*(1), 3–6. https://doi.org/10.1177/1420326X211048577

WHO. (2023). Climate change. Accessed March 25, 2025, from https://www.who.int/news-room/fact-sheets/detail/climate-change-and-health

World Bank. (2023). Advancing climate action through an urban lens. In: World Bank. Accessed December 18, 2023, from https://www.worldbank.org/en/topic/urbandevelopment/brief/climate-action-through-an-urban-lens

Yin, J., Yu, D., Yin, Z., Liu, M., & He, Q. (2016). Evaluating the impact and risk of pluvial flash flood on intra-urban road network: A case study in the city center of Shanghai, China. *Journal of Hydrology, 537*, 138–145. https://doi.org/10.1016/j.jhydrol.2016.03.037

Zhou, Q., Mikkelsen, P. S., Halsnæs, K., & Arnbjerg-Nielsen, K. (2012). Framework for economic pluvial flood risk assessment considering climate change effects and adaptation benefits. *Journal of Hydrology, 414–415*, 539–549. https://doi.org/10.1016/j.jhydrol.2011.11.031

Part II

Equity, Planning, and Resilience in Urban Waterscapes

4 Bridging the Gap in Urban Water Accessibility: Exploring the Barriers to Equity in Water Access in Ghana

Maxwell Fobi Kontor,
José Rafael Núñez Collado, and André Brown

Abstract

Access to clean water is recognised by the United Nations as a fundamental human right. However, in urban Ghana, significant disparities exist in access to clean water, particularly in poor communities. This chapter analyses the persistent inequalities in urban water accessibility in Greater Accra to understand the barriers hindering equitable water provision. We do this by examining four interconnected aspects: colonial legacies that shape fragmented water infrastructure, contemporary infrastructural deficits, commodification and privatisation of water services, and fragmented governance and regulatory mechanisms. Using these dimensions as analytical lenses, we demonstrate how historical patterns of urban planning and water infrastructure development under colonial rule established spatial inequalities that persist today. Contemporary infrastructure gaps and poor water governance further deepen these disparities, particularly in low-income and informal urban communities. These areas often rely on costly informal water sources such as tankers and sachets/bottled water. Privatisation efforts promoted under neoliberal policies and commodification practices have further exacerbated these inequities by prioritising profit over equitable distribution. By unpacking these interconnected factors, the chapter calls for a new paradigm of urban water governance, one that explicitly prioritises equity, reclaims water as a social and ecological good, and foregrounds the voices and agency of marginalised communities. This lays the foundation for advancing sustainable water management to drive resilient urban waterscapes.

Keywords

Water equity · Infrastructure, privatisation · Governance · Colonial legacy · Ghana

4.1 Introduction

Over the past decade, notable progress has been made in expanding global access to improved drinking water sources (Ando et al., 2025). The proportion of the global population using safe drinking water services rose from 69% in 2015 to 73% by 2022 (World Health Organisation & United Nations Children's Fund, 2023). These improvements reflect coordinated interventions by international organisations, national govern-

M. Fobi Kontor (✉) · J. R. Núñez Collado · A. Brown
School of Architecture, Victoria University of Wellington, Wellington, New Zealand
e-mail: maxwell.fobikontor@vuw.ac.nz; jose.nunezcollado@vuw.ac.nz; andre.brown@vuw.ac.nz

U. Chatterjee, A. I. Almulhim (eds.), *Urban Waterscapes*, Sustainable Development Goals Series,
https://doi.org/10.1007/978-3-032-20649-7_4

ments, and local stakeholders to address one of humanity's most basic needs. Despite these global advancements, substantial disparities persist, particularly in urban areas across the Global South, where access to safe, reliable water remains uneven (Babuna et al., 2023; Tetteh et al., 2022). In many urban contexts, low-income populations struggle with inadequate supply and poor water quality, exposing them to heightened risks of waterborne diseases and broader health inequities (Jambadu et al., 2023). For these communities, reliance on unsafe water sources is not merely a technical failure but a reflection of more profound structural inequity and vulnerabilities (Leahy et al., 2024). In response to these persistent challenges, multilateral institutions such as the United Nations, the World Health Organisation (WHO), and the United Nations Children's Fund United Nations Children's Fund (UNICEF) have assumed prominent roles in shaping water policy agendas and providing financial and technical support. Through ambitious frameworks such as the Sustainable Development Goals (SDGs), these actors have promoted universal access to safe drinking water, advocating for systemic reforms that extend beyond infrastructure to include regulatory accountability, social equity, and participatory governance (United Nations, 2017). This emphasises the recognition that ensuring access to safe water requires physical delivery systems, political will, institutional capacity, and collaborative governance.

Recognising the vital role of safe water and sanitation in human life, Target 6.1 of Sustainable Development Goal Six (Clean Water and Sanitation) mandates universal and equitable access to safe and affordable drinking water for everyone by 2030. This goal aligns with the United Nations' declaration that access to safe water is a fundamental human right (United Nations, 2010). Despite widespread recognition of the importance of safe water, millions across the globe, particularly in Sub-Saharan Africa (SSA), still lack proper access to it. According to the WHO & UNICEF (2023), only 31% of the SSA population had access to safely managed drinking water, compared to 68% in Central and Southern Asia, 73% in Latin America and the Caribbean, 79% in Eastern and South-Eastern Asia, and 94% in Europe and Northern America. Safely managed water is defined as drinking water sourced from an improved source that is available on-site, accessible when needed, and free from faecal and priority chemical contaminants. By this metric, over 60% of SSA's population remains underserved with safely managed drinking water, with 22% still dependent on unimproved or surface water.

Equally critical is SDG 6.2, which commits to achieving universal access to adequate sanitation and hygiene by 2030. Safe and improved drinking water is inherently linked to enhanced sanitation. Proper sanitation and hygiene are fundamental to maintaining water quality and promoting public health (Alum et al., 2024; Fuhrmeister et al., 2023). The absence of adequate sanitation and hygiene facilities undermines water quality through contamination. Particularly, in urban poor communities and informal settlements, where households often lack both piped water and proper sanitation (Hoelzel, 2024; Sprouse et al., 2024), these issues intersect, leading to increased risks and perpetuating cycles of disease. This undermines both human well-being and urban sustainability (Abdulhadi et al., 2024). By situating this chapter within the context of SDG 6.1 and 6.2, we emphasise that safe and equitable water access should be meaningfully advanced in concurrence with the simultaneous improvements in sanitation and hygiene. Together, these elements lay the groundwork for a resilient urban waterscape and highlight the urgent need for integrated strategies that link equitable water provision and sanitation improvement to foster inclusive and sustainable solutions. Figure 4.1 illustrates the interconnections between SDG 6.1 (safe and affordable drinking water) and SDG 6.2 (adequate sanitation and hygiene) in shaping sustainable urban waterscapes. It highlights how improvements in water access and sanitation are mutually reinforcing, and together provide the foundation for resilient, equitable, and sustainable urban water systems.

The scale and complexity of water access have attracted significant scholarly attention. Scholars

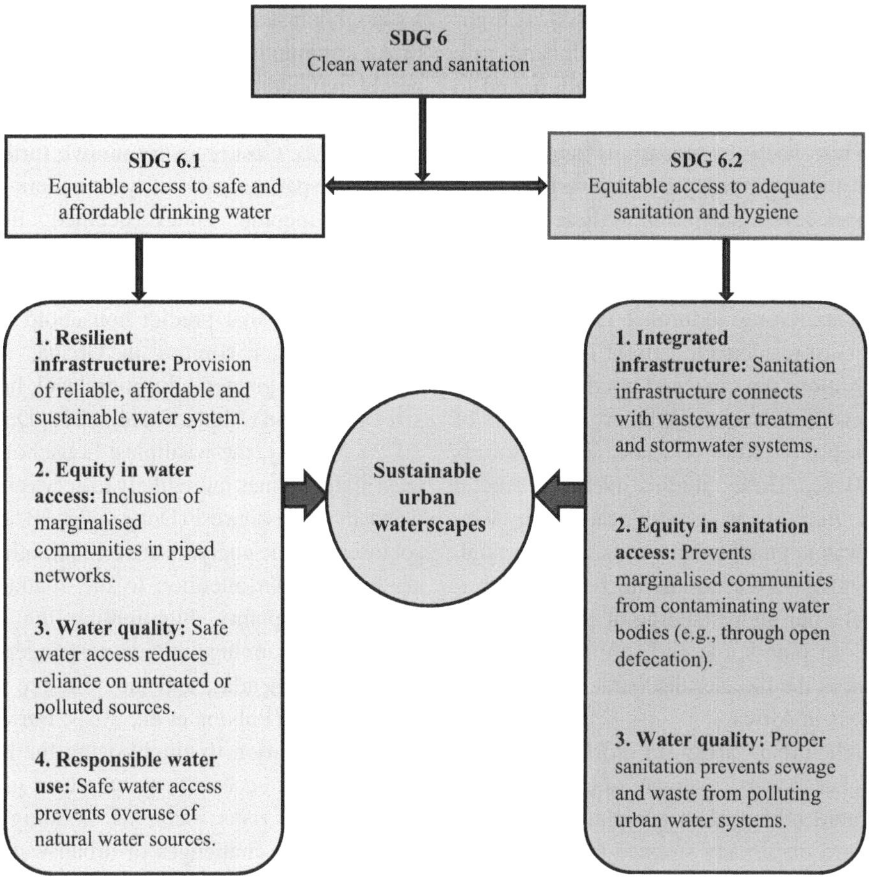

Fig. 4.1 Linkage between SDG targets 6.1 and 6.2 in advancing sustainable urban waterscapes. Source: Authors own construct

have examined multiple dimensions of water access, including water security, scarcity, service delivery quality, water quality, and associated public health outcomes, as well as disparities in access across and within socio-economic groups and urban geographies (Harris, 2021; Spronk, 2010; Uitermark & Tieleman, 2021). In addition to these dimensions, there has been increased attention to understanding challenges to water access as it relates to colonial legacies and socio--spatial justice. In the United States, for instance, environmental justice scholars have investigated the intersection of systemic racism, poverty and water insecurity, prompting an entry point for thinking about water equity (Mueller & Gasteyer, 2021). Other critical perspectives have offered more nuanced framings of equity and fragmentation in urban water access and why it has proven to be more ingrained in cities in the South. In the Global South, water supply and access are emblematic of socio-economic disparity (Thommandru et al., 2025) and a 'splintered city', where affluent households connected to the public water network pay significantly lower rates per unit of water than the low-income and poor residents dependent on informal vendors (Graham & Marvin, 2002; Hoelzel, 2024).

While there is increasing interest in understanding the dimensions affecting water accessibility across the Global South, the situation in informal settlements in SSA, which increasingly house new migrants into expanding urban centres, remains relatively underexplored. These settlements are usually excluded from formal development plans, lack sufficient infrastructure and often struggle with water accessibility. In line with the global patterns, Harris (2021) highlights that limited access to potable and safe water in Sub-Saharan Africa is closely linked to financial constraints and inadequate housing arrangements, such as precarious land tenure. Many low-income residents and informal communities lack legal recognition in these contexts, preventing them from establishing formal utility connections. More recent contributions reinforce these insights, citing additional factors such as the inability of residents to pay for improved services, variations in cost and quality of water delivery, and the compounded marginalisation of poor households (Ablo & Etale, 2024; Bartels et al., 2018). These studies identify crucial dynamics that shape access, including who receives water, under what terms, and through which channels. This expanding body of literature deepens our understanding of contemporary challenges in pipe-borne and safe water supply and advances the broader discourse on equitable water access in Africa.

Although urban areas in Africa typically report higher water coverage rates than rural regions, rapid urbanisation has placed considerable pressure on already strained systems, exacerbating service delivery failures (Dongzagla, 2022). Efforts to extend water infrastructure have not kept pace with urban population growth, particularly in informal settlements, low-income communities and peripheral zones, where exclusion from piped networks remains widespread (Tetteh et al., 2022). These disparities are embedded within the spatial politics of urban governance, wherein formal planning and infrastructure investments often favour affluent neighbourhoods while marginalising low-income and unplanned communities. This reflects historically rooted institutional logics that privilege formality, centralised planning, and cost-recovery models that systematically disadvantage those unable to meet connection criteria or afford commercialised services (Bartels et al., 2018).

Ghana exemplifies these regional dynamics. Like many countries in the Global South, Ghana faces the persistent challenge of delivering equitable water services in rapidly growing urban areas. Water supply in cities often favours wealthier neighbourhoods, while low-income and peri-urban communities, often described as "outskirt neighbourhoods", experience limited access and erratic supply (Owusu & Kofi Teye, 2015; Tetteh et al., 2022). Cost recovery motive further exacerbates disparities, as utility providers deprioritise low-income zones deemed financially nonviable (Ablo & Etale, 2024). Multiple studies have identified various social, economic, and spatial factors that predict household access to improved water sources in Ghana, including wealth status, gender, education level, household size, and region (Akpakli et al., 2018; Dongzagla, 2022). Notably, the wealthiest households are at least sixteen times more likely to access improved water than the poorest (Dongzagla, 2022). While socio-economic analyses are critical, other scholars have drawn attention to the spatial dimensions of inequality, highlighting the enduring influence of planning regimes, post-independence governance agendas, and the structure of urban land markets (Babuna et al., 2023; Bartels et al., 2018). In addition to emphasising the heterogeneity of water access inequity, recent global climate change risks, such as flooding, further compound the challenges of urban water access in Ghana (Dzodzomenyo et al., 2022; Ntajal et al., 2022). This is due to pollution and damage to the water supply infrastructure.

Despite this rich and growing body of work, significant gaps remain in understanding how infrastructural and governance frameworks reproduce water access inequities in urban centres. Much of the literature has focused on socio-economic predictors or informal coping mechanisms, often sidelining the institutional and spatial configurations that shape access at the local scale (Frimpong et al., 2024; Hoelzel, 2024). Intra-urban inequalities in piped water access are particularly pronounced in cities like Accra, where poor urban neighbourhoods consistently report access levels below the national urban average of 30% (Obeng-Odoom, 2012). This landscape of uneven access is often attributed to urban sprawl and the rapid urbanisation trend, which obscures the core underpinning drivers. However, a deeper investigation of water governance failures, infrastructure neglect, and historical legacies remains a critically important area to explore. Addressing these gaps requires a reorientation of research and policy to foreground how governance systems and infrastructure deficits shape water's access and spatial distribution in cities. This chapter responds to this need by critically examining the nexus of water governance and infrastructural inequality in Greater Accra, Ghana's most urbanised region and administrative capital. Through a synthesis of existing literature, policy analysis, and empirical insights, the chapter analyses the persistent inequalities in urban water accessibility.

The primary objective of this chapter is to examine the barriers that hinder equitable water provision in Greater Accra. We do this by examining four interconnected barriers or dimensions: colonial legacies that shape fragmented water infrastructure, contemporary infrastructural deficits, privatisation and commodification of water services, and fragmented governance and regulatory mechanisms. Using these dimensions as analytical lenses, the chapter demonstrates how water inequity is structured, perpetuated and embedded in the fragmented urban landscape. We advance two central arguments. First, we argue that the multipronged dimensions of water inequity cannot be fully understood or addressed without engagement with deeper histories of colonial legacies and systemic fragmented water infrastructure and governance frameworks. Second, we emphasise that the multiple barriers of water access intersect and compound to shape lived realities of water inequity in a way that makes universal access elusive. In doing so, we aim to promote a transformative agenda for both research and practice that foregrounds water equity within the broader lens of structural injustices. Generally, the chapter directly contributes to SDG 6.1 (equitable access to safe and affordable drinking water) by offering an understanding of how structural injustices constrain progress toward equitable water access and thus necessitate integrated strategies for sustainable urban water management.

The chapter is organised as follows: first, we present the methodology of the study. In what follows, we highlight the water supply situation in the Greater Accra region. In the remainder of the paper, we discuss the barriers to water supply and how they structure inequitable water access, suggesting novel strategies for sustainable water management to advance equity in access and resilient urban waterscape.

4.2 Methods and Study Area

This chapter employs a review of the literature to examine water equity issues in urban Ghana (Snyder, 2019). The research process involved identifying and synthesising evidence from peer-reviewed journal articles, historical records, policy documents, institutional reports, and grey literature sources. This approach allowed for a holistic understanding of the complex factors influencing water accessibility and equity in water access in Ghana's urban centres. The literature search was conducted across major academic databases including Web of Science, Scopus, and Google Scholar. Additional materials were drawn from institutional repositories of key stakeholders, including the Ghana Water Company Limited (GWCL), the Ministry of Sanitation and Water Resources, and international development agencies working in Ghana's water sector. Key search terms for the secondary data included (a) urban water access, (b) water equity, (c) water governance, (d) water infrastructure, (e) historical legacies, and (f) water privatisation. The selection focused on publications relevant to urban water access in the Ghanaian context, with particular emphasis on studies in Greater Accra. Special attention was given to empirical studies documenting the lived experiences of marginalised, informal, and low-income communities about water access. Figure 4.2 presents the methodology flow chart for the chapter.

Data extraction focused on identifying key themes related to infrastructure development, governance mechanisms, water privatisation, and historical influences. This approach facilitated the identification of recurring patterns, which were used to analyse the barriers that shape inequitable water access. A thematic analysis of the reviewed literature was conducted based on the

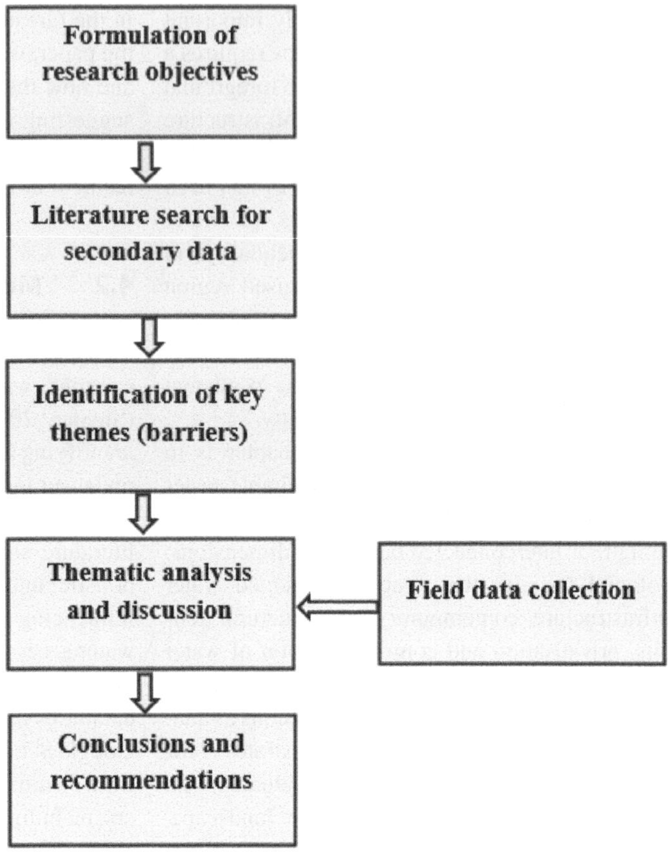

Fig. 4.2 Methodological flow chart

key themes and patterns identified. The thematic analysis uncovered four key barriers: colonial legacies, infrastructural deficits, privatisation, governance/regulatory mechanisms, and their implications on sustainable water management. Additionally, the discussions are also supported by observations in the field through extensive fieldwork in the informal settlement of Old Fadama, where the lead author conducted research as part of a larger research project. While the discussions in this chapter are grounded mostly in the review of academic and policy documents, it is significantly complemented by key observations and lived experiences of the urban poor at the margins.

4.2.1 Study Area: Greater Accra

The Greater Accra Region, located along the southern coast of Ghana, is the geographic focus of this chapter (Fig. 4.3). Some of the major communities in the Greater Accra region include Tema, Accra, Ridge, Ablekuma, Madina, Dodowa, Prampram, Weija, including prominent informal settlements such as Old Fadama, Nima, James Town and Ashaiman. Accra is the regional capital and serves as the administrative, political, and economic hub of Ghana. Spanning approximately 3245 square kilometres, Greater Accra is home to an estimated 5.4 million residents, accounting for roughly 18% of Ghana's total population, while recording one of the highest urbanisation rates in West Africa (Ghana Statistical Service (GSS), 2021). This rapid demographic expansion exerts increasing pressure on public services, particularly water infrastructure, aggravating pre-existing inequalities in access and reliability of supply. The urban landscape of Greater Accra is marked by an uneven juxtaposition of planned, affluent neighbourhoods and sprawling informal settlements, often

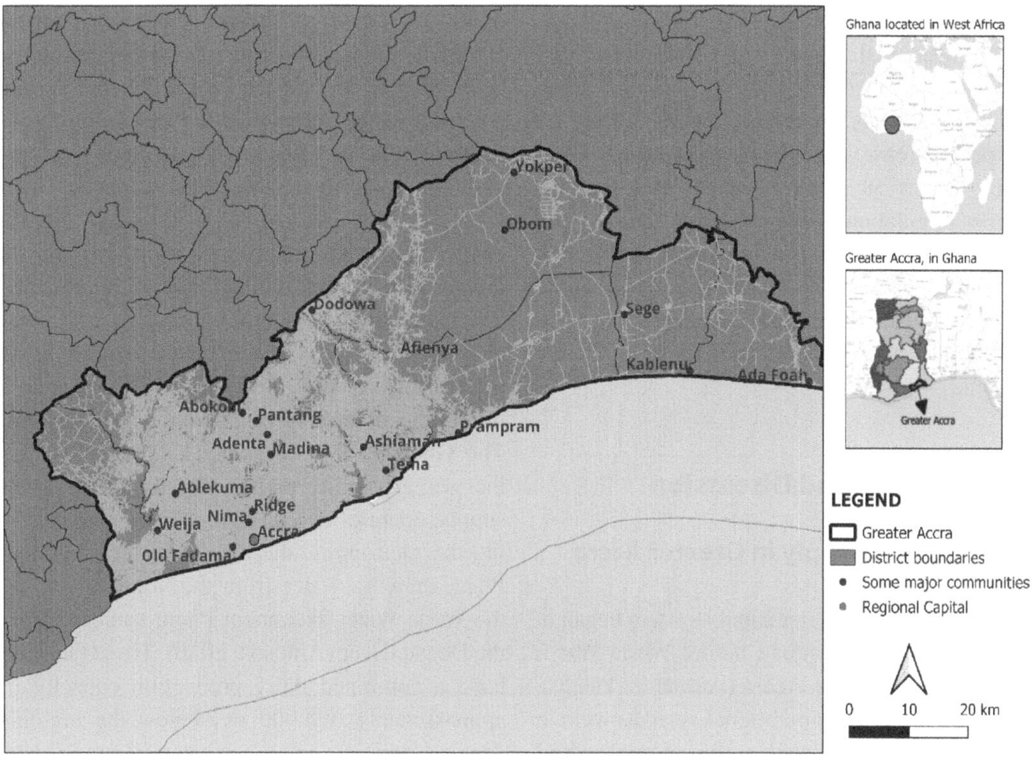

Fig. 4.3 The study area location. Source: Authors own construct

situated on the urban periphery or in interstitial spaces in city centres (Adamtey et al., 2021; Morrison, 2017). Accra is therefore home to a large informal proletariat excluded from formal wage labour and housing markets, thus compelled to establish self-organised systems of provision for survival. The region's urban morphology reflects both historical patterns of colonial planning and post-independence development logics, resulting in a patchwork of serviced and underserved areas. Compact land use, rapid commercial development, and real estate-driven expansion have outpaced infrastructural investments, especially in water supply systems (Frimpong et al., 2024). Consequently, differential access to potable water remains a defining feature of the urban experience across socio-economic and spatial divides.

According to the Ghana Statistical Service, piped water remains the second predominant source of drinking water in Greater Accra (Ghana Statistical Service (GSS), 2022). However, this statistic masks significant disparities in connection rates, service reliability, and affordability between neighbourhoods. While wealthier communities enjoy relatively consistent access to piped water, many low-income and peri-urban areas are characterised by limited or intermittent supply, forcing residents to rely on expensive and often lower-quality alternatives such as sachet water, tanker deliveries, and communal standpipes. The region's water governance structure is shaped by a pluralistic configuration of state agencies, private actors, and community-based initiatives. The Ghana Water Company Limited (GWCL) is the primary organisation responsible for providing and managing urban water services. Yet, its operations are complemented and often substituted by informal and semi-formal providers (Twum & Abubakari, 2020). These include water tanker associations, sachet water manufacturers, neighbourhood vendors, and local water boards. The multiplicity of actors reflects pragmatic responses to service delivery gaps and the institutional fragmentation that complicates regulation, accountability, and equitable service distribution. As a microcosm of Ghana's urban water challenges, Greater Accra offers a compelling case study for examining water equity issues in urban Ghana. The region's diverse population, governance structures, and water access patterns provide rich insights into the dynamics of water inequity and potential pathways toward integrated strategies for sustainable water management and more equitable outcomes.

4.3 Results and Discussion

4.3.1 Water Supply in Greater Accra

Ghana's first public water supply system began in the early twentieth century, before World War I, and was established in Accra (Amoah & Yahaya, 2013). By the 1920s, operational systems were in place in various cities, such as the colonial capital Cape Coast, Sekondi, Winneba, and Kumasi, the capital of the Ashanti kingdom. During this time, water infrastructure, including artificial reservoirs, pumps, and piped supply systems, was developed in commercial and residential areas where British officials lived (GWCL, 2016). These investments largely excluded the indigenous population, who lived in unplanned settlements with little or no access to formal water services. Efforts were made during the post-colonial era to expand the water supply in Accra, which therefore required some investment in infrastructure and service provision.

Following Ghana's independence in 1957, President Kwame Nkrumah's government made concerted efforts to expand water infrastructure across urban centres, seeking to redress colonial-era exclusions. International development agencies, including the World Health Organisation, supported these efforts and collaborated with the Ghanaian government to design a national water sector masterplan in the 1960s (UN-Special Fund/WHO, 1968). Substantial financial support from the World Bank and bilateral donors facilitated the extension of water networks into newly planned areas. Although the inequalities entrenched in the water supply system during the colonial period persisted (areas covered by the colonial network continued to have the best connections), the initiatives of postcolonial governments resulted in notable improvements (Duku et al., 2024). The network's coverage significantly increased, marking a transition toward greater infrastructural integration within the city. The water supply system in Greater Accra presents a complex and fragmented landscape characterised by formal utility services alongside a multitude of alternative provision mechanisms. The Ghana Water Company Limited (GWCL), the primary utility responsible for urban water supply, operates two major treatment plants serving the metropolis: the Kpong Water Treatment Plant, drawing water from the Volta River, and the Weija Water Treatment Plant, sourcing from the Densu River (Ainuson, 2010). These facilities have a combined daily production capacity of approximately 695,000 m^3, below the region's demand, and thus create a substantial supply deficit (Frimpong et al., 2024).

The production shortfall has led to the implementation of a water rationing system locally called "open-close," where different neighbourhoods receive water on alternating schedules. Even residents with piped connections cannot count on a consistent water supply, as interruptions continue due to the rationing of piped water (Stoler et al., 2012). According to Peloso and Morinville (2014) the rationing schedules often disadvantage lower-income neighbourhoods, with some experiencing water deliveries as infrequently as once a week or not at all. On the contrary, high-income neighbourhoods and commercial areas generally benefit from a more dependable supply, highlighting the inequities in service distribution. Premised on these, water users can be categorised into three groups: (a) those with consistent access to piped water (the privileged minority), (b) those with piped connections who face frequent interruptions and need supplementary water from private providers, and (c) those that depend entirely on private water providers due to the lack of piped connections (Amankwaa et al., 2014). The field observa-

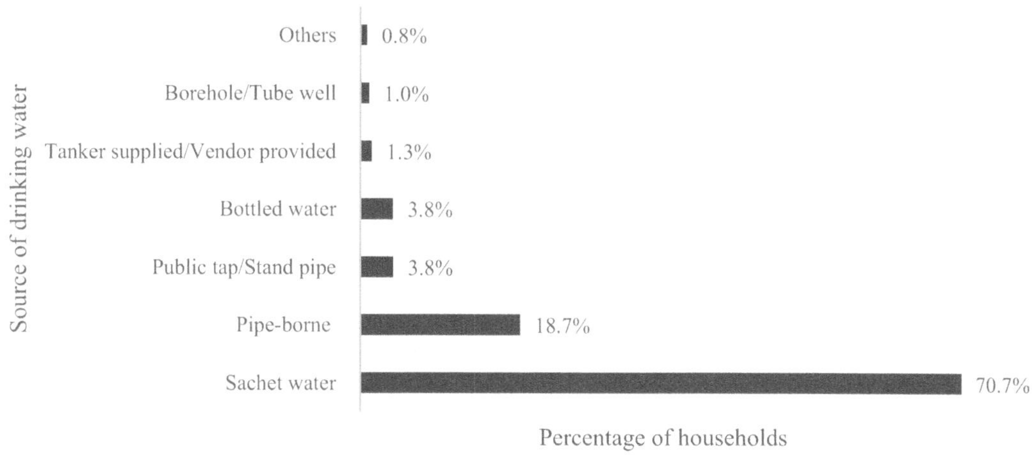

Fig. 4.4 Main source of drinking water in the Greater Accra Region. Source: Authors based on Ghana Statistical Service, 2022

tion in Old Fadama revealed a fourth group of water users: individuals, particularly the homeless, who lack access to any regular water source.

In response to water delivery shortages, residents have developed a range of coping strategies that have created the multiplicity and diversity in drinking water sources in Greater Accra. According to the Ghana Statistical Service (Fig. 4.4), the primary drinking water sources across Greater Accra reflect a diversity of supply mechanisms, including piped connections, tanker deliveries, sachet water, bottled water, boreholes, and communal standpipes. Notably, sachet water accounts for 70.7% of household drinking water sources, while pipe-borne water constitutes 18.7%. together making up over 80% of the primary drinking water supply. For example, the proliferation of sachet water, known locally as "pure water", has become a defining feature of urban water consumption in Accra. While sachet water provides a relatively accessible and portable source of drinking water, it raises serious concerns regarding quality control, cost, and environmental sustainability (Stoler et al., 2012). Additionally, water tanker services are critical to the urban water economy, especially in unserved areas. These services, typically organised through loosely regulated associations, supply water at rates often higher than regulated tariffs, imposing a disproportionate burden on poor households (Twum & Abubakari, 2020). This phenomenon, usually called the "poverty penalty," illustrates how residents with the least financial means end up paying the highest unit cost for water. This pluralistic landscape underscores the functional limits of formal utility provision. It reveals the extent to which informal and commodified alternatives have become institutionalised within the region's everyday practices of water access.

4.3.2 Factors Contributing to Inequity in Urban Water Access

4.3.2.1 Colonial Legacies: The Pathway to Fragmented Water Infrastructure

A critical understanding of equity in water access in Greater Accra necessitates an examination of the colonial legacies embedded within the city's urban infrastructure. Scholars argue that the socio-spatial configuration of water infrastructure and access in Accra bears the enduring imprint of colonial planning logics that instituted, normalised, and perpetuated urban inequality (Bohman, 2012; Harris, 2021). During the British colonial administration, residential and infrastructural segregation was systematically implemented, favouring administrative elites and

European residents while marginalising indigenous populations. The first piped water system in Accra, introduced in 1914, was designed to serve colonial administrative buildings, expatriate residences, and commercial establishments (Bohman, 2012). Private, metered connections were primarily reserved for European quarters, while African residents were relegated to communal standpipes or left to rely on wells and surface water (Shaloff, 1972). This initial setup created a pattern of spatial favouritism that continues in various ways today. As Kooy and Bakker (2008) contend, colonial water delivery systems were inherently "splintered," designed to inscribe privilege through infrastructural inclusion and exclusion spatially. Fanon (1963) conceptualisation of the "compartmentalised city" aptly captures this dynamic: colonial urbanism produced a city divided not only by race and class but also by differential access to basic services.

Records from the colonial Public Works Department show that between 1920 and 1950, investments in water infrastructure were concentrated in elite enclaves such as Ridge, Cantonments, and the Central Business District, areas demarcated for colonial administrators and commercial elites (Acquah, 1958). This early infrastructural footprint laid the groundwork for a stratified urban water system that postcolonial governments would inherit. Post-independence attempts to overcome this colonial infrastructure legacy have achieved mixed results. The immediate post-independence period saw ambitious water infrastructure projects, including the Kpong Water Works (Grönwall & Oduro-Kwarteng, 2018), which aimed to democratise water access. However, subsequent economic challenges and structural adjustment programs in the 1980s and 1990s constrained public investment in water infrastructure. Although Ghana's first independent administration under Kwame Nkrumah sought to expand access and dismantle colonial-era inequalities (Bohman, 2012), the spatial logic of infrastructure provision remained largely intact, favouring planned and economically viable zones.

In post-colonial contexts, infrastructural disparities often evolved from and perpetuated existing racial divides within urban waterscapes (Uitermark & Tieleman, 2021). The patterns of infrastructure development in contemporary settings align with the narrative of previously disjointed structures. Therefore, the historical evolution of the piped water system in Greater Accra should not be viewed as a transition from fragmentation to integration; rather, it represents a continuous lineage of historical inequalities in the city's water infrastructure. Similar critiques have been recently observed in other urban settings across the Global South. Neves-Silva et al. (2024) research in India, for instance, illustrates the role of historical legacies in contemporary water infrastructure, highlighting how the water supply network in colonial India was highly fragmented. During the fragmented water supply regime, the Europeans developed piped systems directly to their homes, while the natives relied on rivers and open channels in the city; legacies that persist today. Mathur and Mulwafu (2018) study also reinforces the impacts of colonial planning legacies on contemporary water accessibility. They found that water struggles and water inequality in postcolonial Malawi traced its roots to colonial legacies. Critical scholars have argued that decades of racialised urban planning, a defining feature in colonial governance, were central to these inequities (Smiley, 2020). Importantly, formally planned and wealthier neighbourhoods have historically and continue to attract disproportionate water infrastructure investment, thereby compounding enduring patterns of water injustices. When urban water supply systems are tied to city bifurcations, they often reinforce racial segregation and systemic inequity in access.

In scrutinising contemporary water access equity, it is vital to consider how historical legacies continue to shape modern practices and policies, reproducing inequality and exclusion in managing and distributing water resources at local scales. Recognising these is key to understanding the current challenges of water scarcity and unequal access to water infrastructure. Therefore, integrating strategies for water equity and sustainable water management in cities demands a critical engagement with past planning approaches, which continues to shape both access and governance in the present. This provides the point of departure for understanding and creating sustainable water management solutions in contemporary dispensations. The following section turns to the contemporary infrastructural deficits facing Greater Accra. We examine how this reflects and reproduces spatial injustice and inequity in the delivery of urban water.

4.3.2.2 Water Infrastructural Deficits in Contemporary Accra

The legacy of spatially uneven infrastructure established during the colonial and postcolonial periods is visibly reproduced in the current water supply architecture of Greater Accra. While historical logics explain the origins of fragmentation, contemporary infrastructural deficits exacerbate and entrench these disparities, particularly in informal settlements and peri-urban zones. Water infrastructure in Accra is characterised by chronic underinvestment, ageing networks, and limited extension into rapidly urbanising peripheral areas (Tetteh et al., 2022). This stems from a combination of factors, including limited public financing and political prioritisation. Investment tends to favour formally planned areas, while infrastructural development in informal settlements is deprioritised due to their lack of legal recognition, a phenomenon we observed in the Old Fadama informal community. This pattern aligns with what Graham and Marvin (2002) conceptualise as "splintered urbanism," where infrastructure is selectively distributed according to profitability, political salience, and spatial visibility. In Accra, this logic privileges central business districts and planned residential enclaves, such as those in the Tema Metropolitan Assembly, while systematically excluding low-income communities and informal neighbourhoods that lack planning approval or legal tenure.

Although formally mandated to provide urban water, the Ghana Water Company Limited (GWCL) operates within a cost-recovery framework that structurally disincentivises investment in marginal and low-income areas. As Frimpong et al. (2024) pointed out, GWCL's operational mandate is primarily technical, focusing on connections to main grids rather than household-level service extensions. Capital-intensive infrastructure development falls to the state, yet public authorities often lack the resources or political will to undertake such investments. In this context, access to infrastructure is mediated by residents' ability to finance their connections or lobby political representatives (Foli & Uitermark, 2024), a dynamic that systematically disadvantages poorer households. This approach differs from the United Nation normative framing of water as a fundamental human right, and limits Ghana's approach towards contributing to SDG six.

Formal connection to the GWCL grid is both bureaucratically and financially burdensome. The observations from Old Fadama informal settlement revealed that households are often required to finance the full suite of infrastructure costs, including pipeline procurement, excavation, backfilling, and administrative charges, before any service provision begins. Beyond these collective expenses, individual households must still pay separate connection fees. Some of the residents further emphasised that, despite bearing these costs, their status as an informal settlement and 'illegal occupants on government land prevents any connection to 'formal' water supplies. As a result, households rely on informal water vendors, who, through negotiations and lobbying with government officials, have secured access to piped water in the community. These layered financial demands and barriers to access render formal water access unattainable for many informal dwellers. Ainuson (2009) observes that nearly two-thirds of low-income residents in Accra lack piped water, compared to only 12% among wealthier households, a striking indicator of the spatial and class-based inequities in water access. As such, water infrastructure in Accra functions less as a universal public good and more as a differentiated commodity, shaped by market logics and uneven state commitments.

The existing piped water infrastructure in Greater Accra illustrates a stark pattern of unevenness. High-income, planned neighbourhoods benefit from relatively consistent service due to sustained investments, long-standing grid integration, and formal tenure arrangements, compared to marginalised and informal settlements, which are often viewed as temporary despite their permanence in the urban fabric (Alba et al., 2022). The case of the Tema Metropolitan Assembly is illustrative: originally conceived as a planned industrial city, Tema benefited from substantial post-independence investment in infrastructure through the Tema Development Corporation (Kirchherr, 1968). The result is a well-connected grid serving Communities 1 to 25, all of which have access to GWCL services. This "planning-before-development" model sharply contrasts with the experience of fast-growing, informal zones such as Ashaiman, James Town, Old Fadama and Nima, where housing has proliferated without corresponding investment in water systems. The spatial expansion of Greater Accra has consistently outstripped the pace of water infrastructure development, producing what Allen et al. (2006) term "peri-urban water poverty." The rapid urbanisation and lack of access to land in central Accra have pushed many to the outskirts. Rapid urbanisation along the city's periphery, particularly in areas such as Pantang, Abokobi and parts of Ashaiman, Adenta, and Madina, has unfolded without corresponding investments in basic water infrastructure (Bartels et al., 2018; Peloso & Morinville, 2014; Tetteh et al., 2022). The case of the Adentan Municipality is illustrative. Although much of its development occurred after the 1980s, the expansion was largely unregulated, with housing constructed outside the existing water network's coverage.

Similar to the peri-urban situation, many homes in informal settlements also lack access to piped water altogether, while others endure erratic supply. For instance, residents in Old Fadama often rely on unregulated and costly private water sources, such as sachet water and tanker services, due to the unavailability of piped water systems in their homes. The phenomenon of widespread disconnection from piped water infrastructure in homes is pervasive in African urban poor communities and informal settlements. Hoelzel (2024) & Smiley (2020) highlight how this plays out in low-income areas in Nigeria and Tanzania. They discuss how fragmented urban waterscapes influenced by poor piped connections in slums and urban poor communities lead to a dependence on informal water provision, such as tanker delivery services. The daily infrastructure provision for a piped connection remains the responsibility of residents. This has created a hegemony of informal water suppliers, often in poorer areas that struggle to connect to the formal piped network. The growing reliance on informal private water vendors, such as tanker services and sachet water, for instance, despite concerns regarding their quality (Ahiabor & Donkor, 2025) underscores the shortcomings of targeted investment in infrastructure to expand water services to underserved areas. Residents in Old Fadama informal settlement elucidated that the infrastructural deficit is further compounded by service rationing, with consumers receiving piped water as infrequently as once a week, or not at all.

The consequences of infrastructural fragmentation and deficits have exacerbated the economic burden on poor urban residents in accessing water. For example, residents often pay significantly higher per-unit prices for water delivered by tankers or sold by vendors in areas such as outskirt/peri-urban communities and low-income/informal settlements disconnected from the formal piped network. This phenomenon, sometimes described as the "poverty premium," means that the urban poor pay more for basic services than their wealthier counterparts. Ainuson (2010) estimated that residents in low-income Accra neighbourhoods pay up to three or four times the regulated price for water. Similar dynamics have been observed in other urban poor settlements in the Global South. Kashem et al. (2023), for example, reported significant disparities in Tejgaon slum in Bangladesh, where residents without a formal piped connection to their homes had to pay more than 17 times the regulated water price. Documenting these infrastructure inequities in water access demonstrates how it is crucial to reconceptualise water infrastructure as a site of justice. This demands governance frameworks and integrated strategies that prioritise equity, recognise the lived realities of marginalised communities, and challenge the market-driven logics that have long shaped urban water delivery.

4.3.2.3 Privatisation and Commodification of Water

Privatising water utilities has become widespread throughout the Global South, reflecting a broader ideological shift towards neoliberal, market-oriented governance. Neoliberalisation of water governance encompasses policies and practices that privilege economic markets and their associated mechanisms, including commodification, commercialisation, marketisation, and privatisation (Twum & Abubakari, 2020; Yates & Harris, 2018). Neoliberalism, by design, encourages the state to withdraw from its role as a direct provider of essential services, reconfiguring it instead as a facilitator of market dynamics. In the water sector, this withdrawal is frequently justified on the grounds of economic efficiency and fiscal discipline, leaving service provision to private actors who operate on the logic of profit rather than need (Rooney, 2024). Although proponents typically justify privatisation for enhanced efficiency and sustainability, empirical evidence suggests a more complex reality: private interests frequently capture benefits, often undermining the broader public good (Thommandru et al., 2025). As neoliberal paradigms deepen their influence on water governance, they intensify pre-existing disparities, depicting water delivery as a contested sphere of privatised authority and economic gain. This undermines the normative ideal of water as a public good and a universal human right.

The application of neoliberal dynamics is at the centre of controversies regarding water delivery and provision in Ghana (Twum & Abubakari, 2020). The core premise is that market mechanisms can deliver public services more efficiently than state-led models. This is evident in how international institutions, notably the World Bank, shape local governance structures by promoting privatisation, cost recovery initiatives, and public–private partnerships. Beginning in the late 1990s, under the pressure from international financial institutions, Ghana embarked on a path toward private sector participation in urban water services as part of broader structural adjustment programs (Whitfield, 2006). This trajectory culminated in the 2006 management contract between Ghana Water Company Limited (GWCL) and Aqua Vitens Rand Limited (AVRL), a joint venture between Dutch and South African companies which operated until 2011 when the contract was not renewed following performance controversies.

The privatisation experience produced mixed outcomes from an equity perspective in water delivery, often exacerbating rather than alleviating inequalities. Proponents anticipated that private sector efficiency would improve service delivery across all customer segments. However, empirical evidence suggests that service improvements disproportionately benefited areas with existing infrastructure and higher revenue potential (Twum & Abubakari, 2020). Amenga-Etego and Grusky (2005) document how performance-based incentives in the management contract encouraged the private operator to prioritise commercial loss reduction and revenue collection over service expansion to underserved areas, effectively marginalising low-income communities in the Greater Accra. Performance-based incentives within management contracts encouraged private operators to focus on cost recovery and revenue optimisation, rather than universal and equitable service delivery.

This model of market-led water governance illustrates what Lesch et al. (2023) critique as the displacement of social justice objectives in favour of economic efficiency. The commodification of water, framed as a tradable good subject to market forces, transforms access into a function of purchasing power rather than need or rights. In such systems, poor households are doubly penalised: excluded from formal infrastructure due to affordability constraints and forced to rely on informal providers who charge significantly higher rates. Evidence from Greater Accra illustrates this reality: water from private tankers or vendors can cost up to 20 times more than regulated tariffs, imposing a disproportionate burden on the poorest residents (Amankwaa et al., 2014). Commodification emerges as an intrinsic consequence of de facto privatisation, enabling private providers to determine prices based solely on market principles, rather than state-regulated tariffs. For example, our conversations with residents from the Old Fadama informal settlement revealed that significant price variability exists among private providers, reflecting individual profit motives rather than collective welfare considerations. We uncovered that significant price variability persists in this community, particularly during periods of piped water shortage, where private tanker delivery services charge exorbitant prices. Such disparities are symptomatic of weak regulatory oversight. Private water vendors operate outside the Public Utility Regulatory Commission (PURC) mandate, allowing water pricing to fluctuate according to market forces and providers' motives (Alba et al., 2022). What emerges is not simply a fragmented system but a form of water delivery injustice in which those (urban poor) unable to connect to formal piped water and least able to pay for water services are charged the most for water access.

The Ghanaian experience mirrors broader regional and global trends. In Johannesburg, South Africa, the implementation of prepaid water meters in impoverished townships following privatisation resulted in widespread service disconnections, protests, and legal actions against private water providers (Spronk, 2010). Similarly, Rooney's (2024) study in Dar es Salaam, Tanzania, demonstrated how privatisation resulted in poor quality of water supply, especially for the urban poor households. Given the poor performance of the private water company, the privatised service model eventually collapsed. Moreover, Bolivia's Cochabamba Water War exemplifies how the commodification and privatisation of municipal water systems, driven by international financial institutions, caused water tariffs to surge beyond the reach of many residents, triggering widespread social unrest (Tomba, 2023). Such cases underscore how commodification and privatisation exacerbate existing socio-economic inequalities and pose significant barriers to achieving sustainable, equitable, and universally accessible drinking water.

Ultimately, privatisation and commodification, though often justified by appeals to efficiency, have reinforced existing inequities in Accra's water landscape. They have institutionalised differential access while eroding the normative commitment to water as a public good and a human right. These findings imply that sustainable water management cannot rely exclusively on market mechanisms. This requires both regulatory reforms and a fundamental rethinking of the principles underpinning urban water governance. Sustainable water management and resilient urban waterscapes demand an integrated governance strategy that combines equitable public investment, community co-management and strong regulatory frameworks to ensure inclusive and affordable services. Reasserting equity, justice, and collective responsibility at the heart of policy can help imagine a more inclusive, equitable and resilient urban waterscape to foster effective solutions and sustainable urban water management.

4.3.2.4 Water Governance and Regulatory Mechanisms

The predominant factors shaping water inequities discussed in the previous sections, whether stemming from colonial legacies, infrastructural fragmentation and deficits, or neoliberal privatisation, are sustained and often exacerbated by Ghana's water governance architecture. In Greater Accra, water governance is marked by institutional complexity, regulatory fragmentation, and operational ambiguities, which together undermine the pursuit of equity and efficiency in service delivery (Frimpong et al., 2021). Despite multiple policy reforms and organisational mandates, the governance system remains beset by coordination failures, blurred institutional boundaries, and weak accountability mechanisms.

The formal governance architecture comprises multiple actors, including the Ministry of Sanitation and Water Resources, Ghana Water Company Limited (GWCL), the Public Utilities Regulatory Commission (PURC), the Water Resources Commission (WRC), and various local government authorities. While each institution is tasked with specific mandates, policy formulation, service delivery, regulation, or resource management, their roles often overlap in practice (Eduful, 2024). For example, tensions exist between the regulatory authority of the PURC and that of the WRC, creating regulatory overlaps that obstruct coherent policy implementation. Additionally, jurisdictional ambiguities persist between the GWCL and the Community Water and Sanitation Agency, particularly in peri-urban contexts where settlement typologies defy precise categorisations (Ainuson, 2010). These institutional overlaps portend administrative inconveniences and translate into complex inefficiencies in connection and delivery. The process is cumbersome and bureaucratically burdensome for residents seeking to connect to the GWCL grid. Application procedures typically require formal documentation, such as land title, building permits, or legally recognised addresses (GWCL, 2023), a requirement that many residents in informal settlements are unable to provide. This procedural barrier operates as a de facto exclusion mechanism, denying access to already structurally marginalised populations.

The regulatory environment itself also fails to address fundamental equity concerns. Public Utility Regulatory Commission, responsible for setting tariffs and protecting consumer interests, has often prioritised financial sustainability over justice considerations. Although its tariff structures, designed with lifeline provisions for low-income consumers, primarily benefit households already connected to the grid. Those outside the formal network, who often rely on informal vendors, remain subject to unregulated, volatile, and frequently exploitative pricing regimes. This regulatory gap reinforces inequity in water access by institutionalising a two-tiered water economy: one governed by public standards, and another shaped by market opportunism.

Policy frameworks have not been entirely silent on equity. The 2007 National Water Policy acknowledges water as a human right and commits to improving service provision in underserved areas. However, political expediencies and fiscal constraints have limited the policy's operationalisation. A persistent gap exists between policy commitments and the actual implementation of inclusive water access strategies (Morinville, 2017). In practice, cost recovery and service efficiency often override considerations of affordability and universal access, particularly under conditions of constrained public financing. The result has been a governance landscape in which responsibility is dispersed, accountability is diluted, and equity remains subordinated to fiscal discipline. For many urban residents, especially those in informal or peri-urban areas, navigating this landscape involves negotiating multiple actors, uncertain procedures, and inconsistent pricing standards. As a result, water governance becomes not a means of guaranteeing access, but a site of everyday struggle, exclusion, and conditional belonging.

To improve water equity across Greater Accra, governance reform must move beyond superficial coordination to embrace more profound critical changes for effective and sustainable water management. This includes strengthening regulatory oversight of informal providers and embedding participatory mechanisms that elevate the voices of marginalised communities. This is of particular importance in the current water management conjuncture, in which water supply governance have been blamed for inequity of access and supply. Without such reforms, urban water governance will continue to reproduce rather than resolve the inequities it is ostensibly designed to overcome.

4.4 Conclusions

This chapter has examined the multifaceted dimensions of water access equity in urban Ghana, focusing on the Greater Accra. The analysis reveals that water access disparities are manifestations of deeper socio-political dynamics that systematically disadvantage low-income and informal settlements. The enduring colonial blueprint of infrastructure segregation, the selective expansion under postcolonial regimes, and the recent turn toward neoliberal water governance have all produced a deeply stratified urban waterscape, where spatial, economic, and social lines increasingly shape access. At the heart of these dynamics lies a failure of water governance. Rather than acting as a redistributive force, Ghana's water governance system has often reinforced existing inequities. Regulatory bodies have struggled to reconcile cost-recovery imperatives with social equity goals. At the same time, the overlapping mandates of multiple institutions have produced a landscape of weak accountability and regulatory oversight. This governance vacuum has allowed informal and private actors to proliferate, many of whom operate with minimal supervision and pass exorbitant costs onto the most vulnerable populations.

Moreover, water commodification, fuelled by privatisation and infrastructural neglect, has transformed water from a public good into a market-controlled asset. In this dual economy, those with secure tenure and formally planned high-income areas access subsidised piped water, while low-income and peripheral communities navigate a precarious patchwork of vendors, tankers, and sachets, paying higher prices for lower quality (increased risk of disease) and unreliable supply. The irony is acute: those least able to afford water are compelled to pay the most, while formal governance mechanisms remain structurally indifferent to their plight.

The implications of these analysis are both empirical and conceptual. Empirically, it highlights the need to go beyond service delivery metrics and interrogate the political economies of infrastructure, policy, and spatial planning that produce unequal outcomes. Conceptually, it demands a shift from technocratic understandings of water access to an integrated strategies rooted in justice, rights, and citizenship to foster resilient urban waterscapes and a sustainable water management. Unequal access to water is not just an infrastructural deficit issue; it is a denial of urban belonging, the praxis of spatial exclusion, and a reflection of whose needs are recognised by the state. The findings underscore that addressing water governance and infrastructural inequities is crucial for realising the transformative agenda of equitable urban water access. This affirms that achieving universal access to safe drinking water, along with adequate sanitation and hygiene is essential for a resilient and sustainable urban waterscapes and directly contributes to SDG 6.1 and 6.2.

In light of these findings, a new paradigm of urban water governance is needed, one that explicitly centres equity, reclaims water as a social good, and foregrounds the voices and agency of marginalised communities. This involves transitioning from narrow cost-efficiency models to hybrid governance arrangements that combine public accountability with community-led solutions. Thommandru et al. (2025), for instance, propose participatory and locally informed strategies that prioritise community empowerment, equity and resilience in addressing water insecurity, while Rooney (2024) advocates for state-led development models that prioritise public collaboration and equitable resource distribution. Successful examples across the Global South, such as remunicipalisation in Jakarta, Indonesia (Marwa, 2019), community engagement in water management in Bangkachao, Thailand (Kittiprapas & Daniere, 2025) and participatory and community voices on water quality and access in Makhanda, South Africa (Khumalo et al., 2024), demonstrate that decentralised and rights-based approaches are not only viable but necessary for integrating effective strategies into sustainable water management.

These shifts would also require regulatory reform to incorporate affordability into tariff structures, public investment in neglected areas, formal recognition of informal settlements, and the integration of grassroots knowledge into planning processes. Equally important is the need to historicise reform efforts: recognising how colonial and postcolonial urbanisms have shaped contemporary disparities and ensuring that current interventions do not inadvertently reproduce these patterns. The challenge and the opportunity lie in transforming the water governance from a system that dismantles privilege and fragmentation to one that guarantees dignity, equity, and shared resilience. As research in this field continues to advance, there is significant value in the engagement of urban water policy scholars with frameworks such as historical legacies, political ecologies of water infrastructure and the fragmented water governance. This integration helps situate contemporary urban water issues within critical theoretical thoughts and broadens the analytical perspective to encompass the deeper histories and underlying causes of water equity. Such engagements, which accommodate the overlapping drivers of water inequity, lay a strong foundation for intersectional and integrated approaches to urban water governance and sustainable water management. They also provide an empirical basis for advancing more equitable and sustainable urban water futures, to drive resilient urban waterscapes.

Chapter Abbreviations The following abbreviations were used in this chapter:

AVRL	Aqua Vitens Rand Limited
GoG	Government of Ghana
GSS	Ghana Statistical Service
GWCL	Ghana Water Company Limited
PURC	Public Utility Regulatory Commission
SDG	Sustainable Development Goals
SSA	Sub-Saharan Africa
UN	United Nations
UNDP	United Nations Development Programme
UNICEF	United Nations Children's Fund
WRC	Water Resources Commission
WHO	World Health Organisation

Funding The research was supported by the Wellington Doctoral Scholarship at the Victoria University of Wellington, Faculty of Architecture and Design Innovation, New Zealand.

Potential Conflicts of Interest The authors declare that there is no conflict of interest.

Informed Consent Informed consent was obtained from all individual participants included in the study.

Authors' Contribution All authors have contributed equally to this work.

Data Availability Statement No new data were created.

References

Abdulhadi, R., Bailey, A., & Van Noorloos, F. (2024). Access inequalities to WASH and housing in slums in low- and middle-income countries (LMICs): A scoping review. *Global Public Health, 19*(1), 1–20. https://doi.org/10.1080/17441692.2024.2369099

Ablo, A. D., & Etale, A. (2024). Beyond technical: Delineating factors influencing recycled water acceptability. *Urban Water Journal, 21*(8), 1014–1021. https://doi.org/10.1080/1573062X.2022.2155847

Acquah, I. (1958). *Accra survey: A social survey of the Capital of Ghana, formerly called the Gold Coast, undertaken for the West African Institute of Social and Economic Research, 1953–1956*. University of London Press.

Adamtey, R., Mensah, J. V., & Obeng, G. (2021). Making cities resilient in Ghana: The realities of slum dwellers that confront the Accra metropolitan assembly. *Journal of Sustainable Development, 14*(1), 70–83. https://doi.org/10.5539/jsd.v14n1p70

Ahiabor, W. K., & Donkor, E. S. (2025). Microbial safety of sachet water in Ghana: A systematic review. *EnvironmentalHealthInsights, 19*, 11786302241307830. https://doi.org/10.1177/11786302241307830

Ainuson, K. (2009). Ensuring adequate water supply to disadvantaged urban communities in Ghana. https://tigerprints.clemson.edu/all_dissertations.

Ainuson, K. (2010). Urban water politics and water security in disadvantaged urban communities in Ghana. *African Studies Quarterly, 11*, 60–82. http://www.africa.ufl.edu/asq/v11/v11i4a4.pdf

Akpakli, D. E., Manyeh, A. K., Akpakli, J. K., Kukula, V., & Gyapong, M. (2018). Determinants of access to improved sanitation facilities in rural districts of southern Ghana: Evidence from Dodowa health and demographic surveillance site. *BMC Research Notes, 11*(1), 473. https://doi.org/10.1186/s13104-018-3572-6

Alba, R., Kooy, M., & Bruns, A. (2022). Conflicts, cooperation and experimentation: Analysing the politics of urban water through Accra's heterogeneous water supply infrastructure. *Environment and Planning E: Nature and Space, 5*(1), 250–271. https://doi.org/10.1177/2514848620975342

Allen, A., Dávila, J. D., & Hofmann, P. (2006). The peri-urban water poor: Citizens or consumers? *Environment and Urbanization, 18*(2), 333–351. https://doi.org/10.1177/0956247806069608

Alum, E. U., Obeagu, E. I., & Ugwu, O. P. C. (2024). Enhancing quality water, good sanitation, and proper hygiene is the panacea to diarrhea control and the attainment of some related sustainable development goals: A review. *Medicine, 103*(38), e39578. https://doi.org/10.1097/MD.0000000000039578

Amankwaa, F. E., Owusu, A. B., Owusu, G., & Eshun, F. (2014). Accra's poverty trap: Analysing water provision in urban Ghana. *Journal of Social Science for Policy, 2*(2), 69–89.

Amenga-Etego, R. N., & Grusky, S. (2005). The new face of conditionalities: The World Bank and water privatization in Ghana. In D. McDonald & G. Ruiters (Eds.), *The age of commodity: Water privatization in Southern Africa* (pp. 275–290). Earthscan.

Amoah, S. T., & Yahaya, A.-K. (2013). Evolution of water systems and its challenges in the Wa municipal of Ghana. *Journal of Environment and Earth Science, 3*(7), 15–24. https://www.researchgate.net/publication/260105667

Ando, H., Kitajima, M., Oki, T., & Murakami, M. (2025). Advancements in global water and sanitation access (2000–2020). *Scientific Reports, 15*(1), 6399. https://doi.org/10.1038/s41598-025-90980-7

Babuna, P., Yang, X., Tulcan, R. X. S., Dehui, B., Takase, M., Guba, B. Y., Han, C., Awudi, D. A., & Li, M. (2023). Modeling water inequality and water security: The role of water governance. *Journal of Environmental Management, 326*, 116815. https://doi.org/10.1016/j.jenvman.2022.116815

Bartels, L. E., Bruns, A., & Alba, R. (2018). The production of uneven access to land and water in peri-urban spaces: De facto privatisation in Greater Accra. *Local Environment, 23*(12), 1172–1189. https://doi.org/10.1080/13549839.2018.1533932

Bohman, A. (2012). The presence of the past: A retrospective view of the politics of urban water management in Accra, Ghana. *Water History, 4*(2), 137–154. https://doi.org/10.1007/s12685-011-0047-2

Dongzagla, A. (2022). Socioeconomic and demographic factors affecting urban household access to improved water and sanitation in Ghana. *GeoJournal, 87*(6), 4763–4773. https://doi.org/10.1007/s10708-021-10534-y

Duku, G. A., Appiah-Effah, E., Gyamfi, C., & Nyarko, K. B. (2024). Ghana's water safety journey: A review of efforts toward a risked-based water quality management. *Journal of Water and Health, 22*(12), 2370–2384. https://doi.org/10.2166/wh.2024.371

Dzodzomenyo, M., Asamoah, M., Li, C., Kichana, E., & Wright, J. (2022). Impact of flooding on microbiological contamination of domestic water sources: A longitudinal study in northern Ghana. *Applied Water Science, 12*(10), 1–10. https://doi.org/10.1007/S13201-022-01757-6

Eduful, M. (2024). Toward good governance in water resources management in Ghana. *Natural Resources Forum, 48*(2), 485–507. https://doi.org/10.1111/1477-8947.12324

Fanon, F. (1963). *The wretched of the earth*. Grove Press. https://monoskop.org/images/6/6b/Fanon_Frantz_The_Wretched_of_the_Earth_1963.pdf

Foli, A., & Uitermark, J. (2024). The politics of drains: Everyday negotiations of infrastructure imaginaries in Accra. *Urban Studies, 61*(11), 2099–2117. https://doi.org/10.1177/00420980241230432

Frimpong, J., Adamtey, R., Pedersen, A. B., Wahaga, E., Jensen, A., Obuobie, E., & Ampomah, B. (2021). A review of the design and implementation of Ghana's National Water Policy (2007). *Water Policy, 00*(02), 1170–1188. https://doi.org/10.2166/wp.2021.042

Frimpong, L. K., Mensah, S. L., & Ablo, A. D. (2024). Households' access and expenditure on water services: Examining intra-urban differences in the Accra metropolis, Ghana. *Urban Governance, 4*, 222–231. https://doi.org/10.1016/j.ugj.2024.07.002

Fuhrmeister, E. R., Harvey, A. P., Nadimpalli, M. L., Gallandat, K., Ambelu, A., Arnold, B. F., Brown, J., Cumming, O., Earl, A. M., Kang, G., Kariuki, S., Levy, K., Pinto Jimenez, C. E., Swarthout, J. M., Trueba, G., Tsukayama, P., Worby, C. J., & Pickering, A. J. (2023). Evaluating the relationship between community water and sanitation access and the global burden of antibiotic resistance: An ecological study. *The Lancet Microbe, 4*(8), e591–e600. https://doi.org/10.1016/S2666-5247(23)00137-4

Ghana Statistical Service (GSS). (2021). *2021 population and housing census publications*. Ghana Statistical Service.

Ghana Statistical Service (GSS). (2022). *Ghana 2021 population and housing census, general report volume 3M: Water and sanitation*. Ghana Statistical Service.

Graham, S., & Marvin, S. (2002). *Splintering urbanism*. Routledge. https://doi.org/10.4324/9780203452202

Grönwall, J., & Oduro-Kwarteng, S. (2018). Groundwater as a strategic resource for improved resilience: A case study from peri-urban Accra. *Environmental Earth Sciences, 77*(1), 6. https://doi.org/10.1007/s12665-017-7181-9

GWCL. (2016). *History of water supply in Ghana*. Ghana Water Company Limited.

GWCL. (2023). *Standard operating procedure for distribution operations*. Ghana Water Company Limited.

Harris, L. M. (2021). Everyday experiences of water insecurity: Insights from underserved areas of Accra, Ghana. *Daedalus, 150*(4), 64–84. https://doi.org/10.1162/daed_a_01873

Hoelzel, F. (2024). Water supply systems in urban slum communities in Lagos, Nigeria: Between self-supply and co-production. *The International Journal of Community and Social Development, 6*(1), 27–55. https://doi.org/10.1177/25166026241231434

Jambadu, L., Dongzagla, A., & Kabange, I. (2023). Understanding intra-urban inequality in networked water supply in Wa, Ghana. *GeoJournal, 88*(1), 841–857. https://doi.org/10.1007/s10708-022-10662-z

Kashem, S., Tahsin, N., Subah, Z., Murshed, S. B., Nowreen, S., & Mondal, M. S. (2023). Assessing the right to water of the urban poor in Dhaka city. *GeoJournal, 88*(3), 3183–3204. https://doi.org/10.1007/S10708-022-10804-3

Khumalo, L., Mickelsson, M., Fogel, R., Mutingwende, N., Madikiza, L., & Limson, J. (2024). Progressing from science communication to engagement: Community voices on water quality and access in Makhanda. *Sustainability (Switzerland), 16*(1), 459. https://doi.org/10.3390/SU16010459

Kirchherr, E. C. (1968). Tema 1951-1962: The evolution of a Planned City in West Africa.

Urban Studies, 5(2), 207–217. https://doi.org/10.1080/00420986820080281
Kittiprapas, S., & Daniere, A. (2025). Community engagement in water management to enhance sustainability: A case study of Bangkachao, Thailand. *Local Development and Society, 6*(2), 384–415. https://doi.org/10.1080/26883597.2025.2480592
Kooy, M., & Bakker, K. (2008). Splintered networks: The colonial and contemporary waters of Jakarta. *Geoforum, 39*(6), 1843–1858. https://doi.org/10.1016/j.geoforum.2008.07.012
Leahy, W., Achore, M., & Dery, F. (2024). "In the dry season, our suffering doubles": Barriers to water access in poor urban settlements in Ghana. *PLOS Water, 3*(9), e0000265. https://doi.org/10.1371/journal.pwat.0000265
Lesch, D., Miörner, J., & Binz, C. (2023). The role of global actors in sustainability transitions–Tracing the emergence of a novel infrastructure paradigm in the sanitation sector. *Environmental Innovation and Societal Transitions, 49*, 100787. https://doi.org/10.1016/j.eist.2023.100787
Marwa, M. (2019). Changing water regime: Remunicipalization in Indonesia under the global hegemony of privatization. *Global South Review, 1*(1), 6. https://doi.org/10.22146/globalsouth.33131
Mathur, C., & Mulwafu, W. (2018). Colonialism and its legacies, as reflected in water, incorporating a view from Malawi. *WIREs Water, 5*(4), e1287. https://doi.org/10.1002/wat2.1287
Morinville, C. (2017). Sachet water: Regulation and implications for access and equity in Accra, Ghana. *WIREs Water, 4*(6), e1244. https://doi.org/10.1002/wat2.1244
Morrison, N. (2017). Struggling for the right to be recognized: The informal settlement of old fadama, Accra, Ghana. In *Geographies of forced eviction: Dispossession, violence, resistance* (pp. 25–45). Palgrave Macmillan. https://doi.org/10.1057/978-1-137-51127-0_2
Mueller, J. T., & Gasteyer, S. (2021). The widespread and unjust drinking water and clean water crisis in the United States. *Nature Communications, 12*(1), 1–8. https://doi.org/10.1038/s41467-021-23898-z
Neves-Silva, P., Aires Gomes, M. C., & Heller, L. (2024). Talks about privatization of water and sanitation: A critical discourse analysis of contributions to a UN report. *Water Security, 23*, 100179. https://doi.org/10.1016/j.wasec.2024.100179
Ntajal, J., Höllermann, B., Falkenberg, T., Kistemann, T., & Evers, M. (2022). Water and health nexus—Land use dynamics, flooding, and water-borne diseases in the Odaw River basin, Ghana. *Water, 14*(3), 461. https://doi.org/10.3390/W14030461
Obeng-Odoom, F. (2012). Beyond access to water. *Development in Practice, 22*(8), 1135–1146. https://doi.org/10.1080/09614524.2012.714744
Owusu, K., & Kofi Teye, J. (2015). Supplementing urban water supply with rainwater harvesting in Accra, Ghana. *International Journal of Water Resources Development, 31*(4), 630–639. https://doi.org/10.1080/07900627.2014.927752
Peloso, M., & Morinville, C. (2014). Chasing for water': Everyday practices of water access in Peri-urban Ashaiman, Ghana. *Water Alternatives, 7*(1), 121–139.
Rooney, E. (2024). Failed privatisation in urban water utilities: Can PuPs pick up the pieces? Reviewing evidence from Dar es Salaam, 2005–2018. *Geoforum, 152*, 103998. https://doi.org/10.1016/J.GEOFORUM.2024.103998
Shaloff, S. (1972). The Gold Coast water rate controversy 1909-1938. *Institute of African Studies Research Review, 8*(3), 21–34.
Smiley, S. L. (2020). Heterogeneous water provision in Dar es Salaam: The role of networked infrastructures and alternative systems in informal areas. *Environment and Planning E: Nature and Space, 3*(4), 1215–1231. https://doi.org/10.1177/2514848620908194
Snyder, H. (2019). Literature review as a research methodology: An overview and guidelines. *Journal of Business Research, 104*, 333–339. https://doi.org/10.1016/j.jbusres.2019.07.039
Spronk, S. (2010). Water and sanitation utilities in the global south: Re-centering the debate on "efficiency". *Review of Radical Political Economics, 42*(2), 156–174. https://doi.org/10.1177/0486613410368389
Sprouse, L., Lebu, S., Nguyen, J., Muoghalu, C., Uwase, A., Guo, J., Baldwin-SoRelle, C., Anthonj, C., Simiyu, S. N., Akudago, J. A., & Manga, M. (2024). Shared sanitation in informal settlements: A systematic review and meta-analysis of prevalence, preferences, and quality. *International Journal of Hygiene and Environmental Health, 260*, 114392. https://doi.org/10.1016/J.IJHEH.2024.114392
Stoler, J., Weeks, J. R., & Fink, G. (2012). Sachet drinking water in Ghana's Accra-Tema metropolitan area: Past, present, and future. *Journal of Water, Sanitation and Hygiene for Development, 2*(4), 223–240. https://doi.org/10.2166/washdev.2012.104
Tetteh, J. D., Templeton, M. R., Cavanaugh, A., Bixby, H., Owusu, G., Yidana, S. M., Moulds, S., Robinson, B., Baumgartner, J., Annim, S. K., Quartey, R., Mintah, S. E., Bawah, A. A., Arku, R. E., Ezzati, M., & Agyei-Mensah, S. (2022). Spatial heterogeneity in drinking water sources in the Greater Accra Metropolitan Area (GAMA), Ghana. *Population and Environment, 44*(1–2), 46–76. https://doi.org/10.1007/s11111-022-00407-y
Thommandru, A., Turdialiev, M. A., & Mone, V. (2025). Hydro-hegemony in the Anthropocene: Neoliberal paradigms and global south marginalization in water scarcity governance. *Journal of Developing Societies, 41*(3), 383–405. https://doi.org/10.1177/0169796X251326720
Tomba, M. (2023). Social property in the Cochabamba water war, Bolivia 2000. *Angelaki, 28*(1), 73–86. https://doi.org/10.1080/0969725X.2023.2167785
Twum, K. O., & Abubakari, M. (2020). Drops in the city: The puzzle of water privatization and consump-

tion deficiencies in urban Ghana. *Water Policy, 22*(3), 417–434. https://doi.org/10.2166/wp.2020.175

Uitermark, J., & Tieleman, J. (2021). From fragmentation to integration and back again: The politics of water infrastructure in Accra's peripheral neighbourhoods. *Transactions of the Institute of British Geographers, 46*(2), 347–362. https://doi.org/10.1111/tran.12420

Un-Special Fund, & WHO (1968). Design phase of the water supply and sewerage works for the Accra-Tema Metropolitan Area, Ghana. https://www.ircwash.org/sites/default/files/824GH.AC68-539-I.pdf.

United Nations. (2010). Resolution adopted by the general assembly on 28 July 2010: 64/292. The human right to water and sanitation. In *Treaty series: Vol. I* (No. 9464).

United Nations. (2017). The sustainable development goals report. United Nations, New York, USA.

Whitfield, L. (2006). The politics of urban water reform in Ghana. *Review of African Political Economy, 33*(109), 425–448. https://doi.org/10.1080/03056240601000812

WHO, & UNICEF. (2023). *Progress on household drinking water, sanitation and hygiene 2000–2022: Special focus on gender*. https://washdata.org

Yates, J. S., & Harris, L. M. (2018). Hybrid regulatory landscapes: The human right to water, variegated neoliberal water governance, and policy transfer in Cape Town, South Africa, and Accra, Ghana. *World Development, 110*, 75–87. https://doi.org/10.1016/j.worlddev.2018.05.021

5 Integrating Wetlands into Urban Planning for Resilient Cities—Towards an Urban Autopoiesis

Kemo Usto and Lineker Max Goulart Coelho

Abstract

Given the climate change pressure on the cities, water infrastructure resilience is crucial in ensuring sustainable development. While there are known benefits to Nature-based Solutions (NbS), further research is needed to explore the multifunctionality of NbS and make them more desirable in urban planning. If cities are to become resilient, they need to approach an ecosystem-like dynamic of autopoiesis where the city can reconstitute its components in a balanced relation with its environment. This chapter explores how the NbS strategy of constructed wetlands can be planned in the city to induce a biomass yield that can supply the construction industry with the purpose of establishing a working framework for future work and research. This exploration is a multi-method abductive research approach that combines urban and architectural design methodology with water management science to explore how a different approach to urban planning could manifest if cities are to become resilient through self-making of building components. The chapter provides an experimental exploration of a spatial-temporal design strategy for how to plan and design a constructed wetland area for cultivating materials, increasing human livability, increasing biodiversity, other cultivation etc. Said experiment allows the development of a framework from which one can pursue the goals of Urban Autopoiesis through wetlands design and planning. The framework indicates that the many parameters are systemically entangled, and much focus should be put on using such a framework and design prototyping to establish a dialog with relevant stakeholders through a systems thinking approach.

Keywords

Autopoiesis · Urban design · Constructed wetlands · Built environments · Systems design

5.1 Introduction

Climate change is putting cities under pressure to deal with new emerging conditions which are to a large degree caused by the same urban areas and their material and energy demands and subsequent consumption. Cities are under pressure to not only develop design mechanisms that take these changes into account as to continue business as usual but also need to develop a resilient

K. Usto (✉) · L. M. G. Coelho
Department of Engineering Technology and Didactics, Technical University of Denmark, Ballerup, Denmark
e-mail: kemu@dtu.dk; linco@dtu.dk

U. Chatterjee, A. I. Almulhim (eds.), *Urban Waterscapes*, Sustainable Development Goals Series,
https://doi.org/10.1007/978-3-032-20649-7_5

transformative approach (Datola, 2023) towards envisioning innovative ecological ways of urban living with green and blue infrastructure. Urban Resilience is a complex notion and there are many definitions of resilience in urban contexts (Kapucu et al., 2024). The literature tends to highlight that resilience is the capacity to withstand perturbation and return to a previous state (Büyüközkan et al., 2022). While the conventional engineering approach is still functional, we need a more complex and interconnected understanding of resilience to design more complex networks with surplus benefits (Datola, 2023). Particularly, water is reported in the literature as a significant aspect of urban resilience (Du et al., 2020; Ebrahimi et al., 2022). In this context, considering SDG 6.5, which addresses an Integrated Water Resource Management (IWRM), it could be noted that this approach is crucial to strengthen sustainable urban water governance, as it enables at the same time having a systemic overview of existing issues as well as facilitates harmonized solutions. Moreover, IWRM also supports achieving SDG 6.6, as IWRM also helps to be more effective in terms of actions for the protection and restoration of water-related ecosystems such as rivers, lakes, and wetlands. Indeed, IWRM is essential to ensuring sustainability and the multi-dimensional benefits of urban water infrastructure, which includes resilience and climate adaptation.

The literature emphasizes flooding and excessive rainwater management but also highlights that urban waterscapes can provide wellbeing and health benefits for people. In extension of this, Nature-based Solutions (NbS) are a significant and relevant strand of urban resilience research and constructed wetlands are a well-suited solution for urban areas. We are familiar with the established benefits of constructed wetlands; recharge or enrichment of groundwater aquifers, control/amendment of flood incidents (protective buffers), trapping of sediments and other substances, absorption of carbon dioxide, heat storage and release, absorption of solar radiation and respective support to food chains (Dotro et al., 2017; Stefanakis et al., 2014; MEA, 2005; De Groot et al., 2006; Ghermandi et al., 2010). Furthermore, leading literature on NbS highlights the need for further exploring of multifunctionality of NbS systems (Tzoulas et al., 2021; Kabisch et al., 2022) as to further increase their desirability and increase their application in urban areas. Given that constructed wetlands can yield biomass which can be used as construction material (i.e. thatched roofs etc.), this chapter considers it pertinent to explore how urban wetlands, while providing recreative and wellbeing benefits, can also influence construction material supply chain by diversifying material procurement. Procurement and manufacturing of materials was not directly mentioned in the urban resilience literature reviews (Datola, 2023; Büyüközkan et al., 2022) but there was mentioned of urban resilience in relation to possible disruption of supply chains in case of extreme events but not explicitly regarding construction materials. Considering urban waterscapes, the challenge is not only to mitigate changes in the conventional engineering sense but also to develop a systemic design thinking approach to allow waterscapes to become regenerative and provide a variety of benefits for the city. Given that even green growth and circular design efforts in the building industry can cause rebound effects (Castro et al., 2022; Mavropoulos & Nilsen 2020; Andrew & Pigosso, 2024), there is an urgency to radically explore new solutions by cross-pollinating thinking from different fields and design the city's anatomies as a complex interrelated network of both material and immaterial considerations.

The theoretical background for this is the ecosystems research and in particular the concept of autopoiesis. Autopoiesis (Maturana & Varela, 1980) is any system's or organism's ability to reconstitute its own component. While it can be a "blind" mechanism that leads to growth, more recent variants of this idea (Upreti, 2024) entail that an organisms or systems must have an understanding of the environment they find themselves in as to be able to reconstitute their own components in the long term as well as short term. The challenge is thus to translate the ecosystemic concept to social city systems. Cities are large consumers of materials (Mavropoulos & Nilsen, 2020; Krausmann et al., 2018; Miljøstyrelsen, 2019) whose unbalanced dynamics between society and the hinterland pose challenges for biodiversity and ultimately challenge humanity's ability to sustain themselves in the future. Nevertheless, in approaching such challenges, there is a need for a systems understanding grounded in ecosystem principles (Kabisch, et al., 2022; Upreti, 2024). If we observe the city as a complex dynamic system, we can see that growth is a key dynamic that necessitates material consumption increase (Bettencourt, 2022) which causes emission rise and biodiversity loss, leading to poor and unsustainable land use schemes which in the end necessitate increasingly more mitigating measures for urban areas to be resilient. While this is the case of societal systems, all-natural physical systems demonstrate Negative feedback in their dynamics (Upreti, 2024)—this means that, despite any perturbations, it finds a meaningful balance between itself and its environment without permitting any exponential over-consumption and exploitation. And incidentally, when in balance, an ecosystem provides significant benefits (or ecosystem services) to human existence—ranging from aesthetic to utilitarian and physical (De Groot et al., 2002; Upreti, 2024). In extension, autopoiesis (Maturana & Varela, 1980; Upreti, 2024) is a key concept that is characteristic of natural systems which have the capacity to demonstrate a balanced and resilient relation between a particular organism or system with its environment. More precisely, the system and its environment co-evolve where the system in question develops behaviors that perpetuate the health of its environment so that the system itself can sustain its self-replicable existence within said environment (Fig. 5.1). This potentially forms the basis of a given system's capacity of sustainable resilience in wake of climate change, material shortages and overconsumption, continuing Greenhouse gases (GHG) emissions (Keeling curve) etc.

A constructed wetlands (CW) is a Nature-based Solution (NbS) and consists in an engineering system designed to replicate the process that naturally occurs in wetland environments (Vymazal, 2014; Santos et al., 2023). It is a consolidated technology used for decades in rural areas and small communities for pollutant removal from stormwater and wastewater (Yang et al., 2023; Coelho, 2024). Apart from technical and material benefits from NbS, there is likewise potential for significant immaterial and social benefits like well-being and quality of life dimension to introducing NbS in urban contexts (Cohen-Shacham et al., 2019; Kabisch et al., 2020; 2022; Buijs & Jacobs, 2021; Rao et al., 2022). In extension, given the complexity of differing parameters when dealing with NbS, the NbS literature highlights the need for systems approaches (Kabisch et al., 2022). This aspect is likewise an important consideration through the insight from ecosystemic thinking and research (Upreti, 2024)—particularly, if we are to link urban design and water management with autopoiesis. This entails that the manufacturing of components is made by the city itself, bringing production back to the city. Figure 5.2 presents a diagram in which the interfaces and relations of different fields of water ecosystems are represented, and how IWRM and urban design are related to them acting as an enabler towards SDG 6.5 and 6.6.

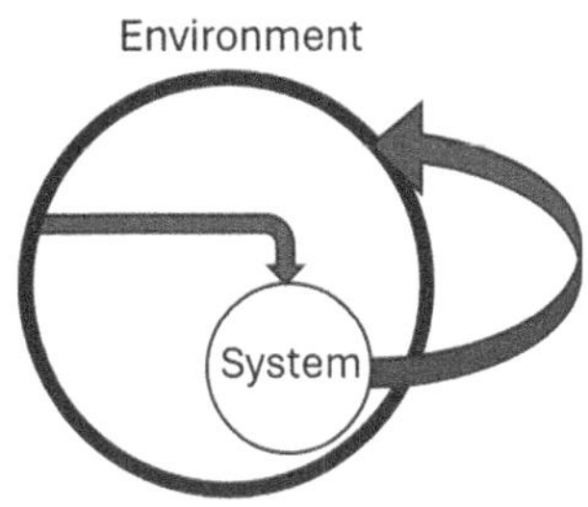

Fig. 5.1 Autopoietic relation between system and its environment

Building physics research and architectural technology research (CINARK, 2025) has come a long way and still developing the research and practice of sustainable organic construction materials. However, the production, procurement and acquiring of said materials is not integral to

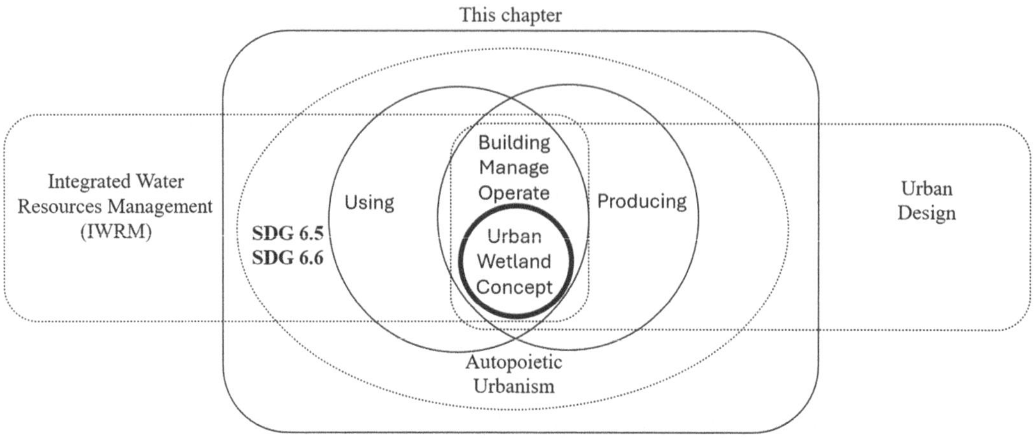

Fig. 5.2 Overall scope of the paper in relation to key fields in dealing with water ecosystems (SDG 6.5 and 6.6)

said research yet. In this regard, urban waterscapes can demonstrate and potentially increase the resilience of not only waterscapes themselves, but also the urban areas by proving construction materials alongside the well-established climate change mitigation and wellbeing affordances of urban wetlands. City planners and urban designers as well as water management engineers are not used to planning urban wetlands with the purpose of biomass harvesting. In this study, we will attempt to construct an appropriate framework for how to start thinking systematically about urban autopoiesis through constructed wetlands. This will be done qua an abductive a theoretical experiment (though related to real-life physical conditions) where a few key aspects will be tested with regard to the physical challenges and potential of designing constructed wetlands in urban areas through an (approximate or increasingly) autopoietic logic. So, we have the following Research question: can we create a circular autopoietic dynamic of an urban system? While the final realization of such an idealized design is far away, the main objective is to develop an approach and an appropriate framework in this trajectory. The chapter will explore the physical feasibility and potential plausibility of the proposed innovation as to map what considerations, design parameters, challenges and potentials need to be considered when exploring this trajectory.

5.2 Methods

5.2.1 Case Study Description and Experiment Conditions

The study of this chapter is situated in Denmark, more precisely in the municipality of Frederiksberg in Copenhagen, DK. The Municipality of Frederiksberg is already very ambitious with implementation of sustainable measures and strategies towards both mitigating climate effects and enhancing quality of life for its citizens and visitors. The municipality likewise harbors a significant archival dataset of urban conditions (Frederiksberg Kommune, 2024a) which would aid the design process of the urban wetlands of this chapter. Figure 5.3 shows a map of Frederiksberg Municipality.

As the purpose of the experiment of this chapter is to explore the approach and the potential of the urban wetlands as an autopoietic system, the experiment will limit its scope to exploring spatial design schemes in relation to the introduction of wetlands and biomass cultivation in urban areas in a smaller scale chosen spatial boundary and not include the whole municipality. Frederiksberg is a dense urban area and it is relatively difficult to simply propose the place of cultivation sites, even with the study being a desktop study. In choosing the site to conduct the experiment, several aspects have been considered.

Fig. 5.3 Frederiksberg municipality. Source: KlimaDatastyrelsen

Given that this is a desktop experiment, the choice of the specific site is less important compared to establishing a methodology approach and exploring the design potentials and limitations in the ambition of exploring and potentially developing a systems way of thinking about urban settlements and their autopoietic trajectory. For a practical reason, more generally, Frederiksberg was chosen not only because of their ambitions about sustainability, but also due to the fact that the municipality has much publicly available data of the municipality. This meant that we got the 3D of the building stock of the whole municipality and likewise the web-based archive of up to date very different data layer of the municipality (Frederiksberg Kommune, 2024b). This was considered important very early in the process given the serendipitous nature of the experiment of this chapter. Frederiksberg is situated in the proximity of a large lake (Damhussøen) and the fjord, which separates mainland Copenhagen with Amager island. The green area indicates the watershed, i.e. the extension of a real streaming towards the chosen point. While the above provided general considerations, a more specific design site had to be chosen for the distribution 0.4 hectare of cultivation area had to be composed in the municipality. To this end, a very rudimentary consideration was that the location of the area could be defined strategically to be supplied by water. Figure 5.4 shows where the chosen site is, and that it in fact is located downstream of the general water flow.

The green area (Fig. 5.4) indicates the area which collects water which eventually flows through the southern part in the vicinity of the design site before entering a canal to then run off into the main sink (fjord). The site for the experiment is chosen from the basis of wanting to relieve the existing sinks, but at the same time use the existing water flow from larger areas to be able to provide stable water level for cultivation purposes of biomass in the constructed wetlands. The Urban Wetlands Concept can as such be used to slow water from flowing too quickly to the main sinks, and at the same time provide water for cultivation.

Having to choose the specific area, a simple consideration contingent to this experiment was done. Given that the outset of the exploration was to organize 0.4 hectare area meant for cultivation meant that we had to choose an area which is

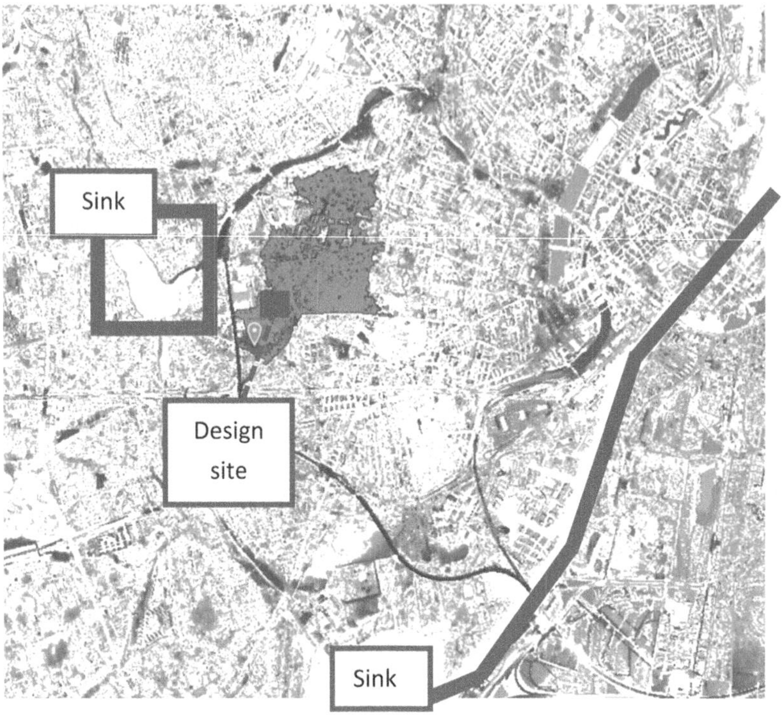

Fig. 5.4 Local of site in relation to water pattern and the relation to main sinks

Fig. 5.5 Specific area chosen in the case study (design site)

large enough, but which contained a certain variation and posed a minimal challenge to the design of the urban space. The choice was Rådmand Steins Alley in near proximity to Roskildevej (Fig. 5.5). While a specific area has been chosen, certain liberties have been taken to explore the experiment. The building placed on the chosen site (hereafter referred as design site) has its own current conventional conditions, but in the experiment of this chapter, certain physical and material conditions have been altered to fit the purpose of the experiment. One of them being that the outdoor spaces do change in function (there are currently playgrounds). The physical characteristics of the roof are changed to fit the requirements of 45 degree angle crucial for thatched roofs.

Another founding condition of this experiment was to design from a feasible problem condition. In part, this entails that the experiment chooses how much wetland for cultivation is to be planned and designed within the urban setting. In doing so, we chose that the

Fig. 5.6 Representation of the geometrical dimensions of the case study

land area we must integrate for wetlands (to cultivate biomass) should correspond to the volume of harvested reeds for a thatched roof area of 141,5 m^2 (45 degree angle) which is placed on a building which is 10m×10m×10m (roof reaching height of 15 m). Figure 5.6 provides a schematic visualization of the geometry used in this case.

On the basis of other studies and grey literature on roof thatching and cultivation of biomass for construction purposes (Vandkunsten, 2020, p. 4; Bondgaard et al., 2022) the wetlands area demand is estimated for said roof size. To cultivate the needed volume of reeds for a roof area of 141,5 m^2 which is 300 mm thick requires an area 3846 m^2 of wetlands. This study increases this to 4000 m^2 to mitigate risks and dangers of possible damage done to the cultivation site (whether by weather or people).

The experiment of this book chapter adheres to the guidelines for the design of constructed wetlands as per sources (Wallace et al., 2001, Brix et al., 2007, Brix & Arias, 2005) which affords implementation strategies taking into account local climate and seasonal changes in the Danish climate. In extension these guides and prescriptions will be used for the mind-map assignment of exploring affordance in the technical application of wetlands to bring forth creative potential for urban areas. There are two sorts of reeds (dunhammer) in Danish climate, and those are the so-called wide-leaf variant (*Typha latifolia*) and the thin-leafed variant (*Typha angustifolia*) (Bondgaard et al., 2022). Based on the aforementioned grey literature, it seems that there is more focus on the wide-leaf variant as it permits an output of 10–20 tons per hectare, with few examples of cultivation sites showing output between 20–30 tons (ibid.). The literature goes on to highlight that dunhammer is a good sink for carbon and phosphorus (ibid.). The experiment of this chapter tries to maximize material output of biomass production (for construction) and thus uses the wide-leaf variant of dunhammer. According to aforementioned grey literature (ibid.), the reeds require a minimum of 3 seasons to establish roots, and followingly, it is possible to harvest fully grown reeds once a year.

5.2.2 Conceptual Research Approach

This chapter approaches the prospect of Urban Wetland Concept through an abductive mixed methodology. The reason for this is that there is a hypothesis of a future scenario which is to be explored in terms of its plausibility—and this followingly requires a systemic interdisciplinary approach as to explore the physical possibilities of the increasing biomass production in urban context and what spatial and social affordance it requires/evokes as a grounding condition and what new ones will emerge. Since we are working in abductive manor, the following hypothesis is going to be explored: the increased cultivation of biomass of constructed wetlands in urban context poses considerable affordances and benefits for social, economic and environmental sustainability and wellbeing. The exploration of such will be structured and systematized as presented in Fig. 5.7. There are of course many aspects to explore in such a hypothesis, but the goal of this chapter is to explore what a framework for approaching urban autopoiesis could consist of and how it could be structured.

The overall methodological approach is represented in Fig. 5.8 of this study is that of Systems Thinking which permits the researchers

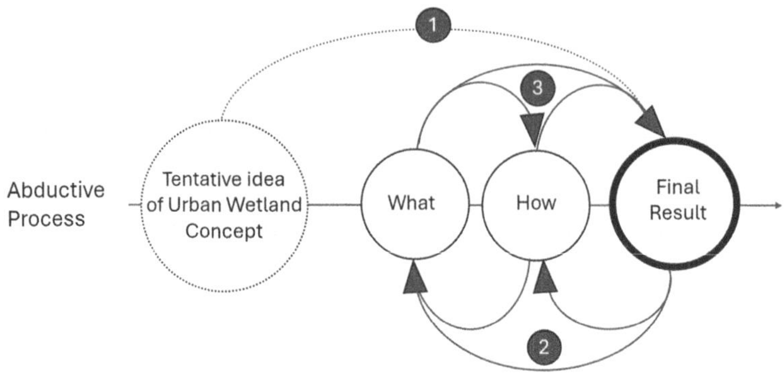

Fig. 5.7 Structure and systematization of the abductive process of iterative back-and-forth between developing the framework and figuring out how and with what. A tentative idea (1) of a potential end-product demands then an iterative testing of concepts and tools (2) and to find approaches and methods (3) for how to arrive and the end-product

Fig. 5.8 Schematics representing the research design used in this chapter

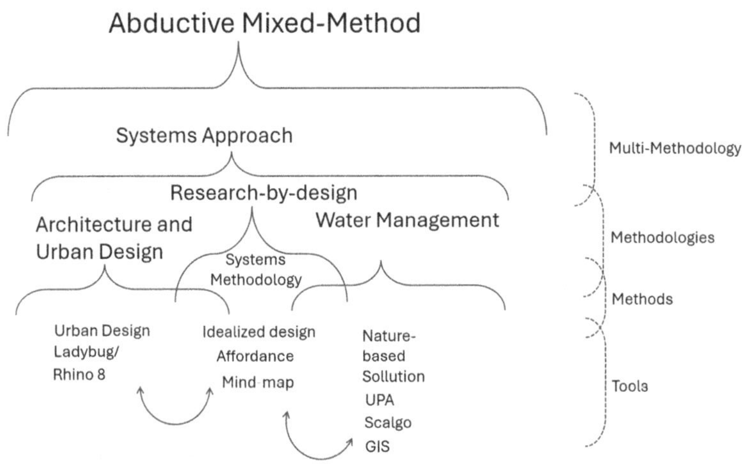

to approach and deal with the problem situation from an interdisciplinary perspective with respect to material conditions and affordance as well as the ones linked with the social system as well. The particular variant of systems thinking of that of soft variant of Systems Methodology where an "idealized design" is developed and followingly explore for its realistic feasibility and applicability in a given social system (Ackoff et al., 2006, Jackson, 2019). This is likewise backed by a Research-by-Design approach and Design Thinking linked with Systems Thinking (GIGAMAPS) based on Sevaldson (2022). Mind-map (ibid.) was used to explore the possible connection of affordance between urban recreative activities and nature-based solutions. Some creative and pragmatic liberty has been taken when having to apply the methods in the iterative creative endeavour of this chapter.

The chapter will concern itself with a specific site context which will lay the founding for the interdisciplinary experiment. The choice of site was intended to be generalizable with a varying urban and architectural/spatial characteristic of which there are many similar occurrences in a Danish context. The study explicitly did not want to take up specific sites with extreme challenges since this would not permit the generalization for the sake of scalability. In choosing the specific site to work with, the study mobilized the software SCALGO (SCALGO, 2024) and QGIS (QGIS, 2024) to map for pertinent criteria which would highlight the "generalizable" conditions

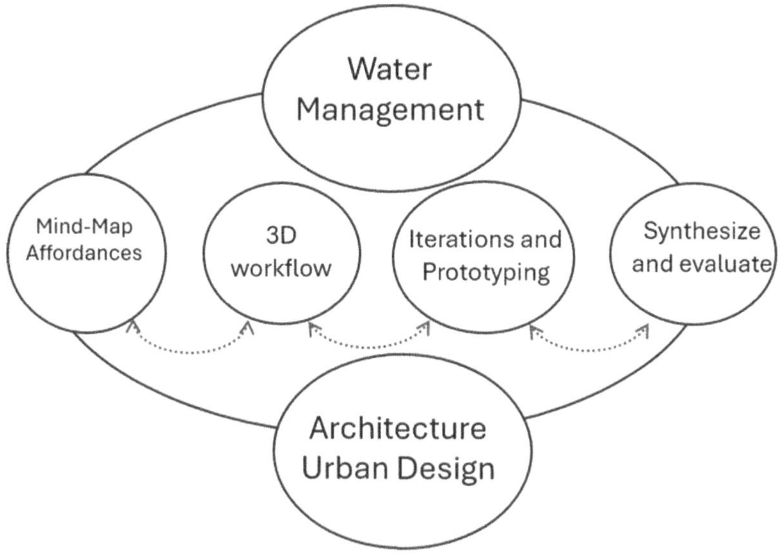

Fig. 5.9 Logic of the process for the implementation of the screening approach

from both a technical water management perspective as well as architectural, social and geographical ones (Fig. 5.9).

5.2.3 Constraints, Potentials and Affordances

This subsection will show the process for the implementation of the proposed framework for this particular experiment. It is initiated by the mapping of affordances and potential links followed by establishing of 3D- modeling environment and the subsequent exploring of how urban areas are to be screened for wetland application. Using the methodology to iterate design proposals in linking affordances with urban context conditions which amounts to the presentation of urban design scheme of the urban wetland concept. On the basis of the chosen site (see Sect. 5.2 of this chapter), it is important to further analyze and map out the micro, meso and macroclimatic and spatial conditions at the given location and the inherent potentials and affordances in crossing of urbanism and constructed wetland. Urban design conditions, pedestrian and traffic flows, building stock, use/function, traffic, sun, shadows will be considered below in synch with the potentiality of urban wetlands. The purpose is to extract the urban affordances and characteristics (which can act as design parameters and design considerations) which can be linked with constructed wetlands.

How to break down the notion of autopoietic systems? Firstly, this aspect is very complex, and the attempt to translate such a mechanism (or structuring logic) into urban and social systems is a great challenge as not to manifest it merely as a biomimicry metaphor without any actual ecosystems potential and autopoiesis as in ecosystems. Autopoiesis entails at a most fundamental level that a system is able to reconstitute itself (meaning production must be central to the conception of the urban design). The challenge is then how each of the three corners of the diagram presented in Fig. 5.10; using, building and producing gain evermore affordances for each other. While the experiment is conceived to explore innovative ways to bring together new social practices while living within planetary boundaries, there is likewise an economic potential in the concept as it would allow the particular stakeholders to provide construction material for themselves and potentially becoming partially independent in terms of material demand and future material scarcities, but at the same time surplus material volumes of biomass could sold to other entire for potential monetary profit gain. Though the economic aspect has not been a driving force in this experiment.

Fig. 5.10 Abstract conceptualizing of the urban autopoiesis

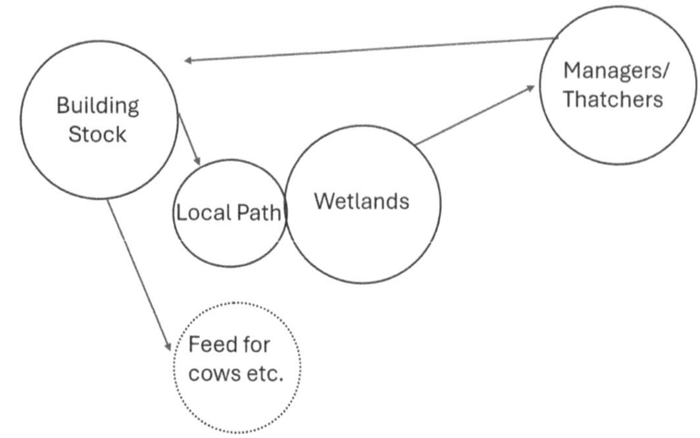

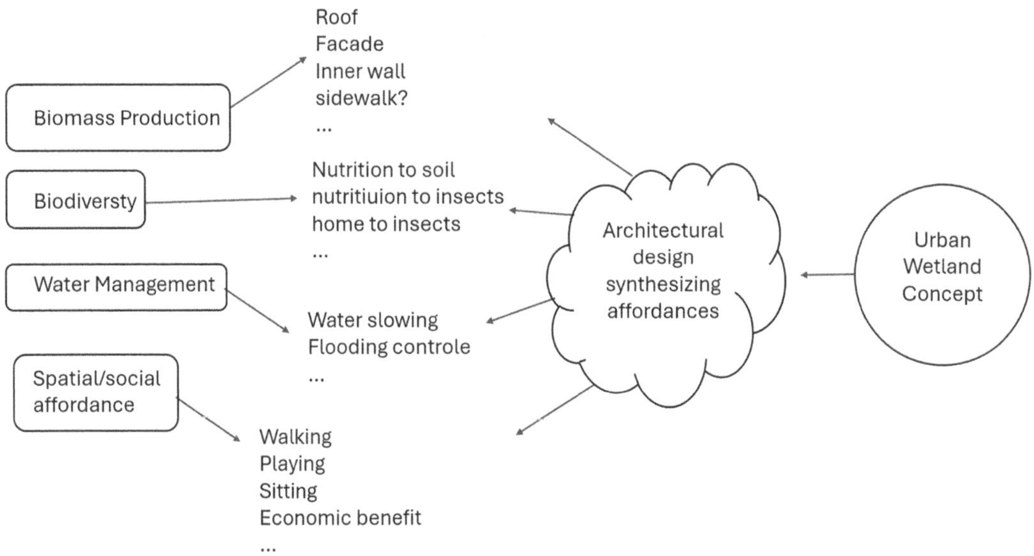

Fig. 5.11 Mind-map with the affordances from urban wetland concept

Towards the exploration of implementing relevant urban and architectural affordances from NbS, a mind-map was used to both map the current properties of the systems of interest and their potential links (Fig. 5.11). There are four main categories of performance when juxtaposing urban dwelling and constructed wetlands; those being biomass production, potential of biodiversity, management of water and social functionality of urban spaces. Properties of these are extrapolated and brought out to be synthesized through an urban design scheme where technical necessities can potentially overlap (to lesser or greater degrees) with architectural and recreative niceties of urban living.

Throughout and in the end of this mapping exercise it becomes evident that there are both material and technical and likewise social and "immaterial" properties (Usto et al., 2022; Usto, 2023) and potentials associated with the trajectory of juxtaposing urban dwelling with implementation of constructed wetlands. The spectrum diagram for this approach is conceptualized in Fig. 5.12.

Going forward, such a diagrammatic understanding of NbS and in particular constructed wetlands will be used when having to design a scheme for the urban spaces. The use of this is to integrate multi-functional dimensions in the

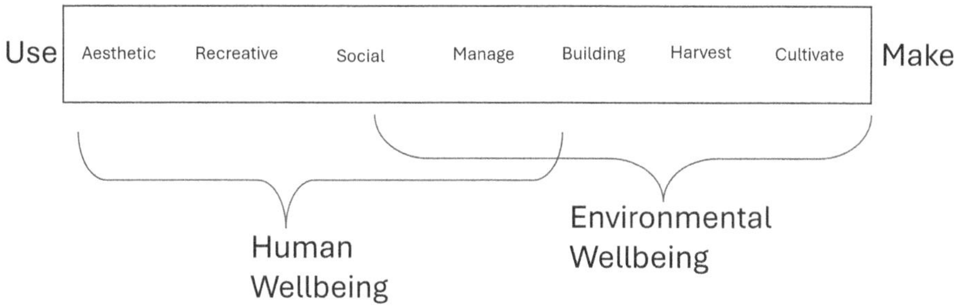

Fig. 5.12 The spectrum of affordances relevant to consider when integrating constructed wetlands into urban contexts

design and spatial organization of constructed wetlands where the technical needs of mitigating urban heat islands and contributing to handle water and cultivating biomass for construction purposes are accompanied with surplus affordances of introducing more biodiversity while also having aesthetic and mental health potential as well.

5.2.4 Method Used for the Site Screening Framework

This section presents the method used to develop the screening framework. When posing the challenge of designing urban spaces where cultivation overlaps with recreative urban dwelling, a way of working is required to increase the likelihood of the feasibility of any such design. The intention was thus to set up a methodological scheme for both developing prototypes/iterations of design but also be able to develop such iterations from an informed set of data. At the same time, the same 3D-modelling environment would also permit the evaluation and simulation of said design iterations. Initially it was important to understand the diversity of tools that architects/urban designers use as well as the tools used in water management. Given that the experiment is conducted in a Danish context, the tools of QGIS was important since Denmark has wide set of data available—and since QGIS is open source and easy to export data from it becomes an important tools to use. Along with this, the SCALGO web based blue spot analysis tool is likewise of great relevance for this experiment as it provides data and simulation potential of precipitation. From the architects' perspective, Rhino 3D is an important 3D-tool where it is easy to model sketches but likewise allows for much precision. This too is a software which allows the import of many different kinds of data while also having multiple plugin possibilities which can simulate indoor and outdoor climate conditions of buildings and urban areas and allows of custom programming of tools and components within the software. Figure 5.13 shows a diagram with the methodological flowchart used for the screening process and Fig. 5.14 presents a visual representation of the multilayer integration procedure.

There was much going back and forth between what to include and what not to include. The framework for screening potential sites for the installation of the wetlands was based on Rhino 3D modeling software which is to integrate many layers of data to be used in the relevant plugins with the intention of prototype development and evaluation. Given that it was possible to acquire a 3D model of the building stock of the whole municipality for Frederiksberg Municipality, this was likewise implemented. Next in setting up the 3D-modelling environment, a weather data file of Roskilde was used and linked to the Rhino plugin called Ladybug. Having then the 3D-modelling environment ready it became possible to gain an overview and accordingly start developing iterations on the design scheme. With this tool at hand, several solar radiation analyses were conducted for the

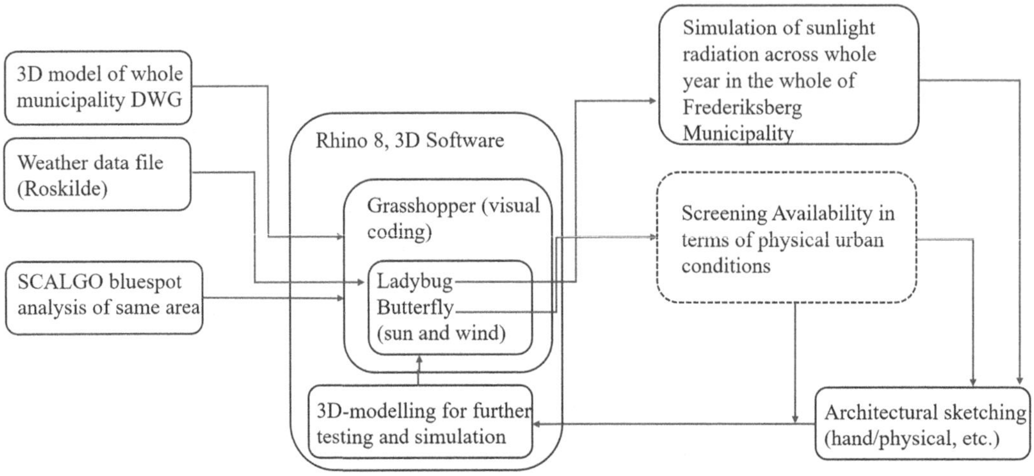

Fig. 5.13 Diagram of methodological flowchart for the screening process and for the application of the tools

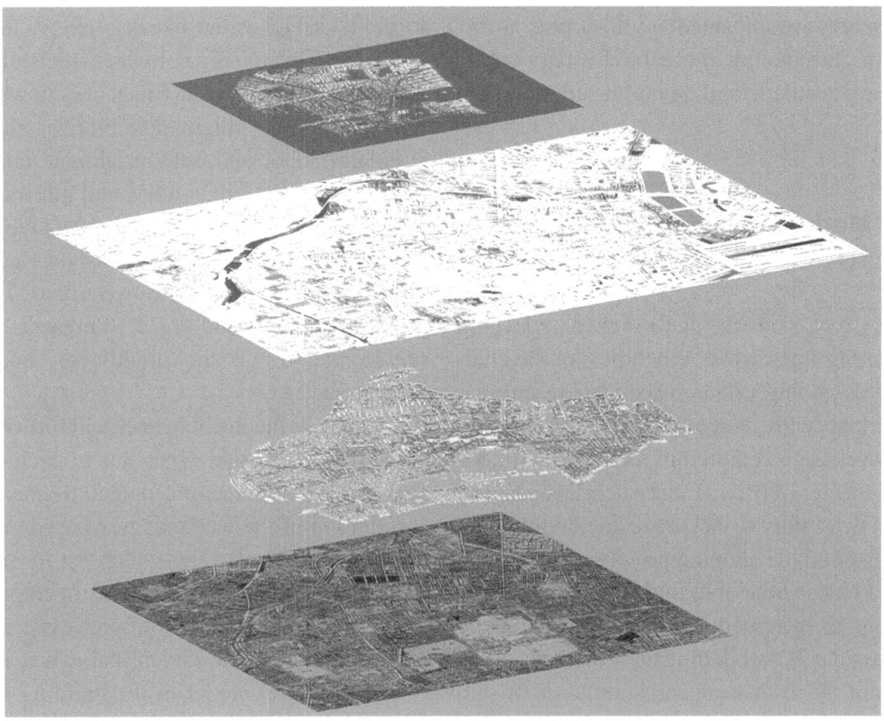

Fig. 5.14 Data layers in 3D modelling environment Rhino 8. Bottom layer 1 is the DWG-drawing of the municipality. Layer 2 is the physical building stock, layer 3 is the Scalgo Bluespot analysis, and layer 4 is the (partial) solar radiation derived from ladybug solar study

whole municipality, but it was necessary to limit the output amount. Another aspect of error margin refers to the terrain. The solar study was conducted on a flat surface. While the terrain is relatively flat in the municipality, there are some variations nevertheless and a complex imported 3d terrain geometry did not permit useful solar simulation.

It is worth mentioning this was the methodology that was set up for the experiment of this chapter. While it is possible to increase the complexity by adding more the layers of data in the modelling environment—we had to limit the scope of the amount of data implemented in this experiment as to develop the key structuring logic of the framework to come.

5.3 Results

5.3.1 Results of the Screening Approach for Site Selection

Having set up the 3D-modelling environment and a general idea of design site, it is not a given where constructed wetlands can be placed. It then becomes possible to use the above modelling environment to start screening for area availability for the placement and design urban wetlands. The initial starting point is an arbitrarily chosen spatial boundary surrounding the design site (Fig. 5.15). The screening approach relies on above described 3d modelling environment but will require a supplementary graphical treatment of data layers in a graphical software as this is not available directly in the 3d modelling environment during the course of the experiment of this chapter. The logic is structured as elaborated below.

The initial staring point is the QGIS-derived city geometry. The screening progresses by mapping the physical conditions that can influence the physical placement of urban wetlands. The logic behind it is the accumulation of the colored area which takes away the white, and the resulting white area shows the availability of the planning and designing urban wetlands. In Fig. 5.16 the first layer is mobility infrastructure, highlighting the main car roads and including main bike routes and pedestrian paths. Layer 2 is the Urban Solar Radiation derived from the 3D modelling environment, where a solar radiation simulation was conducted on the whole municipality. The output was then made into black and white while it otherwise had shades from yellow to blue indicating the intensity and spatial arrangement of the shading of a yearly average from 6 am to 8 pm. Layer 3 is the outdoor func-

Fig. 5.15 Area of interest with surroundings. Source: DataForsyningen

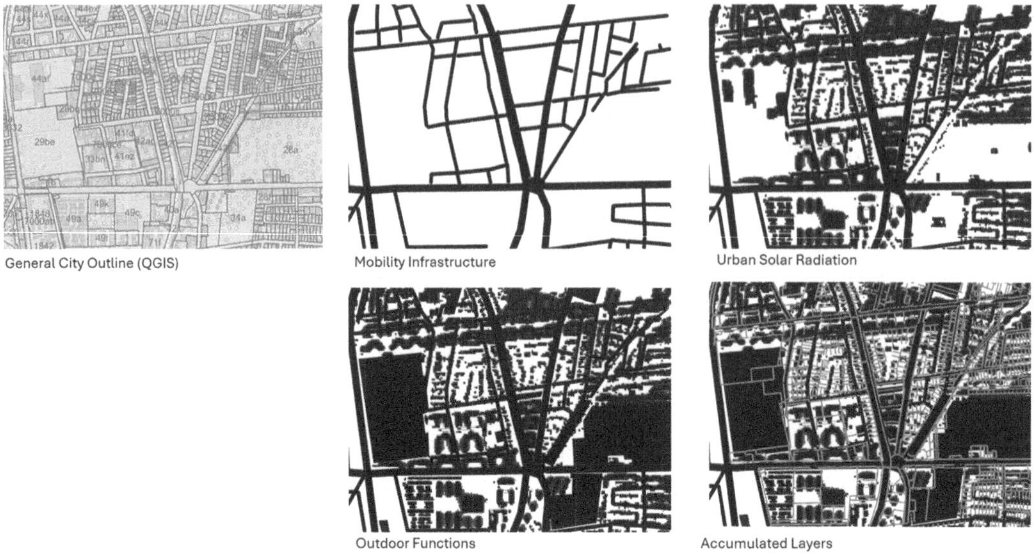

Fig. 5.16 Going from White to Black, the remaining white areas indicate physical availability for planning and designing of urban wetlands. The layers are; (1). Mobility Infrastructure, (2). Urban Solar Radiation, (3). Outdoor Functions and (4). Accumulated layer (including plotlines)

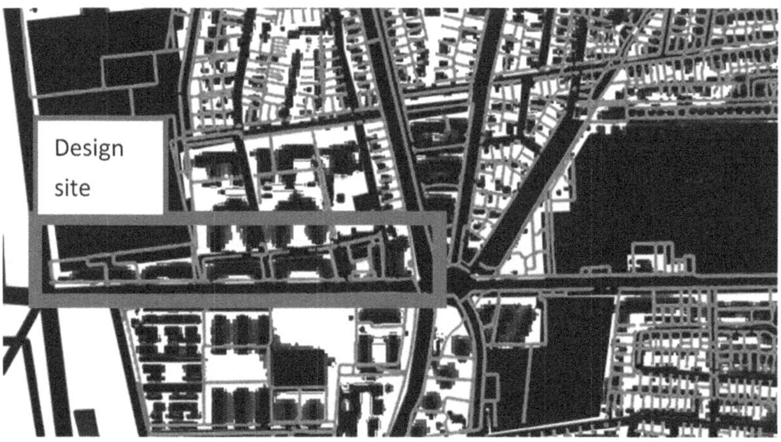

Fig. 5.17 Results of the image filter for the design site

tion derived from the municipal interactive archives (Frederiksberg Kommune, 2024a) outdoor parks and other outdoor activities (football fields, gardening areas etc.). Layer 4 indicates the accumulation of all layers (black) where the remaining white area allows for urban wetlands planning.

After applying the screening process in the case study, we will focus on the region designated as the design site (Fig. 5.5). So, Fig. 5.17 shows the results of the image filter applied to the design site. It is possible. As highlighted in previous works (Jato-Espino et al., 2016, Kuller et al., 2019, Coelho, 2025), GIS-based selection is a decision-making supporting approach, i.e. they can support the decision process by facilitating data integration and visualization, but do not give a final decision, which will depend on further discussions and considerations for the decision makers. Indeed, the resulting remainder of the white areas presented in Fig. 5.17 constitutes physical possibility for the installation of wetlands but does not take into account local plan law and politicians' agendas as this can both be very sturdy but can also be open to dialogue given the shift in cultural and political landscape. Even so, it brings important information together, enabling a fast screening based on physical aspects that narrows the potential sites that can be considered for implementing urban wetlands as will be demonstrated in subsection 5.3.2.

Furthermore, a participatory social design approach (among other relevant ones) will also be necessary to further inform the placement and design of urban wetlands to allow it to be better integrated into existing patterns of use of the outdoor areas but also give rise to new practices and uses. Indeed, Skrydstrup et al. (2020) provided an interesting approach to incorporating stakeholders viewpoints in urban water management. This approach would be extremely useful in this framework for the urban wetlands to facilitate the co-creation process. For now this permits a rudimentary design experiment, which will be useful for the development of the framework, using the 3D modelling environment as design and screening tool.

5.3.2 Results of Urban Wetlands Implementation

After the screening process, for the design site, with a 3d modelling and prototyping environment containing relevant data sets and tools of evaluation and simulation, it became possible to develop designs. The cultivation area is first priority for the experiment of this chapter. Thus much thought has been put into the organization of them (Fig. 5.18a). The reason behind their particular shape is based on the solar study within the 3D-modelling environment which informed very clearly the final concept (Fig. 5.18b). The surrounding pathways, streets and passage are untouched, and the design did not require any major changes in the overall masterplan of the area in terms of urban design (Fig. 5.18c). The recreative space adjacent to the constructed wetlands area design with varying character while the people living there have the possibility to influence how it should be used (Fig. 5.18d).

Since we now have a broad overview of how the wetlands are composed into an urban context,

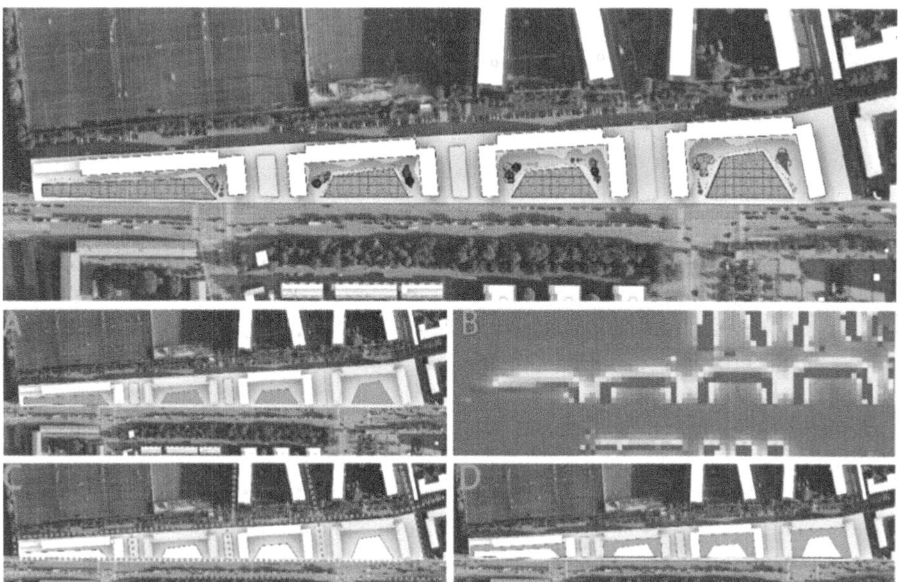

Fig. 5.18 Multiple graphics illustrating the final design and its layers. Sub-graphic (**a**) illustrates the initiating zones of interest and the arrangement of wetland scheme for the cultivation of reeds. Sub-graphic (**b**) illustrates the solar analysis which laid the logic of the shape of the constructed wetlands design. Sub-graphic (**c**) illustrates the surrounding paths and roads -both for bikes, pedestrians and cars. Sub-graphic (**d**) illustrates the recreative/biodiversity zone

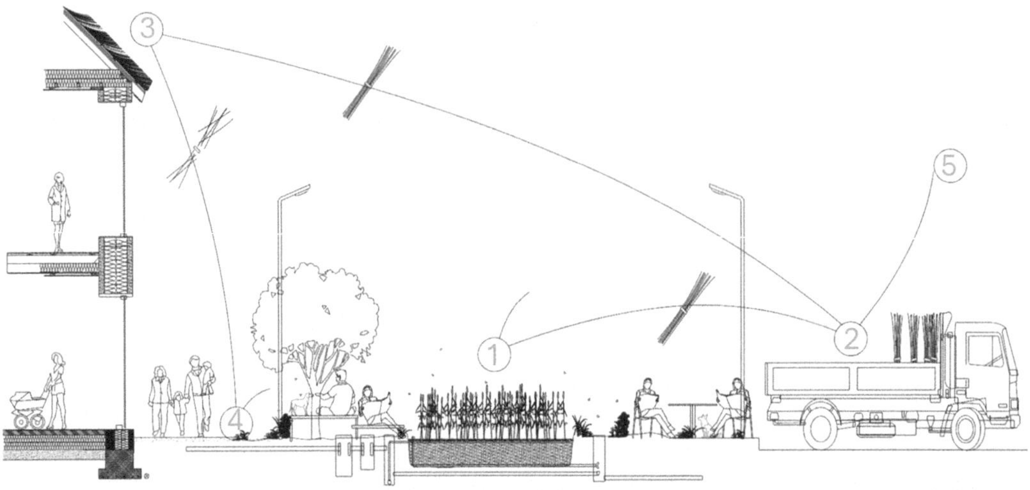

Fig. 5.19 Conceptual (principle) diagrams: Building, recreative space and urban production scheme i.e. the Systems Design concept of Urban Wetlands. Point 1 is where the reeds are cultivated. They are followingly harvested (Point 2) for preparations for construction application and/or being sold elsewhere. Point 3 shows the material applied to construction. Once the materials are broken down from wear and tear, the biomass is shredded and used as pavement (Point 4)—possibly in combination with chipped wood and gravel. This gets broken further down into nutrients from people walking, while then aids the local biodiversity. All of this takes place while the system simultaneously acts as recreative urban space—thus creating a strong urban spatial character which is very different from generic and conventional urban spaces. Point 5 offers a possibility for storage, treatment and preparations or selling of harvest elsewhere

it would be pertinent to look closer at the design scheme. Point 1 (Fig. 5.19) illustrates the placement of the constructed wetlands. They are designed based on the guidelines on Brix et al. (2007, Brix & Arias, 2005). While maintaining the integrity of the basics of a constructed wetlands design, in this experiment we try to implement the sediment tank as part of the urban area (and possibly as seating opportunity). The wetlands are surrounded by urban activity and recreative functions. Point 2 (Fig. 5.19) illustrates that once the reeds are grown they will get harvested locally, driven to local facilities for either preparation for local construction, or even sold for income for elsewhere outside the municipality or other interested parties. Once harvested and prepared the reeds can be applied as roof (Fig. 5.19, point 3)—not necessarily on the chose design sites but other places as well or at least within a close proximity to minimize transportation. After many decades, the roof biomass can be taken down (Fig. 5.19, Point 4), possibly shredded and combined with chipped wood and small gravel to form a nature-like path and/or recreative area and thus storing the carbon with recreative purposes. As people walk the street and area, the carbon will further break down and this could potentially aid local biodiversity and possibly create a relation with the biodiversity potential of the adjacent constructed wetlands. On this aspect, scholars are researching the importance and potential of carbon presence in soil.

There can be a variety of uses and functions which can provide urban and recreative niceties, as illustrated in Fig. 5.20. The design scheme is not uniform and ought to adapt to the needs of providing recreative and meaningful urban functions while also receiving the appropriate amount of solar radiation for cultivation purposes. The space is open to very different activities which can be hard to plan for architects, urban designers and engineers. As part of the quest to meet climate goals, we likewise need to rethink many social and cultural aspects to fit new and sustainable social practices. Research on urban agriculture highlights the notion of "active leisure" (Rao et al., 2022). This could be potentially significant where the local community is engaged in active recreational activity such as gardening, which is both an enjoyment in eating the food grown and

Fig. 5.20 Conceptual (principle) diagrams showing spatial arrangements in urban space. Different spatial setups for recreative activities

the work (active leisure) that went into it (Pungas, 2019). Differing urban spaces can provide the grounds for any such activity while also affording other necessities such as transit, bike parking etc.

5.4 Discussion

With the above experiment and setting up of workflow, it becomes possible to explicate a systemic framework for how to approach the design and planning of urban wetlands as systemic multi-methodology for architects and engineers towards inter- and transdisciplinary work and collaboration, as represented in Fig. 5.21. It can and should be used to both initiate design and planning activities but also use the same to start dialogue with the right legal and social agencies.

This case study was able to effectively attest the feasibility of the proposed method and resulted in an integrated framework that can speed up the process for site selection. Indeed, Inamdar et al. (2013), developed a GIS based mapping for screening sites for the installation of stormwater harvesting, and also highlighted the versatility of the use of digital tools to support multi factor analysis. Similarly, our framework showed the potential for applying the methodology presented here to support the implementation of the urban wetlands concept. There are many dimensions to the design and planning of urban wetlands and on account of the tentative experiment of this chapter, it is possible to speculatively create a methodological framework. The implementation of urban wetlands entails several things simultaneously, all of which need their own methods and tools to assess. As an example, an outdoor space can change in its recreative quality and character when wetlands are introduced, especially with the purpose of cultivating biomass. Said areas cannot be used and walked on as this would ruin the biomass volume and quality for construction application, but on the other hand these influences local biodiversity—though it is not known how much and what it exactly means for quality of life in urban areas. Given that urban heat islands are becoming increasingly a challenge, outdoor comfort as well as local micro-climatic conditions need to be assessed given the introduction of wetlands in the given urban area. All of which, depending on any given design proposal can have varying emis-

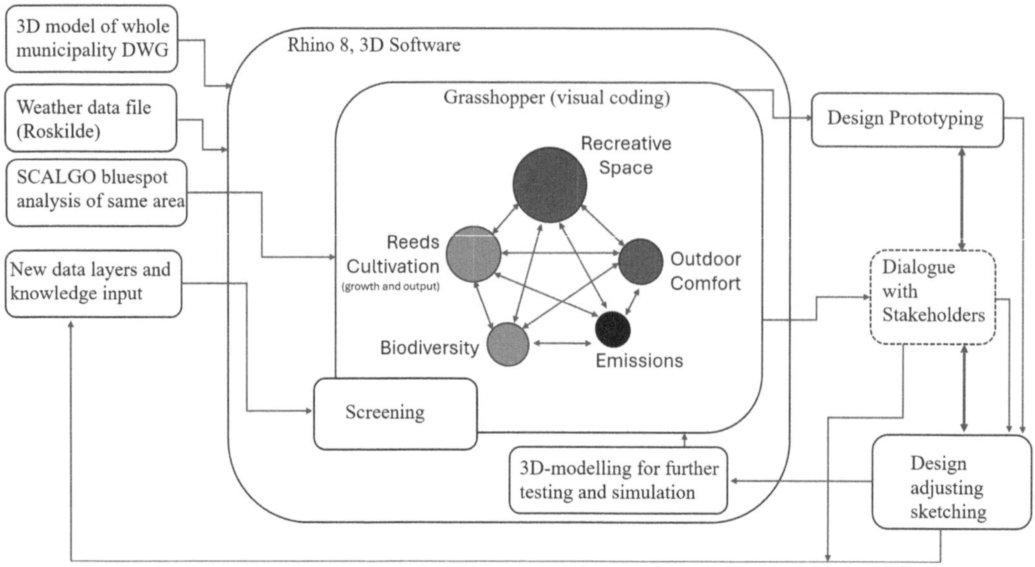

Fig. 5.21 Interconnectivity between factors illustrating the complexity of NbS urban design

sions. The framework is thus an open-ended dynamic system of a high variety of kinds of parameters and data where agility, adaptations and innovative exploration and integration are encouraged.

It is important to highlight that this chapter is a tentative methodology to arrive at a framework for the design and planning of urban wetlands, provided the benefits of such to enhance planning process. Kuller et al. (2019), who developed a planning tool for spatial suitability of stormwater infrastructure also mentioned that the use of GIS-based tools facilitates collaborative planning and a deeper comprehension of the connection concerning complex contexts and urban planning.

Given the explorative, abductive nature of this chapter, the final framework of a speculative nature, which is open-ended and thus necessitates further input and adjustment to any given design task. Other aspects also ought to be considered to further legitimize the research trajectory of autopoiesis qua urban wetlands. There would be a need for more testing and further application of the framework in larger spatial and temporal scopes. Another aspect to consider is that the design prototyping could be more explorative, and it is more general, and does not provide specifics where exactly the harvested biomass will be applied. As such, given the nature of the serendipitous exploration of this chapter of the notion of Autopoiesis, is may at this point in time seem that self-replicability can only be partial and that the biomass can only be applied as roof or facade (possibly also as partition walls) but will not be able to replace load-bearing walls and structure—as different structural performance require different material strategies. Since urban stocks are made up of different materials, an proper autopoiesis will entail different material strategies (e.g. self-healing concrete, selective demolition or use from crushed aggregates from waste/demolition and generally prioritize reuse among many other stratgies) towards minimizing urban material use while having the technical and recreative benefits of green and water infrastructures.

The chapter used a minimum amount of data input to be able to make decisions on design and planning of urban wetlands as to be able to establish initial framing of this kind of research. Needless to say, more data layers should be included and will be even further complicated and made complex when both social and economic aspects are taken into account i.e. Co-creation and co-learning with local population and local politics and planning legislation. So, if we are to furthermore qualify the research of urban autopoiesis, more simulation and evaluation tools for estimating the quality of biodiversity and outdoor comfort. Moreover, another point that was not explored here, but it should be further implemented is the inclusion of stakeholders' preferences during the decision process. In fact, stakeholder engagement is crucial to the development of long-lasting and sustainable practices and participatory methods like the one described in Fagerholm et al. (2021) and Skrydstrup et al. (2020), are something that should be included in the developments of the proposed screening process presented in this chapter.

In extension, there is a temporal dimension that requires management. Given that reeds can be harvested once a year after establishment there is then need for a strategy for the relations between cultivation sites, harvest output, storage and construction application. In relation to the scaling of this practice, it could be necessary to renovate old conventional roof and followingly use a thatched roof and façades where it may be possible. At a more fundamental level, physical tests of shading conditions and output of biomass quality and volume are required—this could allow to streamline the logic of planning wetlands in differing physical conditions (sun, wind etc.). This will be a very interdisciplinary way to approach the design of urban areas, architects and urban designers will be a key part of such efforts. They will both need to be part of the process from the beginning but also be given few key parameters from which they can design without necessarily having the vast technical knowledge of engineers and environmental scientist ideally through systemic methodological approach. Another important aspect is to consider to what extent urban wetlands will impede the already established benefit of wetlands? Will there be a meaningful trade-off between benefits? Will there autopoietic rebound effects? And is it meaningful to continue exploration of reeds production in urban areas as a means of ecological urban resilience? This is pertinent to explore.

While the framework emphasizes that the different dimensions are interconnected it has not been possible or within the scope of this paper to attempt to foresee what kind of emergent dynamics may evolve in the interplay of said dimensions in a real life social and urban context. In this regard, any specific design will be very sensitive to local contextual conditions where small differences in use and incentives can have more or less influence on the autopoietic function of any proposed design scenario in a specific urban context. Once more research and development of the urban autopoiesis and urban wetlands has been done, it will be relevant to conceptualize a more complex and possibly more comprehensive management framework, of the design and the harvest and procurement/application strategies within a temporal and spatial boundary given that there could potentially be a management challenge in organizing the logistics of where the harvest should go, where it is stored or sold among other relevant considerations. In order to do it, the application of principles from Industry 4.0 (Mavropoulos & Nilsen, 2020) could be of relevance as it may hold significant approaches to the potential management of such complex urban schemes where use and production overlap. Furthermore, if urban areas area to become autopoietic other materials and energy aspects should also be considered which ought to have or approximate relevant self-replicating strategies (like urban mining, crushed waste into new agrees etc.).

5.5 Conclusions and Recommendations

The very outset of this chapter is the fact that IWRM as preconized in SDG 6.5 and the protection and restoration of water-related ecosystems as aimed by SDG 6.6 are crucial dimensions of sustainable urban water management and thus the chapter explores how introduction of blue infrastructure into urban areas can be made even further attractive. This chapter explored and demonstrated the partial feasibility of the Urban Wetland Concept through an IWRM *qua* an interdisciplinary approach of urban design and water management. The chapter does this with the development of a screening tool to support the identification of suitable locations based on a multicriteria GIS based approach. This chapter sets itself the goal of exploring how NbS can help cities become increasing autopoietic with self-reconstructing capacities. This is of course, only partially possible as different construction layers have differing functions and can thus be better suited with other materials and accompanying strategies. However, given the novelty of this experiment, the chapter shows how an approach and framework can be set up to establish such a research sub-field of NbS. The chapter shows how to link different aspects of architectural design with technical water management aspects. Followingly, a set of tools are joined to establish a methodical way of working with iterating and optimizing and evaluating design solutions. The result shows an urban design scheme showing a principle that contains multifunctional affordances across social dimensions as well as technical performance. The experiment of this chapter shows the difficulty and the interconnectedness of parameters when wanting to explore the above trajectory. And while there are lessons learned from this experiment, the final design is of less importance than some of the overarching mapping of challenges and potentials observed. The experiment thus shows it could be useful to pursue this trajectory of regenerative NbS, even if there are many challenges linked with it. It is important to also highlight that the tradeoff between useful areas, biodiversity effect and output of biomass must be balanced. Furthermore, a different assignment lies in testing the characteristics and quality of such cultivated biomass were not detailed and will it be of high enough quality for construction purposes.

Moreover, while the experiment was done in an existing urban context, the new concept of autopoiesis should require co-evolution in the design of conventional building stock arrangements to give further thought to autopoiesis. This means that entirely new planning strategies will be needed, and once they are implemented, they could have even more elegant potential for autopoiesis. In order to go further, future research trajectories on this topic includes addressing the following questions:

- If we cultivate reeds in dense urban areas with a lot of traffic (cars etc.) does said cultivation pose potential to decrease CO_2 in urban areas (where the CO_2 is emitted) and what is the measure of cultivated wetlands that will see the minimization of CO_2 presence in city areas?
- What is the ideal spatial setup of building stock and urban elements to increase favorable conditions for the cultivation of biomass in constructed wetlands?
- What social systems conditions would be viable to integrate such practices into the everyday life of urban dwellings? Should people manage something like this themselves, or should their municipal managers be assigned to such task? Depending on ownership conditions, this could vary, but a new business model could be developed where private companies could be made to manage the cultivation as well as the recreative conditions of the wetland concept.
- What percentage of urban facades are feasible to have in reeds/straw in relation to fire challenges? While single buildings have fire regulations, what are the urban regulations that should relate to fire safety in relation to reeds/straw facades/roofs?

While this study is located in Denmark, globally, an autopoietic design approach to urban waterscapes and urban design in general should be very sensitive to local climatic, material and socio-economic conditions. The design of an autopoietic urban system in the global south or any other bioregional context would be different and should consider the climate and adapt to a particular material (or multiple) which can be grown and harvested there. At this point it is difficult to propose any specific design and planning recommendations given the nascent nature of this study but also due to very differing urban and socio-economic conditions of different counties worldwide. Instead, it would be feasible to use the approach and framework to initiate dialogue and collaboration towards relevant context-specific autopoietic design and planning policy.

Funding The research was not supported by any funding.

Potential Conflicts of Interest On behalf of all authors, the corresponding author states that there is no conflict of interest.

Informed Consent Informed consent was obtained from all individual participants included in the study.

Authors Contribution All authors have contributed equally to this work.

Data Availability Statement The datasets during and/or analyzed during the current study available from the corresponding author on reasonable request.

A.I. Discloser Statement This chapter did not use A.I. other than occasionally checking for grammar and punctuation and even still, A.I. text was not pasted into this chapter.

References

Ackoff, R. L., Magidson, J., & Addison, H. J. (2006). Idealized design: Creating an organization's future. Philadelphia, PA: Wharton School.

Andrew, E., & Pigosso, D. C. A. (2024). Multidisciplinary perspectives on rebound effects in sustainability: A systematic review. *Journal of Cleaner Production, 470*, 143366. https://doi.org/10.1016/j.jclepro.2024.143366

Bettencourt, L. (2022). *Introduction to urban science: Evidence and theory of cities as complex systems*. MIT Press.

Bondgaard, F., Toft, V., Lars, O., & Ditte, S. (2022). *DYRKNING AF PALUDIKULTURER EFFEKT PÅ KLIMA, MILJØ OG NATUR*. Seges Innovation Aarhus.

Brix, H., & Arias, C. A. (2005). Danish guidelines for small-scale constructed wetland systems for onsite treatment of domestic sewage. *Water Science and Technology, 51*(9), 1–9. https://doi.org/10.2166/wst.2005.0275

Brix, H., Schierup, H. H., & Arias, C. A. (2007). Twenty years experience with constructed wetland systems in Denmark–what did we learn? *Water Science and Technology, 56*(3), 63–68. https://doi.org/10.2166/wst.2007.522

Buijs, A., & Jacobs, M. (2021). Avoiding negativity bias: Towards a positive psychology of human–wildlife relationships. *Ambio, 50*, 281–288. https://doi.org/10.1007/s13280-020-01394-w

Büyüközkan, G., Ilıcak, Ö., & Feyzioğlu, O. (2022). A review of urban resilience literature. *Sustainable Cities and Society, 77*, 103579. https://doi.org/10.1016/j.scs.2021.103579. ISSN 2210-6707.

Castro, C. G., Trevisan, A. H., Pigosso, D. C. A., & Mascarenhas, J. (2022). The rebound effect or circular economy: Definitions, mechanisms and a research agenda. *Journal of Cleaner Production, 345*, 131136. https://doi.org/10.1016/j.jclepro.2022.131136

CINARK. (2025). Biogene Materialer Tektonik og Arkitektur. ISBN: 978.87-7830-907-5. https://realdania.dk/projekter/veje-til-biobaseret-byggeri

Coelho, L. M. G. (2024). Assessing the effect of local characteristics on environmental impacts of constructed wetlands by regionalized life cycle assessment. *Ecological Engineering, 209*, 107423. https://doi.org/10.1016/j.ecoleng.2024.107423

Coelho, L. M. G. (2025). Nature-based solutions applied in urban drainage systems: A case study using GIS-based hydrological Modeling. *Environmental and Earth Sciences Proceedings, 32*(1), 2. https://doi.org/10.3390/eesp2025032002

Cohen-Shacham, E., Andrade, A., Dalton, J., Dudley, N., Jones, M., Kumar, C., Maginnis, S., Maynard, S., et al. (2019). Core principles for successfully implementing and upscaling nature-based solutions. *Environmental Science and Policy, 98*, 20–29. https://doi.org/10.1016/j.envsci.2019.04.014

Datola, G. (2023). Implementing urban resilience in urban planning: A comprehensive framework for urban resilience evaluation. *Sustainable Cities and Society, 98*, 104821. ISSN 2210-6707. https://doi.org/10.1016/j.scs.2023.104821

De Groot, R., Stuip, M., Finlayson, M., & Davidson, N. (2006). Valuing wetlands: Guidance for valuing the benefits derived from wetland ecosystem services. Ramsar Technical Report No. 3 CBD Technical Series No. 27. Ramsar Convention Secretariat, Gland, Switzerland.

De Groot, R. S., Wilson, M. A., & Boumans, R. M. J. (2002). A typology for the classification, description, and evaluation of ecosystem functions, goods and services. *Ecological Economics, 4*, 393–408.

Dotro, G., Langergraber, G., Molle, P., Nivala, J., Puigagut, J., Stein, O., & Von Sperling, M. (2017). *Treatment wetlands* (p. 172). IWA Publishing.

Du, T. L., Bui, D. D., Buurman, J., & Quach, X. T. (2020). Towards adaptive governance for urban drought resilience: The case of Da Nang, Vietnam. In *Urban resilience to droughts and floods* (pp. 125–143). Routledge.

Ebrahimi, A. H., Mortaheb, M. M., Hassani, N., & Taghizadeh-yazdi, M. (2022). A resilience-based practical platform and novel index for rapid evaluation of urban water distribution network using hybrid simulation. *Sustainable Cities and Society, 82*, 103884.

Fagerholm, N., Raymond, C. M., Olafsson, A. S., Brown, G., Rinne, T., Hasanzadeh, K., et al. (2021). A methodological framework for analysis of participatory mapping data in research, planning, and management. *International Journal of Geographical Information Science, 35*(9), 1848–1875.

Frederiksberg Kommune. (2024a). Fakta m Kommunen. Accessed December 13, 2024, from https://www.frederiksberg.dk/kommunen/forvaltning-og-fakta-om-kommunen/fakta-om-kommunen#frederiksbergs-digitale-bykort-43

Frederiksberg Kommune. (2024b). Fakta om Frederiksberg Kommune—Frederkisberg digital bykort. Accessed December 12, 2024, from https://www.frederiksberg.dk/kommunen/forvaltning-og-fakta-om-kommunen/fakta-om-kommunen#frederiksbergs-digitale-bykort-43

Ghermandi, A., Van den Bergh, J. C. J. M., Brander, L. M., De Groot, H. L. F., & Nunes, P. (2010). Values of natural and human-made wetlands: A meta-analysis. *Water Resources Research, 46*(W12516), 12.

Inamdar, P. M., Cook, S., Sharma, A. K., Corby, N., O'Connor, J., & Perera, B. J. C. (2013). A GIS based screening tool for locating and ranking of suitable stormwater harvesting sites in urban areas. *Journal of Environmental Management, 128*, 363–370.

Jackson, M. C. (2019). *Critical systems thinking and the management of complexity*. John Wiley & Sons.

Jato-Espino, D., Sillanpää, N., Charlesworth, S. M., & Andrés-Doménech, I. (2016). Coupling GIS with Stormwater modelling for the location prioritization and hydrological simulation of permeable pavements in urban catchments. *Water, 8*(10), 451. https://doi.org/10.3390/w8100451

Kabisch, N., Frantzeskaki, N., & Hansen, R. (2022). Principles for urban nature-based solutions. *Ambio, 51*, 1388–1401. https://doi.org/10.1007/s13280-021-01685-w

Kabisch, N., Kraemer, R., & Brenck, M. (2020). Physical activity patterns in two differently characterised urban parks under conditions of summer heat [Environ. Sci. Policy 107 May (2020) 56–65]. *Environmental Science & Policy, 114*, 216. https://doi.org/10.1016/j.envsci.2020.08.004

Kapucu, N., Ge, Y., Rott, E., & Isgandar, H. (2024). Urban resilience: Multidimensional perspectives, challenges and prospects for future research. *Urban Governance, 4*(3), 162–179.

Krausmann, F., Lauk, C., Hass, W., & Wiedenhofer, D. (2018). From resource extraction to outflows of wastes and emissions: The socioeconomic metabolism of the global economy, 1900–2015. *Global Environmental Change, 52*, 131–140.

Kuller, M., Bach, P. M., Roberts, S., Browne, D., & Deletic, A. (2019). A planning-support tool for spatial suitability assessment of green urban stormwater infrastructure. *Science of the Total Environment, 686*, 856–868.

Maturana, H. R., & Varela, F. J. (1980). *Autopoiesis and cognition: The realization of the living* (Vol. 42). Springer.

Mavropoulos, A., & Nilsen, A. W. (2020). *Industry 4.0 and circular economy: Towards a wasteless future or a wasteful planet?* Wiley. ISBN: 978-1-119-69927-9.

MEA (Millennium Ecosystem Assessment). (2005). *Ecosystems and human Well-being: Wetlands and water synthesis*. MEA, World Resources Institute. www.millenniumassessment.org

Miljøstyrelsen. (2019). Afffaldsrapport, p. 42. Accessed April 28, 2025, from https://www2.mst.dk/Udgiv/publikationer/2020/12/978-87-7038-249-6.pdf

Pungas, L. (2019). Food self-provisioning as an answer to the metabolic rift: The case of 'Dacha resilience' in Estonia. *Journal of Rural Studies, 68*, 75–86. ISSN 0743-0167. https://doi.org/10.1016/j.jrurstud.2019.02.010

QGIS.org. (2024). QGIS geographic information system. Open Source Geospatial Foundation Project. http://qgis.org

Rao, N., Patil, S., Singh, C., Roy, P., Pryor, C., Poonacha, P., & Genes, M. (2022). Cultivating sustainable and healthy cities: A systematic literature review of the outcomes of urban and peri-urban agriculture. *Sustainable Cities and Society, 85*, 104063.

Santos, K. A., Coelho, L. M. G., & Santos, H. A. (2023). Proposal of a new approach to perform advanced flow modeling for subsurface flow treatment wetlands. *Water Science & Technology, 88*(5), 1155–1177. https://doi.org/10.2166/wst.2023.219

Scalgo. (2024). Scalgo Software: Introduction. https://scalgo.com/

Sevaldson, B. (2022). *Designing complexity: The methodology and practice of systems oriented design*. Common Ground Research Networks. https://doi.org/10.18848/978-1-86335-262-8/CGP

Skrydstrup, J., Madsen, H. M., Löwe, R., Gregersen, I. B., Pedersen, A. N., & Arnbjerg-Nielsen, K. (2020). Incorporating objectives of stakeholders in strategic planning of urban water management. *Urban Water Journal, 17*(2), 87–99.

Stefanakis, A., Akratos, C., & Tsihrintzis, V. (2014). *Vertical flow constructed wetlands—Eco-engineering systems for wastewater and sludge treatment*. Elsevier.

Tzoulas, K., Galan, J., Venn, S., et al. (2021). A conceptual model of the social–ecological system of nature-based solutions in urban environments. *Ambio, 50*, 335–345. https://doi.org/10.1007/s13280-020-01380-2

Upreti, G. (2024). *Ecosciocentrism—The earth first paradigm for sustainable living*. Springer. https://doi.org/10.1007/978-3-031-41754-2

Usto, K. (2023). *Safe sink tectonics: Towards a metabolism of the built environment within planetary capacities*. Aalborg Universitetsforlag. https://doi.org/10.54337/aau548869912

Usto, K., Hvejsel, M. F., & Brunsgaard, C. (2022). An urban sink—Case study of an (im)material metabolism of buildings. In M. F. Hvejsel & P. J. S. Cruz (Eds.), *Structures & architecture: A viable urban perspective?: Proceedings of the Fifth International Conference on Structures and Architecture*. CRC Press.

Vandkunsten. (2020). LCA-Rapport—Straatagets Kontor. https://www.stråtags-info.dk/wp-content/uploads/2024/09/LCA2020-09-29Inkl-review.pdf

Vymazal, J. (2014). Constructed wetlands for treatment of industrial wastewaters: A review. *Ecological Engineering, 73*, 724–751. https://doi.org/10.1016/j.ecoleng.2014.09.034

Wallace, S., Parkin, G., & Cross, C. (2001). Cold climate wetlands: Design and performance. *Water Science and Technology, 44*(11–12), 259–265. https://doi.org/10.2166/wst.2001.0838

Yang, X., Arias, M. E., & Ergas, S. J. (2023). Hybrid constructed wetlands amended with zeolite/biochar for enhanced landfill leachate treatment. *Ecological Engineering, 192*, 106990. https://doi.org/10.1016/j.ecoleng.2023.106990

6 Water Logging Vulnerability Assessment Through Participatory Rural Appraisal (PRA) Method in Chaktai Commercial Area, Chattogram

Md. Habibun Nabi, Md. Shah Jalal Misuk, Maheea Emtiaz Chowdhury, and Jobyda Sultana

Abstract

Chattogram city, known as the commercial capital of Bangladesh, faces extensive waterlogging due to rapid and unplanned urbanization. Chaktai, a key commercial and financial hub, is highly vulnerable to waterlogging. This study uses a combined method of geospatial analysis and Participatory Rural Appraisal (PRA) approaches to evaluate the spatial and socio-environmental vulnerability to waterlogging in the Chaktai Commercial Area. Field surveys, PRA, geospatial analysis, and stakeholder interviews provide a comprehensive understanding of the factors contributing to drainage issues. Data on drainage conditions, flood frequency, and inundation depth were collected by a questionnaire survey with 385 randomly selected respondents. Furthermore, ten community members participated in a snowball sampling-based Participatory Rural Appraisal (PRA) to investigate everyday impacts and local knowledge. High-risk areas were identified by applying Principal Component Analysis (PCA) to key spatial factors, including elevation, slope, drainage density, and proximity to water bodies. The causes, effects, and seasonal trends of waterlogging was gathered using PRA methods like as cause-effect diagrams, pairwise ranking, seasonal and historical diagrams, Venn diagram and daily activity schedules. The findings indicate that 1.34 square kilometers of the study area, which includes portions of Khula Para, Barua Para, and Khwaja Road are extremely susceptible to waterlogging. The primary causes were determined to be unplanned development, inadequate waste management, clogged sewers, and a lack of agency collaboration. The study also emphasizes gender-related difficulties, health problems, and seasonal consequences. For better planning, it emphasizes how crucial it is to combine community involvement with scientific analysis. This strategy can promote long-term solutions, enhance drainage, and reduce the danger of flooding in other urban areas that are vulnerable to floods.

M. H. Nabi · M. S. J. Misuk · M. E. Chowdhury · J. Sultana (✉)
Department of Urban and Regional Planning, Chittagong University of Engineering & Technology (CUET), Chittagong, Bangladesh
e-mail: shahjalalmisuk@cuet.ac.bd

U. Chatterjee, A. I. Almulhim (eds.), *Urban Waterscapes*, Sustainable Development Goals Series,
https://doi.org/10.1007/978-3-032-20649-7_6

Keywords

Urban waterlogging · Flood vulnerability · Participatory Rural Appraisal (PRA) · Principal Component Analysis (PCA) · Geographic Information System (GIS)

6.1 Introduction

Geographically, 75% of people on earth live in regions that have experienced an earthquake, tropical storm, flood, drought, or water logging at least once between 1980 and 2000, making natural hazards a concerning global issue (Iglesias et al., 2021; Peduzzi et al., 2009). Waterlogging is one of them that is becoming a more significant urban problem, especially in the Global South's quickly growing cities (Awal, 2014). Urban waterlogging is frequently confined but chronic, in contrast to riverine or coastal flooding. It is caused by inadequate drainage, encroachment on natural water retention zones, and unplanned land use system (Atauzzaman et al., 2020; Cox & McFarlane, 1995). Climate change makes the problem worse by increasing rainfall, rising sea levels, and extreme weather events (Arnbjerg-Nielsen et al., 2013). Waterlogging is more likely when natural terrain is replaced by hard surfaces like concrete and asphalt due to urbanization (Parvez et al., 2021; Subrina & Chowdhury, 2018).

Urban waterlogging is a common occurrence in low-lying cities including Jakarta, Mumbai, Lagos, and Bangkok, resulting in yearly economic losses exceeding $100 billion (Rentschler et al., 2022; UN-Habitat, 2020). Waterlogging has become the seasonal occurrence in Dhaka, Karachi, and Jakarta, leading to long-term infrastructure damage, public health emergencies, and disruptions to daily life and livelihoods (Alam et al., 2023; Gunandar et al., 2024). Because of inadequate urban drainage and obstructed drainage systems, swampy conditions continue well beyond the dry season in coastal Indian cities like Thoothukudi (Li et al., 2025). According to recent studies, a combination of socioeconomic factors and physical exposure is making medium and large cities throughout Asia more vulnerable to waterlogging (Joe et al., 2025). To improve urban resilience and assess risk, academics and policymakers are turning more to integrated drainage planning, land-use regulation, and GIS approaches (Alam et al., 2023; Chakma, 2017).

In Bangladesh, urban water logging has become worse due to climate change, especially along the coast where mean sea level in the Bay of Bengal is rising at a rate of around 3.4 mm per year and annual rainfall increased by almost 13% between 1990 and 2020 (MoEFCC, 2022). Currently, monsoon rains expose over a million people to waterlogging annually, causing disruptions to agriculture, employment, and economic activities (Papry & Ahmed, 2015). Approximately 20% of urban area in large cities experiences seasonal waterlogging annually, according to a 2023 report by the Bangladesh Water Development Board (BWDB) (BWDB, 2023). Because of canal encroachment, inadequate drainage, and mostly uncontrolled growth, megacities like Dhaka, Khulna, and Chattogram frequently experience flash floods (Hossain et al., 2020). During the 2022 monsoon, Dhaka alone mapped more than 500 high-risk locations (Alam et al., 2023). The main cause of the same increases in flood frequency reported by secondary cities such as Pabna, Tangail, Jessore, and Cox's Bazar is plugged drains (Alam et al., 2023).

The second-largest coastal city in Bangladesh and a significant port and commerce center, Chittagong, is topographically vulnerable because of its low-lying sections and hilly terrain. About 40% of people in Chattogram live less than 5 meters above sea level. The city experienced 142 water logging events in 2018–2022 (CDA, 2022; Islam et al., 2023), and one surge in 2017 flooded Khatunganj–Chaktai to a depth of 0.5–0.9 meters, drowning 80% of warehouses and damaging 30% of roads (Quddusi, 2017). Despite the Chattogram Development Authority's 2017 commitment of US$650 million, which increased to around US$1.05 billion by 2024, the ongoing flood indicates that structural improvements alone are insufficient (Islam & Raja, 2021; The Business Standard, 2024; The

Financial Express, 2025). Economic losses from monsoon water logging are estimated by the Bangladesh Institute of Planners (2023) to be around BDT 1800 crore (about US $170 million) annually on a national level. This highlights the urgent need for integrated drainage, land use regulation, and climate resilient urban design (BIP, 2023).

Chaktai is one of Chattogram's oldest and most economically important zones, significantly affected by urban waterlogging. Historically, illegal encroachments and excessive sediment accumulation have caused the Chaktai Canal, which was previously 61 feet wide according to RS studies, to decrease to just 10 to 30 feet (The Daily Star, 2022). The canal, today referred to as the "sorrow of Chittagong," was formerly a waterway used by freight vessels (Haque & Hoque, 2018). Poor solid waste management, careless dumping, and haphazard development are the main causes of the degradation, which has turned the canal front into an unhealthy and dangerous urban area (Haque & Hoque, 2018). Similar circumstances may be seen in other significant drainage canals, like Rajakhali Khal and Mohesh Khal, where blockage and narrowing have exacerbated flash floods and further weakened the city's drainage capacity (Ashraf et al., 2012). The city's capacity to manage floods and other issues is seriously threatened by this severe condition, which demonstrates that Chattogram's natural water and green spaces are decreasing. Cities throughout the world are adopting nature-based solutions and Integrated Urban Water Management (IUWM) in order to manage waterlogging more effectively and restore damaged public water areas (Hardy et al., 2005; Islam et al., 2021; UN-Habitat, 2020).

According to Willis and Fitton (2016), the spatial distribution of vulnerability is significantly influenced by the variables' selection, weighting, and aggregation, particularly the weighting approach. Numerous studies use weighting techniques like Principal Component Analysis (PCA) because indicators have varying contributions to overall vulnerability (Ajtai et al., 2023; Borbor-Cordova et al., 2020; Chakraborty et al., 2022; Chang et al., 2021; Sung & Liaw, 2020; Török, 2018). By converting correlated variables into uncorrelated principal components, PCA helps simplify data by capturing the most variance in fewer dimensions (Bro & Smilde, 2014; Greenacre, 2007). According to Abson et al. (2012) & Uddin et al. (2019), PCA is frequently used in vulnerability mapping, a tool that graphically identifies high-risk locations and helps guide policy decisions. In Bangladesh, participatory rural appraisal, or PRA, has emerged as a useful method for comprehending the complex socio-environmental aspects of urban waterlogging. Through direct community involvement, PRA techniques reveal site-specific risks, local knowledge, and useful coping mechanisms. Key determinants of waterlogging and its effects on informal settlements were identified in Chattogram using PRA tools such cause-effect diagrams, seasonal calendars, and social mapping (Islam et al., 2015). In a similar vein, Shakil et al. (2013) used PRA approaches in a slum in Dhaka to draw attention to the problems with regular flooding, drainage, and sanitation. Taufiq (2021) highlighted the importance of community input in policy design for drainage development, whereas Haque et al. (2012) incorporated participatory multi-criteria analysis in flood-prone regions of Dhaka to enhance climate adaptation planning. All of these studies show that PRA improves participatory governance and inclusive urban resilience planning, as well as our understanding of waterlogging from the ground up (Mercer et al., 2008; Pain & Francis, 2003).

Integrated Water Resources Management (IWRM), offers a comprehensive strategy to deal with these significant urban water issues. In order to maintain social justice, economic viability, and environmental sustainability, it places a strong emphasis on the coordinated management of water, land, and associated resources (Dirwai et al., 2021). IWRM facilitates the integration of institutional governance, participatory planning, and upstream and downstream processes in an urban setting (Grison et al., 2023). The international community wants to achieve sustainable water management by 2030 by implementing IWRM, as stated in SDG Target 6.5 (UN-Water,

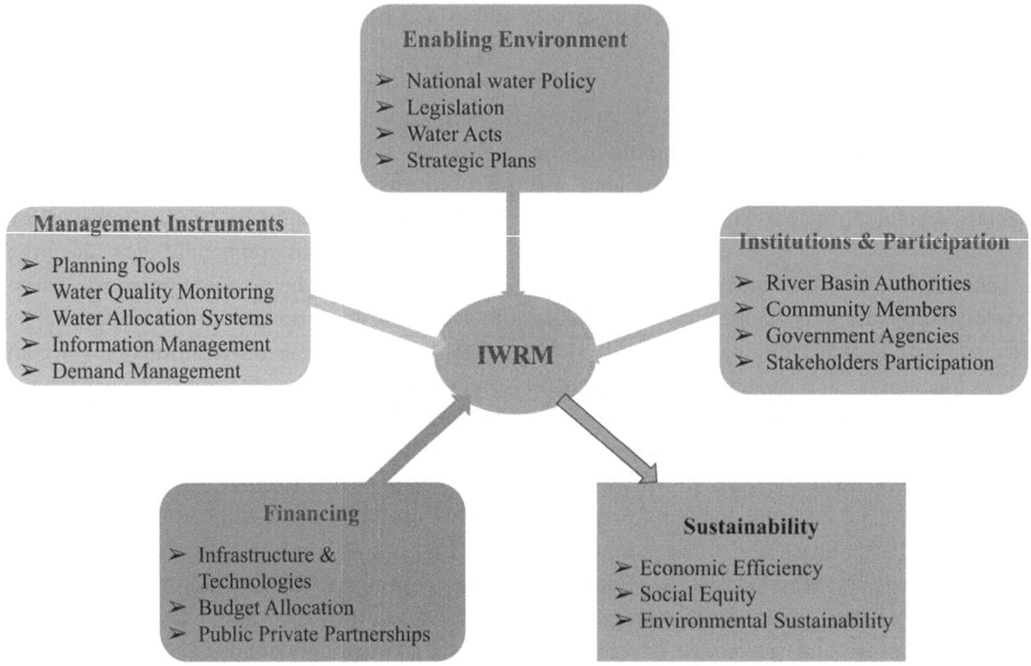

Fig. 6.1 Dimensions of integrated water resource management. *Source: Author, 2025*

2024). Better decisions on drainage, flood risk reduction, and fair water distribution are made possible by the integration of IWRM into urban systems. In the context of the urban water management system, the IWRM flow structure and its connection to SDG 6.5 are shown in the following figure (Jensen & Nair, 2019) (Fig. 6.1).

The four interrelated components of the IWRM framework highlight the cyclical connection between the following: the enabling environment (laws, policies, and plans that support IWRM); institutions and participation (coordination of governance and stakeholder engagement); management tools (tools for monitoring, controlling pollution, and allocating water); and financing (funding and investments for implementation). Together, these components ensure sustainability by economic efficiency, social equity, and environmental sustainability in managing water resources effectively. The framework emphasizes the significance of coordinated, inclusive, and adaptive management for sustainable urban water systems, which is aligned with SDG 6.5.1 and the Global Water Partnership's IWRM planning cycle (UN-Water, 2024).

This chapter aims to assess drainage and waterlogging issues in the Chaktai Commercial Area of Chattogram. It focuses on identifying vulnerable areas using geospatial techniques and analyzing the causes and consequences of waterlogging through Participatory Rural Appraisal (PRA). It includes both technical analysis and community perspectives, in contrast to other studies that just focused on one of these topics. Local knowledge is gathered by PRA techniques such seasonal calendars and group conversations, whereas PCA examines spatial aspects like canal proximity, drainage, and slope. An improved understanding of waterlogging and useful recommendations for enhancing drainage and lowering the region's risk of flooding are provided by this integrated method.

6.2 Materials and Methods

6.2.1 Study Area Profile

Chaktai Commercial Area, located in southeast Chittagong City within Bakalia and Kotwali Thanas (wards 18, 19, and 35), is a historic busi-

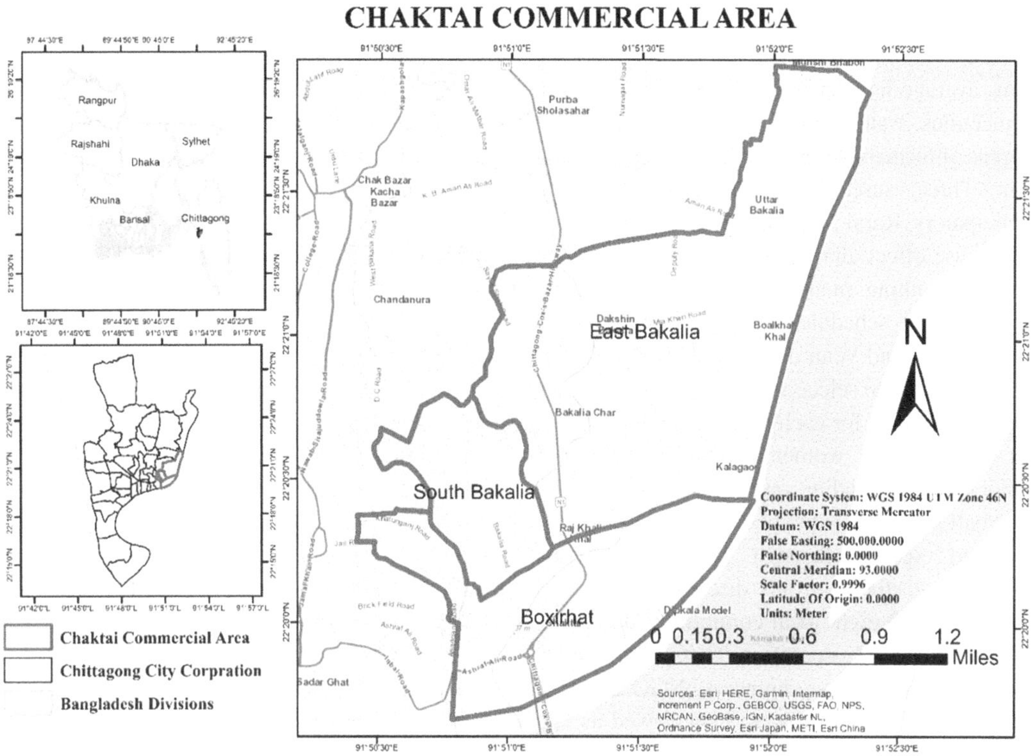

Fig. 6.2 Study area map. Source: Author, 2025

ness hub divided by the Chaktai Canal. The Chaktai Canal originates from Bahaddarhat and drains into the Karnaphuli River at the Chaktai-Chamrar Gudam (Tannery) area, near Shah Amanat Bridge, connecting Chittagong's northern and southern regions. This canal separates the Asad Ganj and Chaktai-Rajakhali commercial zones, while Fishery Ghat and Rajakhali canals border the area on the east and west, respectively. The site falls under DPZ-3 (Sadarghat-Chawkbazar) in Chittagong's Detail Area Plan (DAP) (Haque & Hoque, 2018).

As one of Bangladesh's oldest commercial centers, Chaktai hosts approximately 3000 businesses and 5000 warehouses. The study focuses on a 0.512 km segment of the 6.9 km-long canal, from the end of Khatunganj to the Karnaphuli River. The area is well-connected by roads and waterways, with Asadganj and Khatunganj roads linking to the Marine Drive route. Chaktai is a key supplier of agricultural products such as onions, garlic, rice, lentils, sugar, tea, spices, and herbs. While trade once relied heavily on boats, road expansion has reduced river transport, though canals remain crucial for commerce in this port city (Ashraf et al., 2012) (Fig. 6.2).

6.2.2 Data Collection

Primary data were collected through field surveys, questionnaires, Participatory Rural Appraisal (PRA), reconnaissance surveys, stakeholder interviews, and expert opinions. A questionnaire survey was used to gather primary data, such as waterlogging frequency, drainage condition, inundation depth, and demographic information. Random sampling was used to select 385 respondents from Wards 18, 19, and 35 for a structured questionnaire survey. Between June and September, when waterlogging is at its worst and there have been more rains than usual, the survey was conducted. Because of the timing, the experiences and effects of the community in real time may be accurately captured. The survey asked about road submergence, drainage conditions, average inundation depth, holding-to-drain connections, waterlogging frequency, and demographic information like occupation and education. Three susceptible locations underwent Participatory Rural Appraisal (PRA) using tools like cause-effect diagrams, seasonal diagrams, pairwise ranking matrices, historical diagrams, daily activity schedules of normal and waterlogged days, and Venn diagrams. Snowball sampling was used to select ten community members from three wards for each PRA session, comprising shopkeepers, women, young people, and senior citizens. Finding responders was done by snowball sampling, which began with local leaders and spread through recommendations. Furthermore, focus group discussions (FGDs) were held with eight local community members. Chattogram City Corporation (CCC) ward council members, drainage engineers, NGO employees, and urban planners were interviewed as key informants. Drainage problems, areas of waterlogging, and community concerns were identified with the help of field surveys and reconnaissance visits. Secondary data, such as elevation, slope, drainage density, and distance from waterbodies, were gathered from the Bangladesh Bureau of Statistics, the Chittagong Development Authority's GIS database, and the ASTER GDEM for Digital Elevation Models (DEM). Maps of drainage condition, inundation frequency, and elevation were created by analyzing these datasets using ArcGIS and Excel. Lastly, to determine high-risk waterlogging zones in the study area, survey and spatial data were integrated using Principal Component Analysis (PCA).

6.2.3 Research Design

The methods and techniques utilized in this research are analytical in nature. An overview of the research design is illustrated in the Fig. 6.3:

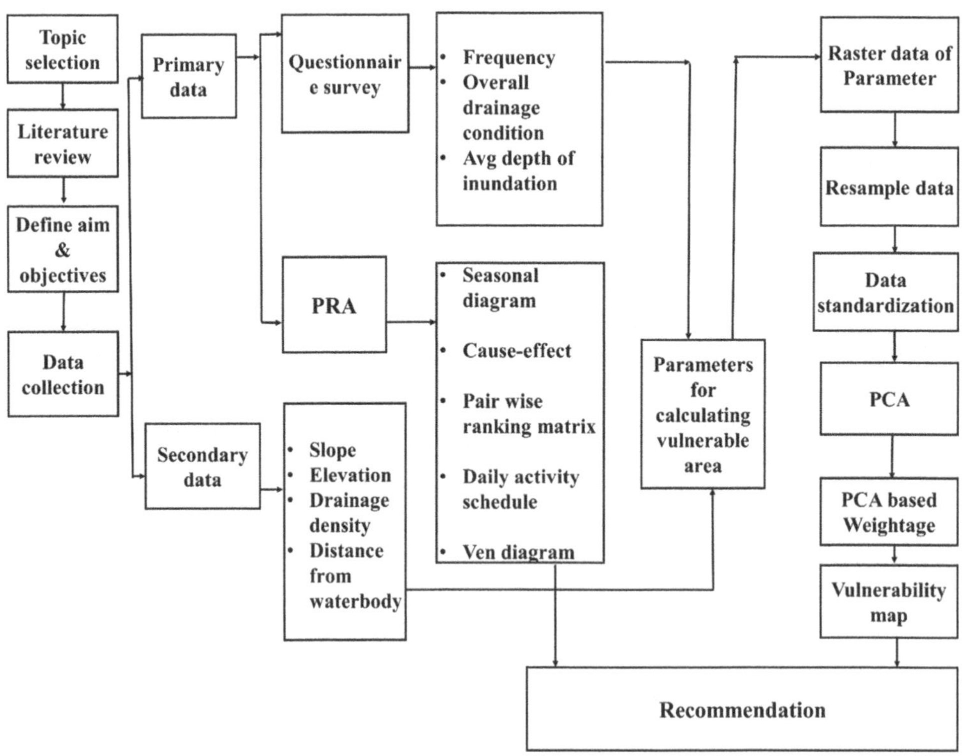

Fig. 6.3 Methodology flow chart of the study. Source: Author, 2025

6.2.4 Investigating Causes and Consequences of Waterlogging

Vulnerability assessment is conducted using Participatory Rural Appraisal (PRA) techniques, after the collection of data from surveys and secondary sources. PRA is a community-based, bottom-up method that empowers locals, regardless of literacy level, to discuss, evaluate, and expand their understanding of local issues and suggest solutions for improved living conditions (Chambers, 1994). To graphically highlight the main environmental, human, and physical variables causing waterlogging in the region, the Cause-Effect Diagram, also known as the Fishbone/Ishikawa diagram, was employed (Chandra, 2010). According to local perspectives, the Pair-Wise Ranking Method assisted in ranking these factors and identified the social groupings and everyday activities that are most impacted (Jones, 1996). Throughout the year, a seasonal diagram was created to monitor variations in rainfall, drainage efficiency, and waterlogging patterns, emphasizing crucial vulnerable times (Chambers, 1994). The Historical Diagram helped to evaluate trends and future hazards by providing a timeline of previous waterlogging episodes, drainage failures, tidal effects, and their long-term implications (Narayanasamy, 2009). According to Narayanasamy (2009). the Daily Activity Schedule (DAS) mapped disturbances to daily living, especially during peak waterlogging hours, illustrating the impact on home and economic operations. The community's perspective of organizational support, stakeholder roles, and institutional links were all depicted using a Venn diagram, which also highlighted the variations in vulnerability between genders and wealth (Upadhyay, 2016). When combined, these PRA resources provided insightful information about the underlying causes, seasonal patterns, social effects, and institutional dynamics of waterlogging in Chaktai, assisting in the creation of more practical, locally based mitigation techniques.

6.2.5 Propose Recommendation

Recommendations for reducing waterlogging in the Chaktai Commercial Area follow the guidelines of Chittagong Storm Water and Drainage Master Plan (1995). with additional proposals that align with its core strategies. These measures aim to improve drainage efficiency and reduce waterlogging. These recommendations were strengthened by information gathered from key informant interviews, PRA tools, and field surveys. In order to better understand the current drainage issues and find workable solutions, experts including ward council members, drainage engineers, and urban planners exchanged useful knowledge. These suggestions are meant to enhance drainage efficiency, reduce the effects of waterlogging, and promote the region's long-term urban resilience.

6.2.6 Result Evaluation and Report Writing

After completing data collection and analysis following the specified methodology, the major findings are summarized to highlight significant outcomes related to the research objectives. The final report is prepared using MS Word for compiling data, MS Excel for statistical calculations, ArcGIS 10.8 for mapping, and ERDAS IMAGINE for satellite image analysis.

6.3 Results

6.3.1 Assessing Water Logging Vulnerability Map

The drainage characteristics in the Chaktai Commercial Area were analyzed using geospatial software like Microsoft Excel and ArcGIS. Key variables such as slope, elevation, drainage density, distance to water bodies, and the frequency and depth of inundations were the focus of the analysis, as shown in Fig. 6.4. The area is classified as low slope (0–2.5 m), medium slope (2.5–6 m), and high slope (6–20 m) on the

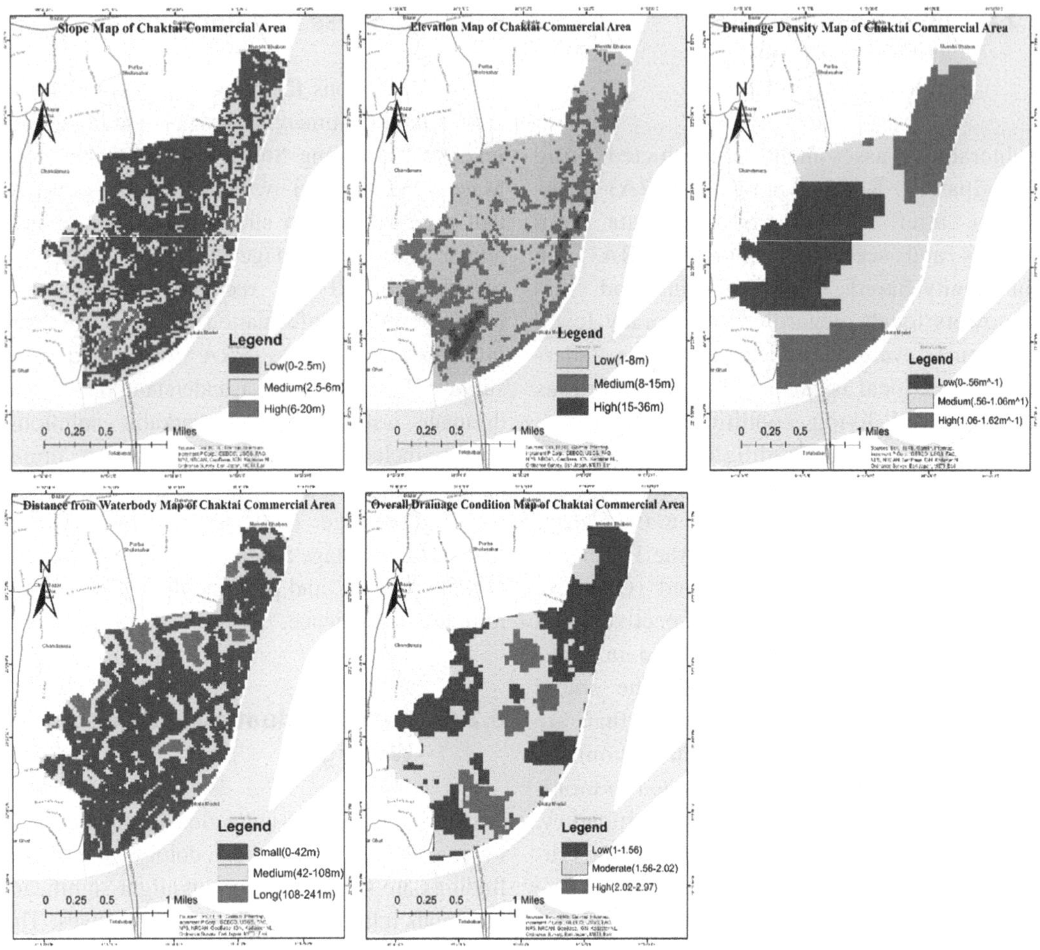

Fig. 6.4 Slope map, elevation map, drainage density map, distance from waterbody, overall drainage condition map, average depth of inundation map, frequency of waterlogging map of Chaktai Commercial Area. Source: Author, 2024

slope map. The ease with which water can drain from the surface is influenced by these slope levels. Land is categorized into low (1–8 m), medium (8–15 m), and high (15–36 m) elevation sections on the elevation map using contour lines. This makes it easier to detect low-lying areas that are more susceptible to waterlogging. Poor, moderate, and good drainage are indicated on the drainage condition map. This classification is predicated on the presence of physical barriers or encroachments, gravity flow, and drain maintenance. The number of drainage lines in a certain area is displayed on the drainage density map. The drainage density is separated into three levels: low (0–0.56 m^{-1}), medium (0.56–1.06 m^{-1}), and high (1.06–1.62 m^{-1}). This makes it easier to spot places that might not have enough drainage. The proximity of various areas to neighboring canals or rivers is displayed on the distance-to-waterbodies map. It is divided into three groups: far (108–241 m), medium (42–108 m), and close (0–42 m). The inundation depth map shows the degree of waterlogging in various zones by classifying water levels as low (6 inches to 1.8 feet), moderate (1.8 to 3.1 feet), and high (3.1 to 5 feet). Localities are categorized as low (3–6 times), moderate (6–9 times), and high (more than 9 times) annually based on the frequency of waterlogging. The Inverse Distance Weighting (IDW) approach, which uses spatial coordinates and field survey data to approximate values in places where direct measurements are unavailable, was

Table 6.1 Weightage and ranking of factors

Factor name	Weightage	Rank
Drainage condition	33.2402	1
Frequency of waterlogging	21.7365	2
Average depth of inundation	18.1261	3
Slope	12.7284	4
Elevation	10.7576	5
Drainage density	3.4096	6
Distance from waterbody	0.0017	7

Source: Author, 2024

used to construct the inundation frequency and depth maps. In order to promote efficient drainage planning and management, these spatial analyses collectively provide a thorough understanding of the environmental and physical elements causing waterlogging in Chaktai.

The most waterlogging vulnerable areas in the Chaktai Commercial Area were determined using Principal Component Analysis (PCA). PCA generated Eigen values for each factor, which show how much of an impact each variable has on the problem as a whole. The percentage weight of each component was determined based on the overall influence. Table 6.1 shows the factors' weights and rankings. The factors were then ranked using these weights to determine their relative relevance. The frequency of waterlogging, average depth of inundation, slope, elevation, drainage density, and distance from water bodies were the next most important factors, after the overall drainage condition. The raster calculator in GIS was utilized to apply the weight of each factor to the associated raster layer.

The Chaktai Commercial Area's waterlogging vulnerability zones are depicted in Fig. 6.5. On the map, low vulnerability zones are shown by green, moderately vulnerable areas by yellow, and extremely vulnerable areas by red. Wards 18, 19, and 35 together make up 6.34 square kilometers. The inside areas of Khwaja Road, Barua Para, and portions of Khula Para are among the 1.34 square kilometers that have been designated as extremely vulnerable. About 3.03 square kilometers of the remaining 5 square kilometers are categorized as moderately vulnerable, while 1.97 square kilometers are categorized as low risk.

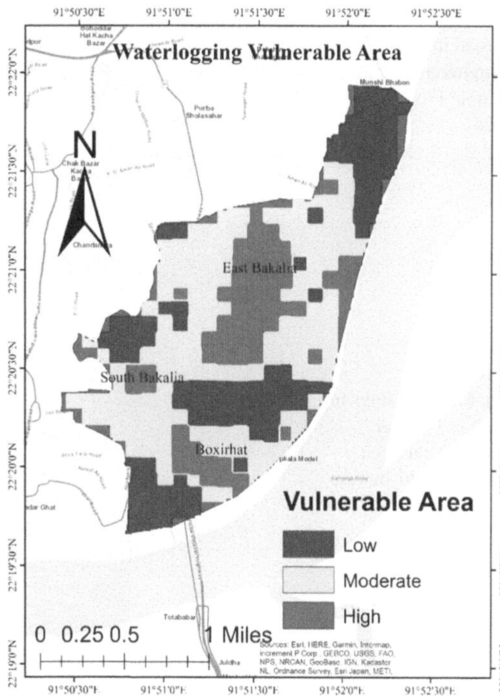

Fig. 6.5 Waterlogging vulnerable area of Chaktai Commercial Area. *Source: Author, 2024*

6.3.2 Existing Condition of Chaktai Commercial Area

Figure 6.6 illustrates road submergence during waterlogging based on the survey. The submergence of road during waterlogging. 75.84% of Main Road faces no submergence as the elevation is quite higher than secondary & tertiary road. 20.78% Main Road faces 1–2 h of submergence. In case of Secondary Road, 13.25% faces more than 2 h of submergence. 82.34% of Secondary Road faces 1–2 h submergence during water logging situation. 69.61% of Mohalla road faces more than 2 h of submergence. 29.09% Mohalla road faces 1–2 h of submergence.

The Fig. 6.7 shows how the respondents drain their dwellings. Approximately 21% of the respondents said that they drained their housing or properties into the main drainage system.

The secondary drain is used by about 33% of people, but the tertiary drainage system is used by 46% of people. Due to inadequate maintenance and low drainage capacity, households

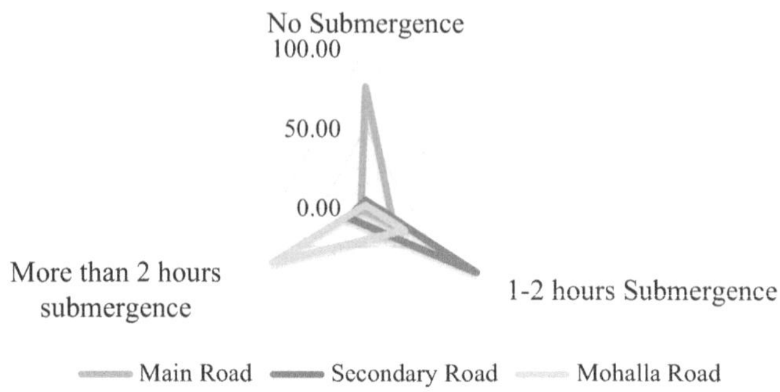

Fig. 6.6 Submergence of road in Chaktai Commercial Area. Source: Field survey, 2024

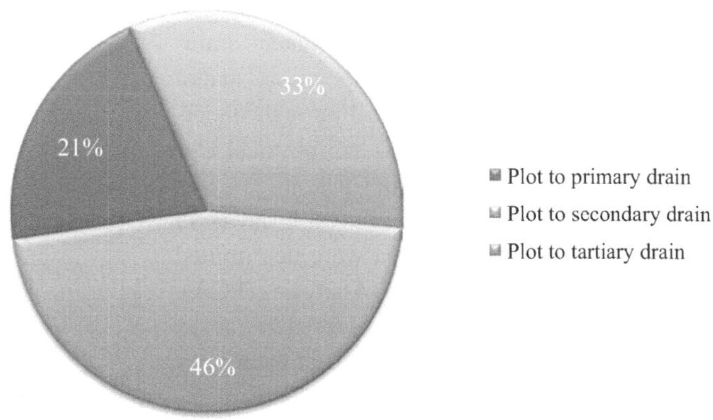

Fig. 6.7 Holdings to drain in Chaktai Commercial Area. Source: Field survey, 2024

with tertiary drains are more vulnerable to waterlogging.

The Fig. 6.8 shows the percentage of drain blockage. 6.75% drains have low blockage that means water trace-passing is very low, 64.68% drains have moderate blockage, 28.57% drains have full blockage that means water flows into the surrounding locality.

6.3.3 Analyzing Causes and Consequences of Waterlogging

In the Chaktai Commercial Area, the PRA approach found both natural and man-made causes of waterlogging. Tidal effects and intense monsoon rain are examples of natural factors that bring in excessive amounts of water that the drainage system is unable to manage. People's actions exacerbate the issue. Unplanned development increases runoff and decreases water absorption by obstructing wetlands and natural drainage pathways. Another big problem is improper solid waste management. The careless disposal of trash and the storage of building supplies close to highways have clogged several drains. Low public awareness and a shortage of trash cans exacerbate the issue. These problems result in water standing still for extended periods of time and reduce drainage capacity. In addition to causing damage to homes and businesses and disruptions to daily life, this also raises the danger of waterborne and mosquito-borne diseases.

The primary causes of waterlogging are shown in Fig. 6.9 and include tidal effects, rainfall, elevation changes, urban sprawl, clogged drains, diminished waterbodies, garbage dumping, and roadside construction storage. Waste dumping and clogged drains are the worst of these. Better garbage disposal systems, better drainage plans that consider the natural slopes of the terrain, and awareness campaigns to alter detrimental public behaviours are all necessary to solve this issue.

The main causes of drainage and waterlogging problems in the Chaktai Commercial Area are

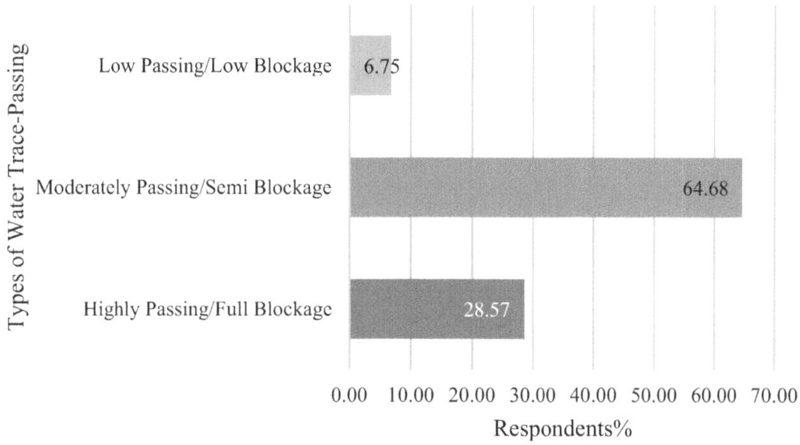

Fig. 6.8 Types of water trace passing or blocked drain in Chaktai Commercial Area. Source: Field survey, 2024

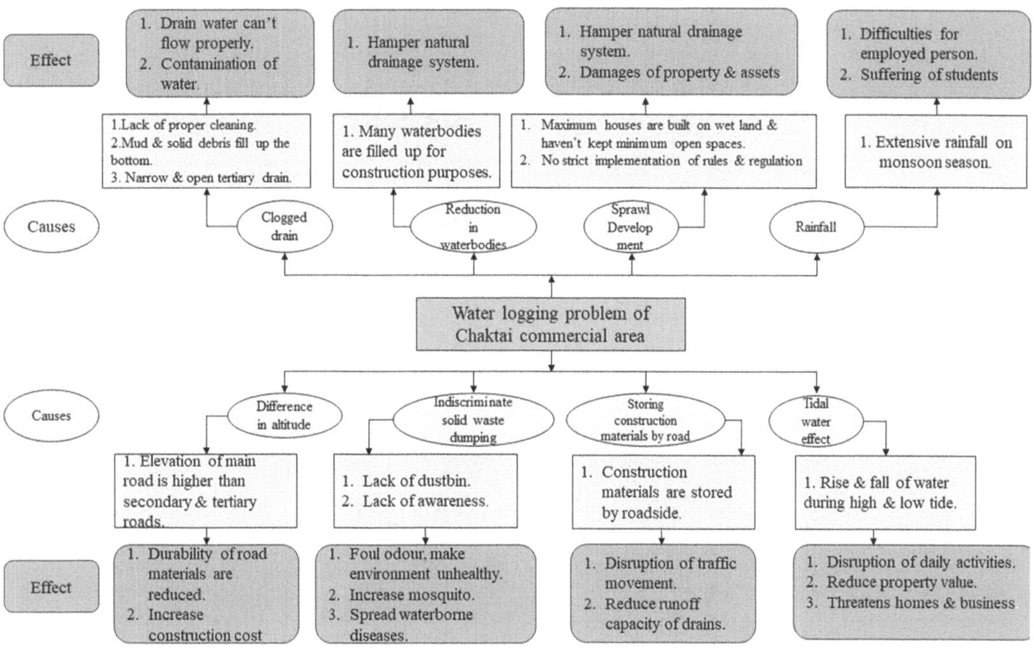

Fig. 6.9 Cause-effect diagram of Chaktai Commercial Area. *Source: PRA, 2024*

determined by the pairwise ranking approach, which is shown in Table 6.2. Each factor is compared against others to determine its relative impact. "Unplanned Development" (6) and "Rainfall" (5) are the most critical contributors, highlighting the effects of rapid urban growth and heavy monsoons. "Difference in Altitude" (4) and "Reduction in Waterbodies" (3) also play major roles, indicating how elevation variations and declining natural reservoirs worsen waterlogging. In contrast, "Tidal Water Effect" and "Storing Construction Materials" received lower rankings, suggesting a lesser but still notable impact on drainage problems.

Table 6.3, Seasonal Diagram, provides a detailed view of how infrastructure, employment, education, health, and waterlogging change seasonally in the Chaktai commercial region. Heavy rainfall (H) occurs from April to August, leading to worsening waterlogging,

Table 6.2 Pair wise ranking matrix of Chaktai Commercial Area

Problems	Drainage problem	Indiscriminate solid waste dumping	Reduction in water bodies	Unplanned development	Rain fall	Difference in altitude	Tidal water effect	Storing cons. materials in road	Total score	Priority frequency
Drainage problem	X	1	1	1	1	1	1	1	7	1
Indiscriminate solid waste dumping	0	X	1	1	1	1	1	1	6	2
Reduction in Waterbodies	0	0	X	1	1	0	1	1	4	4
Unplanned development	0	0	0	X	1	0	1	1	3	5
Rainfall	0	0	0	0	X	0	1	1	2	6
Difference in altitude	0	0	1	1	1	X	1	1	5	3
Tidal water effect	0	0	0	0	0	0	X	0	0	8
Storing construction materials in road	0	0	0	0	0	0	0	X	1	7

Source: PRA, 2024

Table 6.3 Seasonal diagram of Chaktai Commercial Area

Months Criteria	Jan	Feb	Mar	Apr	May	Jun	Jul	Aug	Sep	Oct	Nov	Dec
Rainfall: Little-L Moderate-M Heavy-H			L	M	M	H	H	H	H	H	L	
Water logging situation: Mild-Mi Moderate-M Severe-S			Mi	M	M	S	S	S	S	S	Mi	
Drainage condition: Clean-C Unclean-U	C	C	U	U	U	U	U	U	U	U	C	C
Road condition: Good- G Bad- B	G	G	G	B	B	B	B	B	B	B	G	G
Educational situation: Good-G Fair- F Bad-B	G	G	F	F	F	B	B	B	B	B	F	G
Disease					Diarrhea, Dermatitis, Typhoid	Diarrhea, Dermatitis, Typhoid	Diarrhea, Dermatitis, Typhoid	Diarrhea, Dermatitis, Typhoid				
Employment opportunities: Good-G Bad-B	G	G	G	G	G						G	G

Source: PRA, 2024

which rises from moderate (M) to severe (S) between March and September. Drainage conditions deteriorate from clean (C) to dirty (U) between May and October, while roads degrade from good (G) to bad (B) from June to November. Education quality declines from June to November, shifting from good (G) to fair (F) or bad (B). Common diseases—typhoid, dermatitis, and diarrhea—persist year-round. Employment opportunities remain mostly stable, with minor fluctuations. The monsoon season severely impacts infrastructure and quality of life in the region.

Table 6.4 presents the historical waterlogging data for Chaktai commercial area from 1991 to 2023, highlighting recurring flooding issues. Water levels remain around 4–6 feet, with a slight decrease in flooding frequency in recent years (2022–2023). However, the average flood duration has increased to 5–6 h. Tidal waters remain a major factor, though their impact has slightly improved, leading to a marginal reduction in maximum inundation depth. Despite reduced frequency, worsening flood duration emphasizes the urgent need for better tide control and drainage infrastructure in the Chaktai commercial area.

The daily schedule of activities in Chattogram's Chaktai market region is shown in Table 6.5, with an emphasis on occupational patterns and gender roles. Students, entrepreneurs, day workers, employees, and women are among the groups who are covered. The timetable describes their usual activities, which include breakfast and prayers in the morning, work, housework, and evening relaxation. With time set aside for meals, socializing, and resting throughout the day, it illustrates how people

Table 6.4 Historical diagram of Chaktai Commercial Area

Year	Drainage condition	Tidal water effect	Highest depth of inundation	Frequency of waterlogging (yearly)	Average duration of waterlogging
1991	Bad	High	5–6 feet	25–30	1.5–2 h
1995	Bad	High	5–6 feet	25–30	1.5–2 h
2000	Bad	High	5–6 feet	20–25	1.5–2 h
2005	Bad	High	4–5 feet	25–30	2–3 h
2010	Very bad	High	5–6 feet	20–25	3–4 h
2015	Very bad	High	5–6 feet	25–30	3–4 h
2017	Bad	Moderate	5–6 feet	18–20	4–5 h
2020	Very bad	Moderate	4–5 feet	14–16	5–6 h
2022	Very bad	Low	5–6 feet	12–14	5–6 h
2023	Very bad	Low	4–5 feet	12–14	5–6 h

Source: PRA, 2024

Table 6.5 Daily activity schedule of normal days

Daily activity schedule (normal days)					
Student	Businessman	Day labor	Employee	Women	Time
Sleeping	Sleeping	Sleeping	Sleeping	Sleeping	**12–5 AM**
Sleeping	Sleeping	Prayer & breakfast	Sleeping	Prayer & make breakfast	**6 AM**
Sleeping	Sleeping	Leave for work	Breakfast	Housework	**7 AM**
Breakfast & ready for school	Breakfast	Work	Household work	Eating breakfast	**8 AM**
Tuition	Leave for shop	Work	Leave for work	Household work	**9 AM**
School	Work	Work	Work	Household work & making lunch	**10 AM-1 PM**
Tiffin	Lunch	Lunch	Lunch	Household work & taking bath	**2 PM**
Back to house & take rest	Work	Work	Work	Rest	**3 PM**
Outdoor activities	Work	Snacks	Work	Rest	**4 PM**
Outdoor activities	Prayer & snacks	Work	Work	Gossiping	**5 PM**
Prayer & back to home	Work	Back to house & rest	Back to house & rest	Prayer & leisure time	**6 PM**
Snacks & Reading	Work	Gossiping	Snacks & Leisure time	Help children	**7 PM**
Reading	Work	Gossiping	Gossiping	Help children	**8 PM**
Reading	Work	Watching TV	Watching TV	Making dinner	**9 PM**
Reading	Back to house, take rest &dinner	Dinner	Dinner	Watching TV	**10 PM**
Dinner & Sleeping	Watching TV & sleeping	Sleeping	Sleeping	Watching TV	**11 PM**

Source: PRA, 2024

strike a balance between their jobs, household duties, and downtime.

These daily activities change with waterlogging, as seen in Table 6.6. It draws attention to how inhabitants' activities are affected by flooding, particularly for women, day workers, students, and employees. Important changes include longer sleep durations, postponed work schedules, and lengthier commutes because of flooded roadways. There is a decrease in outside activities and an increase in interior time. Recreational time is also impacted. Normal and flood-affected

Table 6.6 Daily activity schedule of waterlogged days

Daily activity schedule (waterlogged days)					
Student	Businessman	Day labor	Employee	Women	Time
Sleeping	Sleeping	Sleeping	Sleeping	Sleeping	**12–4 AM**
Sleeping	Sleeping	Prayer & breakfast	Sleeping	Sleeping	**5 AM**
Prayer, breakfast & ready for school	Prayer & breakfast	Leave for work	Sleeping	Prayer & make breakfast	**6 AM**
Journey to school	Ready for work	Journey to work	Breakfast	Housework	**7 AM**
Journey to school	Leave for shop	Journey to work	Leave for work	Breakfast	**8 AM**
Tuition	Work	Work	Journey to work	Household work	**9 AM**
School	Work	Work	Work	Household work & making lunch	**10 AM-1 PM**
Tiffin	Lunch	Lunch	Lunch	Household work & taking bath	**2 PM**
Back to house	Work	Work	Work	Rest	**3 PM**
Rest	Work	Snacks	Work	Rest	**4 PM**
Indoor games & gossiping	Prayer & snacks	Work	Work	Gossiping	**5 PM**
Prayer & leisure time	Work	Back to house	Journey to home	Prayer & leisure time	**6 PM**
Snacks & Reading	Work	Take rest & gossiping	Journey to home	Help children	**7 PM**
Reading	Work	Gossiping	Snacks & Leisure time	Help children	**8 PM**
Reading	Back to house	Watching TV	Gossiping & watching TV	Making dinner	**9 PM**
Reading	Take rest & dinner	Dinner	Dinner	Watching TV	**10 PM**
Dinner & Sleeping	Watching TV & sleeping	Sleeping	Sleeping	Watching TV	**11 PM**

Source: PRA, 2024

days are compared in the two tables. Students experience delays in getting to school, and businesses, employees, and day laborers modify their work schedules. Women, on the other hand, continue their regular activities mainly unaltered, even in the event of flooding or intense rain.

Venn Diagram, Fig. 6.10, visualizes key stakeholders in Chaktai's waterlogging management. CDA designs and implements drainage, roads, and infrastructure. CCC coordinates agencies, maintains drainage, and provides emergency services. BWDB manages waterways, dredges channels, and implements flood control. Chaktai Merchant Association advocates for infrastructure improvements and supports drainage maintenance. Arothdar Samity & Khatunganj Warehouse Association raise awareness and collaborate on infrastructure projects. Chittagong Youth Forum mobilizes youth for clean-up drives and advocacy. Mohila Kalyan Samiti organizes women-led resilience initiatives and supports vulnerable groups. Survey results highlight failures in stakeholder responsibilities as a major cause of waterlogging, stressing the need for strong collaboration to mitigate risks and enhance resilience.

6.4 Major Findings

This chapter used a mixed-method approach combining field surveys and GIS analysis to achieve the first objective, which was to identify

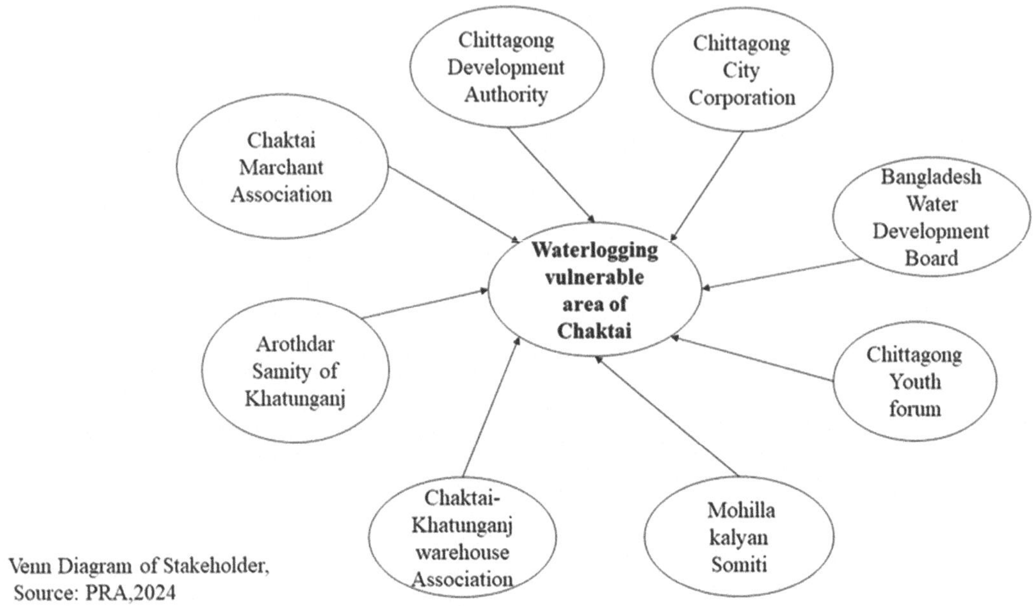

Fig. 6.10 Venn diagram of stakeholders in Chaktai Commercial Area. Source: PRA, 2024

the waterlogging vulnerable areas in the Chaktai Commercial Area. A structured questionnaire survey was used to collect primary data, such as the average depth of inundation, drainage condition, and frequency of waterlogging. Using random sample, 385 respondents in all were chosen from Wards 18, 19, and 35. Secondary sources, such as ASTER GDEM, BBS, and the CDA GIS database, were used to gather spatial data, such as elevation, slope, drainage density, and distance from waterbodies. Excel and ArcGIS were used to process the data. Weights were assigned, variables were standardized, and a vulnerability map was produced using Principal Component Analysis (PCA). The findings revealed that 1.34 square kilometers of the region, particularly in Barua Para and Khula Para, are extremely vulnerable because of their low elevation and inadequate drainage systems.

Field data were gathered using questionnaire surveys, reconnaissance visits, Focus Group Discussions (FGDs), Participatory Rural Appraisal (PRA), and key informant interviews in order to achieve the second objective, which was understanding the causes and consequences of waterlogging. According to the questionnaire survey, secondary and mohalla roads were frequently submerged—up to 69.61% for longer than 2 h—while 75.84% of main highways were unharmed. Primary sewers were used by just 21% of houses, whilst secondary drains were used by 33% and tertiary drains by 46%. 28.57% of drains were completely stopped, and 64.68% were somewhat blocked. Snowball sampling was used to select 10 participants in each PRA sessions in highly vulnerable zones. These participants included women, youths, elderly people, and shopkeepers. Eight people participated in the focus group discussions, and Chattogram City Corporation (CCC) ward councillors, drainage engineers, NGO employees, and planners were among the key informant interviews. Results indicated that tidal effect, canal encroachment, high rainfall, inadequate waste disposal, and uncontrolled urban growth are the main causes of waterlogging. Water levels can rise to 4–6 feet and remain for 5–6 h during peak flooding, which happens between April and September. Static water, broken roads, transportation disruptions, health hazards like diarrhoea, and financial losses for businesses, students, and day laborers are some of the effects. More impacted were tertiary drain users, and gaps in institutional coordination between local actors, CDA, CCC, and BWDB were found.

6.5 Discussion

This chapter used Participatory Rural Appraisal (PRA) and GIS analysis to analyze waterlogging vulnerability and identify its underlying causes in Chaktai, Chattogram. It demonstrated that a combination of natural and man-made factors contributes to the region's susceptibility. The study emphasizes how crucial it is to improve drainage infrastructure and include local people in order to lower the risk of waterlogging and build long-term urban resilience.

According to the findings, drainage obstruction and human-caused problems are the main causes of the 1.34 sq. km of Chaktai that are extremely susceptible to waterlogging. This vulnerability is social, institutional, and physical. Both spatial analysis and community feedback indicated to the inadequate drainage infrastructure as the main problem, with 28.57% of drains completely stopped and 64.68% of drains just partially congested. Failure of the systemic infrastructure is indicated by such high blockage rates. Finding the main areas of vulnerability and the factors that influence them is achieved via these findings.

Human-induced causes include uncontrolled garbage dumping, the shrinkage of the Chaktai Khal, and unplanned development over natural water bodies and wetlands. Because of encroachment, the canal's breadth has decreased from 60 meters to 12 meters, significantly reducing its ability to convey water. Patterns observed in other rapidly urbanizing areas are reflected in these local contexts. Studies in Agartala (Bentekhici et al., 2023) and Dhaka (Karimi et al., 2023), for instance, also link waterlogging to inadequate land use planning and obstructed drainage, demonstrating that Chaktai's issues are a component of a broader urban management issue throughout South Asia. The consequences are severe: commercial areas like Chaktai experience a decline in economic production when daily mobility is interfered and infrastructure is harmed. According to PRA data, tertiary drains—the most vulnerable—are used by 46% of houses. Delays in transportation and decreased customer flow cost businesses money. Similar socioeconomic upheavals were documented in Chinese cities such as Wuhan, where business areas vulnerable to flooding suffered significant financial losses (Cui et al., 2022).

Waterlogging also raises the risks to public health. Water stagnation increases illnesses like typhoid and diarrhoea and provides mosquitoes with nesting grounds. Urban liveability decreases as a result of ecosystems and people being impacted by environmental health degradation. These problems are consistent with studies from other metropolitan centres where inadequate drainage is substantially associated with negative health and economic effects (Migdał et al., 2023). This study emphasizes the necessity of social engagement as well as physical interventions, which directly supports the concept of Resilient Urban Waterscapes. To increase resilience, solutions including implementing sustainable urban drainage systems (SUDS), restoring natural drainage pathways, and enforcing land-use constraints are essential. Cities around the world, like Tokyo and Nanjing, have implemented comparable hybrid systems that combine green and grey infrastructure to lower the danger of urban flooding (Bao et al., 2021).

The resilience paradigm gains local value from the study's participatory PRA approach. Practical knowledge that is sometimes overlooked in strictly technical evaluations, such as everyday activities during waterlogging or reasons for drain clogging, was uncovered via community insights. This is in line with international demands for resilience planning that is driven by citizens (Cabral et al., 2021). Coordination of stakeholders between CDA, CCC, BWDB, and local associations is essential for putting these conclusions into practice and encouraging responsibility and ownership.

Ultimately, the results highlight a complicated trend: inadequately maintained infrastructure, uncontrolled urban growth, and a lack of cooperation among stakeholders all contribute to waterlogging in Chaktai. These trends are similar to those found in other cities around the world, confirming the study's applicability. Systemic infrastructure improvements and inclusive governance are both necessary for the solutions.

6.6 Limitations of the Study

This study contains a number of limitations that could influence the results. First, the analysis's depth and precision are affected by the scarcity of historical data, maps, and thorough drainage records. Observing long-term changes in waterlogging is extremely challenging due to the brief study duration. Furthermore, the frequency and severity of waterlogging can be influenced by weather variations, making it challenging to evaluate the effects of drainage systems with precision.

6.7 Recommendations

Several important suggestions have been observed to reduce waterlogging in the Chaktai Commercial Area based on feedback provided by local responders and in accordance to the Chittagong Storm Water and Drainage Master Plan (1995). The drainage system needs to be maintained because solid waste and mud frequently clog it. Drains must be adequately covered to keep rubbish out, and consistent upkeep is necessary to guarantee smooth water flow. There should be enough dustbins placed all around the place, and they should be emptied frequently to keep the area tidy and stop overflow into the drainage system. In order to restore the natural water retention capacity of previously filled canals and waterbodies, low-lying portions should be appropriately filled and excavated. Construction should be strictly prohibited on waterbodies and important open spaces. Since the main highways are higher than the neighbourhoods around them, when it rains, water frequently floods into the study area. This can be avoided by putting in place drainage systems or protective barriers. Durable materials should be used while building roads to prevent flooding damage. Additionally, to permit appropriate runoff and prevent water collection, tertiary roads should be maintained higher than the linked drains.

6.8 Conclusion

The consequences of urban waterlogging in Chaktai, Chattogram, from a scientific, policy making perspective and community level are carefully considered in this chapter. This chapter provides a comprehensive understanding of waterlogging concerns by combining Participatory Rural Appraisal (PRA) techniques with GIS-based spatial analysis to identify physical vulnerabilities and document lived experiences. The results emphasize that the situation is greatly worsened by socio-institutional variables such as poor solid waste management, encroachment, and low stakeholder coordination, in addition to inadequate infrastructure. It closes a scientific gap by identifying small-scale issues that are frequently ignored in regular studies using technical maps and data from the local community. This strategy fits with the global conversation on water resilience, especially when considering how climate change is affecting the rapidly urbanizing cities of the Global South. It emphasizes the necessity of concerted efforts by planners, community organizations, and local officials to promote integrated and participatory techniques for sustainable water management. This study suggests that in order to increase urban resilience, both natural drainage systems and appropriate land-use planning with significant local involvement should be used. In general, effective urban water management requires the implementation of Integrated Water Resources Management (IWRM). Urban waterlogging can be reduced by an integrated approach that connects environmental sustainability, management tools, funding, and institutional frameworks. Implementing IWRM principles can help achieve SDG 6.5 by 2030 by making water management in cities like Chattogram more resilient, inclusive, and adaptive. Improving IWRM-based planning is essential for attaining sustainable urban water management and constructing long-term climate resilience, despite difficulties in governance and monitoring.

Acknowledgements We are grateful to the local communities, member of the Ward no.18, manager of the local industry for their valuable information and help. We are also grateful to one of our departmental teachers, parents and two students for their help, time, valuable guidance during survey and data collection.

Funding The research was not supported by any funding.

Potential Conflicts of Interest On behalf of all authors, the corresponding author states that there is no conflict of interest.

Informed Consent Informed consent was obtained from all individual participants included in the chapter.

Authors Contribution All authors have contributed equally to this work.

Data Availability Statement The datasets analyzed during the current study available from the field survey and PRA survey from the local community.

References

Abson, D. J., Dougill, A. J., & Stringer, L. C. (2012). Using principal component analysis for information-rich socio-ecological vulnerability mapping in Southern Africa. *Applied Geography, 35*(1), 515–524. https://doi.org/10.1016/j.apgeog.2012.08.004

Ajtai, I., Ştefănie, H., Maloş, C., Botezan, C., Radovici, A., Bizău-Cârstea, M., & Baciu, C. (2023). Mapping social vulnerability to floods. A comprehensive framework using a vulnerability index approach and PCA analysis. *Ecological Indicators, 154*, 110838. https://doi.org/10.1016/j.ecolind.2023.110838

Alam, R., Quayyum, Z., Moulds, S., Radia, M. A., Sara, H. H., Hasan, M. T., & Butler, A. (2023). Dhaka city water logging hazards: Area identification and vulnerability assessment through GIS-remote sensing techniques. *Environmental Monitoring and Assessment, 195*(5), 543. https://doi.org/10.1007/s10661-023-11106-y

Arnbjerg-Nielsen, K., Willems, P., Olsson, J., Beecham, S., Pathirana, A., Bülow Gregersen, I., Madsen, H., & Nguyen, V.-T.-V. (2013). Impacts of climate change on rainfall extremes and urban drainage systems: A review. *Water Science and Technology, 68*(1), 16–28. https://doi.org/10.2166/wst.2013.251

Ashraf, M., Shariful, M., & Chowdhury, A. (2012). Drainage planning in the cities of Bangladesh: Case study of drainage and water logging in Chaktai Commercial area, Chittagong. *Journal of Bangladesh Institute of Planners, 2*, 49–60. https://doi.org/10.3329/jbip.v2i0.9556

Atauzzaman, M., Uddin, M., & Barman, N. (2020). Drainage and water logging in Pabna municipality of Bangladesh: A case study. *Journal of Civil, Construction and Environmental Engineering, 4*, 100–106. https://doi.org/10.11648/j.jccee.20190406.12

Awal, M. (2014). Water logging in southwestern coastal region of Bangladesh: Local adaptation and policy options. *Science Postprint, 1*, e00038.

Bao, Y., Duan, L., Liu, T., Tong, X., Wang, G., Lei, H., Zhang, L., & Singh, V. P. (2021). Simulation of evapotranspiration and its components for the mobile dune using an improved dual-source model in semi-arid regions. *Journal of Hydrology, 592*, 125796. https://doi.org/10.1016/j.jhydrol.2020.125796

Bentekhici, N., Rabehi, W., Bouhlala, M. A., Benharrats, F., Karoui, M. S., Benhamouda, F., & Zegrar, A. (2023). Land cover changes mapping of the west-Algerian territory: A multiscale data analysis for the estimation of the sustainable goal 15.3.1. *Environmental Earth Sciences, 82*(18), 428. https://doi.org/10.1007/s12665-023-11078-9

BIP. (2023). *BIP publications | Bangladesh Institute of Planners | economic impact of monsoon waterlogging in Bangladesh's cities*. BIP. https://www.bip.org.bd/bip-publication?page=1&per-page=12

Borbor-Cordova, M. J., Ger, G., Valdiviezo-Ajila, A. A., Arias-Hidalgo, M., Matamoros, D., Nolivos, I., Menoscal-Aldas, G., Valle, F., Pezzoli, A., & Cornejo-Rodriguez, M. D. P. (2020). An operational framework for urban vulnerability to floods in the Guayas estuary region: The Duran case study. *Sustainability, 12*(24), 24. https://doi.org/10.3390/su122410292

Bro, R., & Smilde, A. K. (2014). Principal component analysis. *Analytical Methods, 6*(9), 2812–2831. https://doi.org/10.1039/C3AY41907J

BWDB. (2023). *Bangladesh Water Development Board | Annual Report*. https://www.bwdb.gov.bd/annual-reports

Cabral, J. B. P., Nogueira, P. F., Becegato, V. A., Becegato, V. R., & Paulino, A. T. (2021). Environmental assessment and toxic metal-contamination level in surface sediment of a water reservoir in the Brazilian Cerrado. *Water, 13*(8), 8. https://doi.org/10.3390/w13081044

CDA. (2022). *Chattogram development Authority-Ministry of housing and public works*. https://cda.gov.bd/

Chakma, M. (2017). Water logging & drainage planning in Bangladesh A case study of Pabna municipality, Pabna. 1st international conference on planning, architecture & Civil Engineering. https://www.academia.edu/34532066/Water_Logging_and_

Drainage_Planning_in_Bangladesh_A_Case_Study_of_Pabna_Municipality_Pabna

Chakraborty, L., Rus, H., Henstra, D., Thistlethwaite, J., Minano, A., & Scott, D. (2022). Exploring spatial heterogeneity and environmental injustices in exposure to flood hazards using geographically weighted regression. *Environmental Research, 210*, 112982. https://doi.org/10.1016/j.envres.2022.112982

Chambers, R. (1994). Participatory rural appraisal (PRA): Challenges, potentials and paradigm. *World Development, 22*(10), 1437–1454. https://doi.org/10.1016/0305-750X(94)90030-2

Chandra, G. (2010). Participatory rural appraisal. In *Issues and tools for social science research in inland fisheries* (pp. 286–302).

Chang, H., Pallathadka, A., Sauer, J., Grimm, N. B., Zimmerman, R., Cheng, C., Iwaniec, D. M., Kim, Y., Lloyd, R., McPhearson, T., Rosenzweig, B., Troxler, T., Welty, C., Brenner, R., & Herreros-Cantis, P. (2021). Assessment of urban flood vulnerability using the social-ecological-technological systems framework in six US cities. *Sustainable Cities and Society, 68*, 102786. https://doi.org/10.1016/j.scs.2021.102786

Cox, J. W., & McFarlane, D. J. (1995). The causes of waterlogging in shallow soils and their drainage in southwestern Australia. *Journal of Hydrology, 167*(1), 175–194. https://doi.org/10.1016/0022-1694(94)02614-H

Cui, P., Ju, X., Liu, Y., & Li, D. (2022). Predicting and improving the waterlogging resilience of urban communities in China—A case study of Nanjing. *Buildings, 12*(7), 7. https://doi.org/10.3390/buildings12070901

Dirwai, T. L., Kanda, E. K., Senzanje, A., & Busari, T. I. (2021). Water resource management: IWRM strategies for improved water management. A systematic review of case studies of East, West and Southern Africa. *PLoS One, 16*(5), e0236903. https://doi.org/10.1371/journal.pone.0236903

Greenacre, M. (2007). *Correspondence analysis in practice* (2nd ed.). Chapman and Hall/CRC. https://doi.org/10.1201/9781420011234

Grison, C., Koop, S., Eisenreich, S., Hofman, J., Chang, I.-S., Wu, J., Savic, D., & van Leeuwen, K. (2023). Integrated water resources Management in Cities in the world: Global challenges. *Water Resources Management, 37*(6), 2787–2803. https://doi.org/10.1007/s11269-023-03475-3

Gunandar, C. M., Hasibuan, H. S., & Tambunan, R. P. (2024). FLOOD VULNERABILITY IN JAKARTA COASTAL SETTLEMENT: A STUDY AT KALIBARU SUBDISTRICT, NORTH JAKARTA, INDONESIA. *PLANNING MALAYSIA, 22*(34), 209–223. https://doi.org/10.21837/pm.v22i34.1584

Haque, A. N., Grafakos, S., & Huijsman, M. (2012). Participatory integrated assessment of flood protection measures for climate adaptation in Dhaka. *Environment & Urbanization, 24*(1), 197–213. https://doi.org/10.1177/0956247811433538

Haque, S. B., & Hoque, M. M. (2018). Design approaches to revitalize a canal front: A case study along Chaktai commercial area of Chittagong, Bangladesh. http://dl.lib.uom.lk/handle/123/16452

Hardy, M. J., Kuczera, G., & Coombes, P. J. (2005). Integrated urban water cycle management: The urban cycle model. *Water Science and Technology, 52*(9), 1–9. https://doi.org/10.2166/wst.2005.0276

Hossain, N., Rahaman, M., Nurain, A., & Sarker, P. (2020). Investigation on causes and effects of waterlogging in the Southern part of Bangladesh. *Management of Sustainable Development, 12*, 4–11. https://doi.org/10.54989/msd-2020-0006

Iglesias, V., Braswell, A. E., Rossi, M. W., Joseph, M. B., McShane, C., Cattau, M., Koontz, M. J., McGlinchy, J., Nagy, R. C., Balch, J., Leyk, S., & Travis, W. R. (2021). Risky development: Increasing exposure to natural hazards in the United States. *Earth's Future, 9*(7), e2020EF001795. https://doi.org/10.1029/2020EF001795

Islam, S., Anjum, F., Kamal, A. B., Akter, M., Rezaie, A. M., & Khan, M. R. (2021). The integration of nature-based solutions into climate adaptation policy and planning in Bangladesh. *Journal of Science Policy & Governance, 18*(2), 1–17. https://www.sciencepolicyjournal.org/article_1038126_jspg180204.html

Islam, K., Chowdhury, S., Chisty, K. U., & Rahman, A. (2015). Community participatory tools for water logging vulnerability assessment in Chittagong City Corporation area. *Journal of Bangladesh Institute of Planners, 8*.

Islam, E., Fiertz, N., Rouleau, T., & Yozell, S. (2023). *CORVI: Measuring multidimensional climate risks in Chattogram, Bangladesh*. Stimson Center. https://www.stimson.org/2023/corvi-measuring-multidimensional-climate-risks-in-chattogram-bangladesh/

Islam, M. R., & Raja, D. R. (2021). Waterlogging risk assessment: An undervalued disaster risk in Coastal Urban Community of Chattogram, Bangladesh. *Earth, 2*(1), 1. https://doi.org/10.3390/earth2010010

Jensen, O., & Nair, S. (2019). Integrated urban water management and water security: A comparison of Singapore and Hong Kong. *Water, 11*(4), 785. https://doi.org/10.3390/w11040785

Joe, R. J. J., Pitchaimani, V. S., Gobinath, R., & Shyamala, G. (2025). Urban waterlogging vulnerability assess using SAR imagery and integrated terrain analysis. *Discover Sustainability, 6*(1), 54. https://doi.org/10.1007/s43621-025-00843-4

Jones, C. (1996). Pairwise ranking: Participatory appraisal methods paper | participatory methods. https://www.participatorymethods.org/resource/pairwise-ranking-participatory-appraisal-methods-paper

Karimi, S., Ahmadi Malek, F., Yaghoubi Farani, A., & Liobikienė, G. (2023). The role of transformational leadership in developing innovative work Behaviors: The mediating role of employees' psychological capital. *Sustainability, 15*(2), 2. https://doi.org/10.3390/su15021267

Li, Y., Huang, T., Li, H., & Li, Y. (2025). Multi-scenario urban waterlogging risk assessment study considering

Hazard and vulnerability. *Water, 17*(6), 6. https://doi.org/10.3390/w17060783

Mercer, J., Kelman, I., Lloyd, K., & Suchet-Pearson, S. (2008). Reflections on use of participatory research for disaster risk reduction. *Area, 40*(2), 172–183.

Migdał, K., Jóźwiakowski, K., Czekała, W., Śliz, P., Tavares, J. M. R., & Almeida, A. (2023). Application of the Monte-Carlo method to assess the operational reliability of a household-constructed wetland with vertical flow: A case study in Poland. *Water, 15*(20), 20. https://doi.org/10.3390/w15203693

MoEFCC. (2022). *Ministry of Environment, Forest and climate change (MoEFCC). Sixth National Communication of Bangladesh to the UNFCCC.* MoEFCC. https://moef.gov.bd/site/press_briefing/66554121-4549-44d0-88d7-fbbdd026365a/front

Narayanasamy, N. (2009). *Participatory rural appraisal: Principles, methods and application.* SAGE Publications Ltd.

Pain, R., & Francis, P. (2003). Reflections on participatory research. *Area, 35*(1), 46–54.

Papry, R., & Ahmed, G. (2015). Drainage condition in water logged areas of central part in Chittagong city corporation. *International Journal of Engineering Science Invention, 4*(1), 24–29.

Parvez, M., Sadat, N., Tasnim, F., & Nejhum, I. J. (2021). Identifying the causes of waterlogging on people's perception towards a resilient community: A case study on Pabna municipality, Bangladesh. *Ecofeminism and Climate Change, 2*(3), 110–126. https://doi.org/10.1108/EFCC-11-2020-0033

Peduzzi, P., Dao, H., Herold, C., & Mouton, F. (2009). Assessing global exposure and vulnerability towards natural hazards: The disaster risk index. *Natural Hazards and Earth System Sciences, 9*(4), 1149–1159. https://doi.org/10.5194/nhess-9-1149-2009

Quddusi, K. S. M. K. A. (2017, August 2). Water-logging in Chittagong: Glimpses of losses | Daily Sun |. Daily Sun. https://www.daily-sun.com/arcprint/details/245013/WaterLogging-in-Chittagong:-Glimpses-of-Losses/2017-08-02

Rentschler, J., Salhab, M., & Jafino, B. A. (2022). Flood exposure and poverty in 188 countries. *Nature Communications, 13*, 3527. https://doi.org/10.1038/s41467-022-30727-4

Shakil, S. H., Ananya, T. H., Islam, I., & Sharmin, N. (2013). Assessment of water supply, sanitation and drainage facilities of south Begunbari slum, Dhaka, Bangladesh through participatory rural appraisal (PRA) method and proposing upgrading schemes. *SSRN Electronic Journal.* https://doi.org/10.2139/ssrn.2344514

Subrina, S., & Chowdhury, F. K. (2018). Urban dynamics: An undervalued issue for water logging disaster risk management in case of Dhaka city, Bangladesh. *Procedia Engineering, 212*, 801–808. https://doi.org/10.1016/j.proeng.2018.01.103

Sung, C.-H., & Liaw, S.-C. (2020). A GIS-based approach for assessing social vulnerability to flood and debris flow hazards. *International Journal of Disaster Risk Reduction, 46*, 101531. https://doi.org/10.1016/j.ijdrr.2020.101531

Taufiq, H. (2021). Dhaka water-logging: Causes, effects and remedial policy options. https://doi.org/10.21203/rs.3.rs-763539/v1.

The Business Standard. (2024, May 28). *CTG waterlogging: Crores spent, deadlines missed, streets still flooded.* The Business Standard. https://www.tbsnews.net/bangladesh/infrastructure/ctg-waterlogging-crores-spent-deadlines-missed-streets-still-flooded

The Daily Star. (2022, June 24). *Demolish illegal structures, recover Chaktai canal.* The daily The Star. https://www.thedailystar.net/views/editorial/news/demolish-illegal-structures-recover-chaktai-canal-3055466

The Financial Express. (2025). Chattogram's costly failure: Waterlogging persists despite mega drainage projects. The Financial Express. https://thefinancialexpress.com.bd/trade/chattograms-costly-failure-waterlogging-persists-despite-mega-drainage-projects

Török, I. (2018). Qualitative assessment of social vulnerability to flood hazards in Romania. *Sustainability, 10*(10), 10. https://doi.org/10.3390/su10103780

Uddin, M. N., Saiful Islam, A. K. M., Bala, S. K., Islam, G. M. T., Adhikary, S., Saha, D., Haque, S., Fahad, M. G. R., & Akter, R. (2019). Mapping of climate vulnerability of the coastal region of Bangladesh using principal component analysis. *Applied Geography, 102*, 47–57. https://doi.org/10.1016/j.apgeog.2018.12.011

UN-Habitat. (2020). *World cities report 2020: The value of sustainable urbanization | UN-Habitat.* https://unhabitat.org/world-cities-report-2020-the-value-of-sustainable-urbanization

UN-Water. (2024). *Progress on integrated water resources management (SDG target 6.5).* https://sdg6data.org/en/indicator/6.5.1

Upadhyay, R. (2016). Uses of Venn diagram in PRA. https://doi.org/10.13140/RG.2.2.34280.44803

Willis, I., & Fitton, J. (2016). A review of multivariate social vulnerability methodologies: A case study of the river Parrett catchment, UK. *Natural Hazards and Earth System Sciences, 16*(6), 1387–1399. https://doi.org/10.5194/nhess-16-1387-2016

7 Adaptive Use of Urban Blue-Green Voids: The Significance of Urban Blue-Green Infrastructure on Mental Well-being, a Case Study of Chennai City

P. Sudharsanamurthy and K. Madhivadhani

Abstract

Rapid urban economic growth often compromises residents' mental well-being due to rising built density and increasing demand for infrastructure, which marginalize natural features such as Blue-Green Infrastructure (BGI). While BGI is widely acknowledged for climate resilience and storm water management, it's potential to enhance mental well-being (MWB) remains underexplored, particularly in high-density urban environments. This study investigates how the spatial presence, quality, and accessibility of BGI influence MWB, including dimensions such as life satisfaction, emotional balance, and social interaction, by focusing on three diverse neighborhoods in Chennai, India—Anna Nagar (Residential), Teynampet (Institutional), and Kilpauk (Commercial and Transit-Oriented). A mixed-methods approach is adopted, integrating structured surveys and spatial indicators through the Community Happiness Index (CHI) to assess user perceptions. Findings reveal that frequency of BGI use proximity, and engagement purpose significantly correlate with higher MWB scores, while spatial inequities and limited access diminish psychological benefits. Notably, the study demonstrates that resilient, community-oriented BGI not only supports mental health but also aligns with the principles of sustainable water management by promoting multi-functional, water-sensitive urban spaces. These findings reinforce the importance of repositioning BGI as dual-purpose infrastructure—delivering both hydrological resilience and psychological restoration. The research contributes to global urban sustainability discourse and offers actionable insights for planners and policymakers aiming to integrate mental well-being into water-resilient urban design.

Keywords

Blue-green infrastructure · Mental well-being · Water-sensitive urban design · Resilient waterscapes · Community happiness index (CHI) · Urban sustainability

7.1 Introduction

World urban population is projected to reach 9.7 billion by 2050 based on United Nations urban population projection, Nations (2014). This acceleration of urbanization impacts land availability, increases the demand for economic estab-

P. Sudharsanamurthy (✉) · K. Madhivadhani
Department of Planning, School of Architecture and Planning, Anna University, Chennai, India
e-mail: sudharsanamurthy@annauniv.edu; vadhani27@annauniv.edu

U. Chatterjee, A. I. Almulhim (eds.), *Urban Waterscapes*, Sustainable Development Goals Series,
https://doi.org/10.1007/978-3-032-20649-7_7

lishments and Industrial setup. This population growth has led to rapid land cover change, directly influencing urban form and social structures Hepat and Khode (2024). This urban expansion places pressure on public life pattern and alters social patterns; and separates human settlements from natural systems Hinshaw (2000). The effect of development activities and change in life pattern have been shown directly to affect their mental health, results in depression and anxiety Harpham (1994). It is difficult to escape from the fully compact built environment, especially on metropolitan cities Bloom (2014).

This population overcrowding has given the pressure on built environment, thus resulted in high built density and reduction on personal spaces. This leads to social tensions and results in mental health issues. Built environment includes the built-up area; emergence of transport networks and its combination with natural environment are made or modified over a period of time by the humans Papas (2007). Concentration of rise in higher density often disconnects the residents from the natural environment, which has played a significant role in the history of evolution Guizzo et al. (2021). The inadequacy and inaccessibility of infrastructure facilities had affected the physical activity, especially the relaxing leisure activities significantly affects the individuals' physical and mental health. Zhong and Liu (2022). The effect of built environment on human mental well-being has eventually overseas an alternate impact, however the potential bias has been correlated with type of activity of the involvement of individual's perspective. Since the maximum percentage of population spend their majority time in the buildings, the relation between the mental health and built environment is considered to be an important factor Xiao et al. (2022), especially when it becomes a matter of assessing their happiness level.

Urbanization, coupled with poor land planning, severely affects Blue-Green Infrastructure (BGI), resulting in ecological degradation and increasing vulnerability to climate change at the micro level Gill et al. (2007). Due to urban sprawl reflects on the rapid expansion of impervious surfaces, which eventually changes a hydrological cycle and resulted in urban flooding Walsh et al. (2005). The destruction of green spaces such as wetlands and forests due to unchecked development reduces biodiversity and weakens the ecosystem services. Elmqvist et al. (2015). On addition to that the fragmentation of green spaces isolated wildlife population, making them more vulnerable to extinction Haddad et al. (2015). Effect of urbansisation was treated to be an important factor on reduction of cooling effect of vegetation, exacerbating the urban heat island effect Bowler et al. (2010). "Resilient urban waterscapes" refer to multifunctional urban ecosystems where blue and green infrastructure is patially and functionally integrated to provide ecological, hydrological, and psychological resilience against urban stresses Van de Ven, (2015). In contrast, well-integrated BGI enhances climate resilience and supports sustainable water management, yet cities with weak planning often overlook its multifunctionality, favoring short-term growth over long-term sustainability Kabisch et al. (2017). This negligence not only affects a built environment, subsequently it has an adverse impact on mental well-being of a society. This interdisciplinary approach combines environmental psychology, spatial planning, and water-sensitive urban design to develop a more inclusive model of climate-resilient cities Zhao, et al. (2022).

The concept of mental health was defined by World Health Organisation (World Mental Health Report, 2022), by describing the well-being condition of an individual, based on their ability to managing stress, responding capacity to the environment and involvement in productive activities. The reduction of blue-green infrastructure (BGI) in urban areas significantly affects mental well-being by limiting access to natural spaces that promote psychological health and stress reduction. Green spaces, such as parks, urban forests and gardens, provide restorative environments that enhance cognitive function, reduce anxiety, and improve mood Bratman et al. (2019). Similarly, blue infrastructure, including rivers, lakes, and wetlands, has been linked to lower stress levels and improved Mental Well-being through its calming effects and opportunities for recreational activities Gascon et al. (2017). However, the decline in BGI due to rapid urbanization exacerbates mental health issues, especially on high-density built environments, where the exposure to natural elements is limited Van den Bosch and Ode Sang (2017).

Mental well-being, as defined by Diener et al. (2010), combines happiness, emotional balance, and a sense of life purpose. It encompasses emotional, social, and psychological well-being, reflected in one's ability to manage stress and contribute meaningfully to their community (Keyes, 2002; WHO, 2004). The Community Happiness Index (CHI) serves as a multidimensional metric for evaluating mental well-being based on environmental, social, and spatial indicators. Spatial planning plays a transformative role in promoting happiness through equitable access to green spaces, public transport, and inclusive urban amenities (Kozaryan & Valente, 2019). Urban plazas, community gardens, and cultural hubs have been linked to improved satisfaction and well-being (Leyden, 2003), while participatory planning enhances attachment and collective identity (Zhang & Gao, 2020).

The theoretical foundation underpinning the relationship between Blue-Green Infrastructure (BGI) and mental well-being is rooted with several key psychological and environmental theories as mentioned in Table 7.1. According to the Attention Restoration Theory (ART) developed by (Kaplan, 1995), natural setting helps the body to replenish to reverse the cognitive, which improves concentration and lessens mental exhaustion. In a similar vein, Ulrich's (1983) Stress Reduction Theory proposes that being in natural environments can reduce stress and encourage relaxation. These presumptions have been supported by empirical research, for example, studies show that residents of greener areas experiences lower rates of mental distress and thus helps in attain higher levels of positive well-beings (White et al. (2021).

Blue-Green Infrastructure (BGI) has emerged as a vital strategy in urban planning for enhancing mental health and well-being across diverse global and local contexts. The integration of natural elements such as urban parks, forests, water bodies, wetlands, and therapeutic gardens into cityscapes consistently demonstrates positive psychological outcomes, including reduced

Table 7.1 Perspectives of BGI as a measure of mental well-being across the world

Reference	Year	Theory/Concept	Description
Ulrich, R. S.	1983	Stress reduction theory	Proposes that exposure to natural environments (including blue and green spaces) reduces psychological stress.
Kalpana, S.	1995	Attention restoration theory (ART)	Suggests that natural environments restore attention and reduce mental fatigue. BGI spaces like parks and waterfronts serve as restorative settings enhancing cognitive function.
Wilson, E. O	2009	Biophilia hypothesis	Argues that humans have an innate affinity for nature. BGI provides necessary contact with nature in urban areas, improving emotional resilience and mental health.
WHO	2014	Urban green spaces and health	Emphasizes the mental health benefits of green spaces, linking well-designed BGI with reduced depression, anxiety, and enhanced life satisfaction.
Gascon, M. et al.,	2017	Outdoor blue spaces and Well-being	Systematic review showing positive mental health outcomes associated with exposure to blue spaces like rivers, lakes, and seas—An essential component of BGI.
Bennett, T. C. & Jones, A.	2018	Dose of nature: Green space and mental health	Exposure to green space is associated with lower stress, depression, and better general health.
White et al.,	2021	Green and blue spaces proximity theory	Consistent associations between proximity to BGI and better mental health across diverse populations and contexts.

Source: Compiled by Author

stress, anxiety, and depression, as well as improved mood and social cohesion. Table 7.2 Indicates the consistent pattern of access to blue-green spaces improves the mental health outcomes across diverse urban setting, regardless of geographic or cultural context. It is evident from the practices of Netherlands canal walks, forested parks in Sweden, and therapeutic landscapes in Singapore, reveals that neighborhood-scale and easy accessibility to the green spaces are more effective measures to attain mental-health of a community. However, studies underscore the need to embed mental health considerations in urban policies and spatial planning, by integrating community oriented BGI interventions that promote emotional resilience and social well-being.

Despite the rich theoretical and international evidence base, existing literature largely treats green and blue spaces in isolation and fails to address their **integrated spatial-perceptual impact** on community mental health, especially in **tropical, high-density cities like Chennai**. Furthermore, the use of **Community Happiness Index (CHI)** as a planning tool remains underexplored. This study investigates the adaptive reuse potential of underutilized urban BGI within the city of Chennai to assess its role in improving residents' mental well-being and to produce a holistic evaluation of urban BGI's psychological impact. Through a spatial-perceptual framework applied in three distinct neighborhoods—Anna Nagar, Teynampet, and Kilpauk—the study integrates urban design evaluation, community perception, and CHI-based analysis to examine how neglected ecological spaces can be transformed into inclusive, restorative, and psychologically supportive urban systems Mell, (2017). While earlier studies addressed the value of green or blue space in isolation, this research identifies the combined influence of integrated BGI as a vital dimension of mental well-being.

The overarching aim is to highlight the importance of embedding CHI as a planning tool to assess well-being outcomes and guide water-sensitive, people-centered design practices. This fills a research gap by offering a replicable framework for linking BGI typologies with perceptual indicators and spatial interventions. The key research questions guiding this study are: (1) does urban Blue-Green Infrastructure enhances community mental health? (2) What aspects of BGI most significantly contribute to mental well-being? And (3) how can spatial planning strategies utilize BGI to support psychological resilience in urban environments? Furthermore, the framework aligns with global sustainable development goals (SDGs), particularly SDG 3 and SDG 11, by promoting cities that prioritize ecological resilience and human well-being. Its replicable structure enables adoption across similar dense urban contexts globally, offering pathways to transform ecological voids into inclusive, restorative, and climate-adaptive urban systems.

7.2 Study Area

The present study was carried out in Greater Chennai City Corporation, India, to evaluate the role of Blue-Green Infrastructure (BGI) across varying urban typologies. Three urban neighbourhoods were selected to represent distinct functional zones—Anna Nagar (residential), Teynampet (institutional), and Kilpauk (commercial and transit-oriented)—based on their land use characteristics, population density, and availability of green spaces. Anna Nagar (13.0875°N, 80.2133°E) is a well-planned residential area with a gridiron street network, low to medium-density housing, and public parks such as Anna Tower Park, Bogainvilla Park, and Corporation Park, as shown in Fig. 7.1(A). In contrast, Teynampet (13.0440°N, 80.2543°E) is centrally located and dominated by institutional land use, including government offices, educational institutions, and healthcare facilities. Green spaces like Semmozhi Poonga and Senganthal Park, as seen in Fig. 7.1(B), serve a diverse demographic and reflect the challenges of preserving BGI in a congested, high-value urban core. Kilpauk (13.0780°N, 80.2356°E), shown in Fig. 7.1(C), lies near major road and rail corridors, repre-

Table 7.2 Perspectives of BGI as a measure of mental well-being across the world

Sl.No.	Reference	Year	Country & City	BGI Measures	Scale of Intervention	Mental Health Benefits	Strategies
1	Mitchell & Popham	2008	United Kingdom	Urban parks, tree-lined streets, river walks	Neighborhood or city	Reduced depression, stress reduction, improved mood	Urban greening in spatial plans, green space access standards
2	Van den Berg et al.	2010	Netherlands Amsterdam	Canal walks, water bodies, pocket parks	Community-level	Enhanced emotional wellbeing, restorative experiences	Design for walkability and proximity to blue-green spaces
3	Grahn & Stigsdotter	2010	Sweden-Stockholm	Forested parks, urban wetlands, trails	City-scale	Cognitive restoration, stress recovery, reduced mental fatigue	Integrating nature reserves within urban development
4	Tan et al.	2013	Singapore	Therapeutic gardens, green corridors, biophilic design	Neighborhood to city	Lower anxiety levels, increased social interaction, emotional resilience	Biophilic city planning, inclusive and sensory-responsive green space design
5	Bhan et al.	2020	India (Delhi, Pune)	Urban forests, riverfronts, green buffers	Neighborhood	Lower psychological distress, community Well-being, relaxation	Inclusion of open space in master plans, blue-green buffer zones
6	Hunter et al.	2019	Australia (Melbourne)	Wetlands, green walls, interactive water features	Building to city scale	Improved mindfulness, enhanced mood	Water-sensitive design and mental health-responsive urban interventions
7	Astell-Burt & Feng	2021	Canada (Toronto)	Waterfront trails, forest parks, rain gardens	Regional to community level	Decreased loneliness, increased feelings of vitality and purpose	Access-based green infrastructure planning, urban social health integration.

Source: Compiled by Author

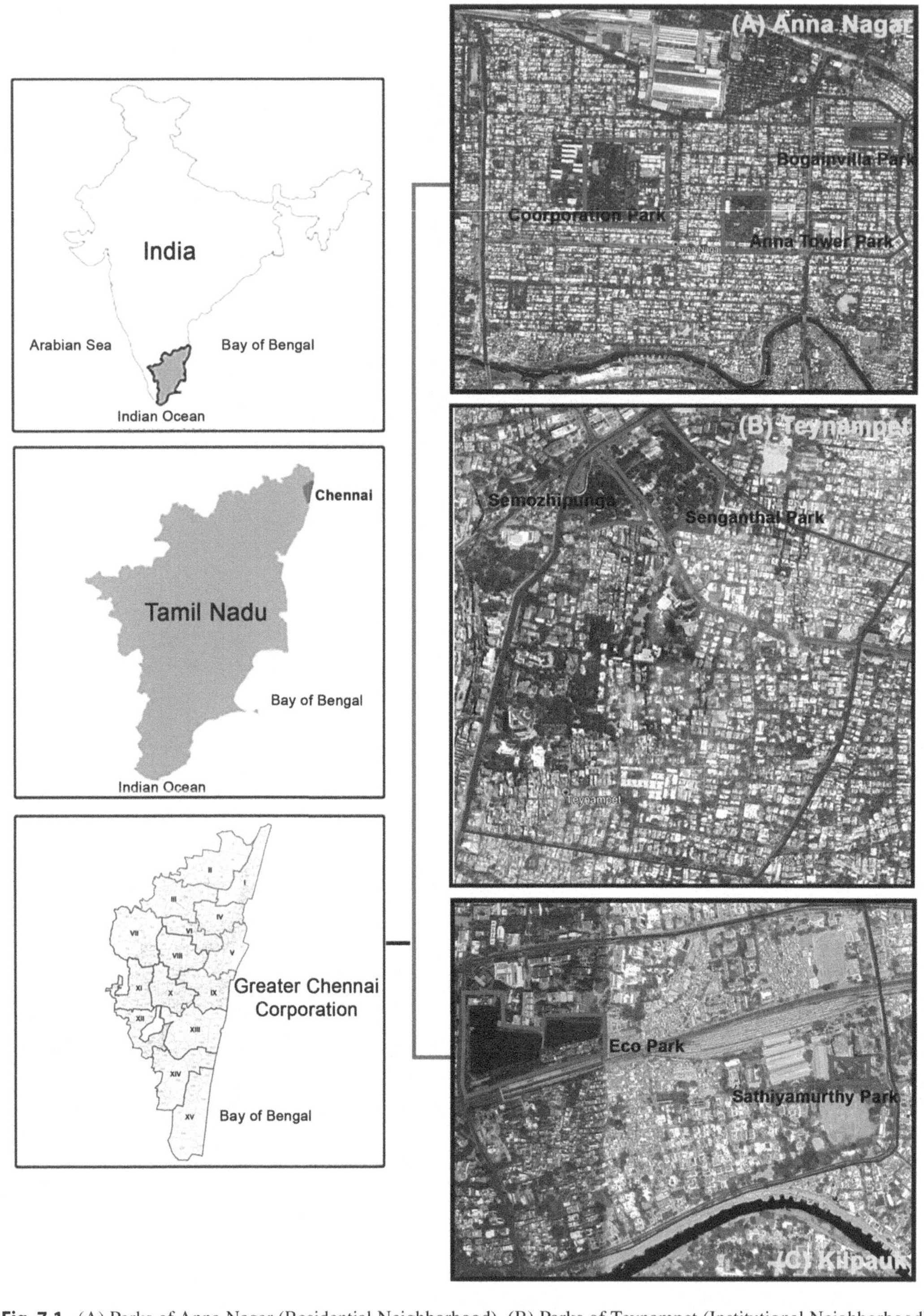

Fig. 7.1 (A) Parks of Anna Nagar (Residential Neighborhood), (B) Parks of Teynampet (Institutional Neighborhood) & (C) Parks of Kilpauk (Commercial & Transit Neighbourhood)

senting a commercial and transit-focused area with heavy pedestrian and vehicular traffic. Parks such as Eco Park and Sathiyamurthy Park in this neighbourhood serve users in a dynamic mixed-use setting. Together, these three neighbourhoods offer a representative cross-section of Chennai's urban landscape and provide a multidimensional framework to examine how different spatial and functional contexts influence the perceived role of BGI in promoting mental health and well-being.

7.3 Material and Methods

The methodology of the present study follows a structured, multi-stage process as illustrated in Fig. 7.2, beginning with a systematic literature review aimed at understanding the relationship between Blue-Green Infrastructure (BGI) and mental well-being. The review covered literature published between 2000 and 2024, retrieved from major academic databases including Scopus, Web of Science, ScienceDirect, Google Scholar,

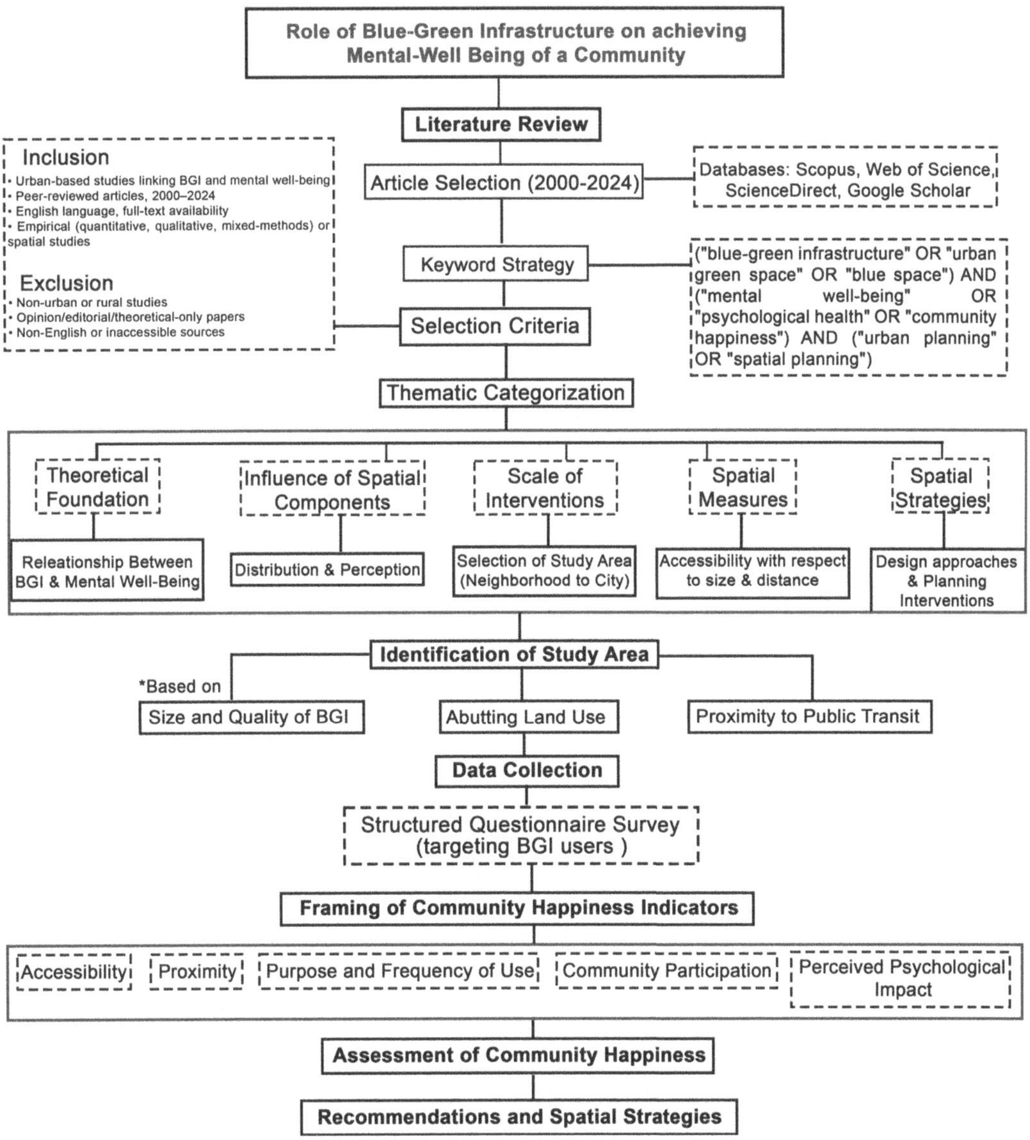

Fig. 7.2 Methodological framework of role of BGI on mental well-being. Source: Compiled by Authors

and PubMed. A Boolean keyword strategy was applied using the combination: ("blue-green infrastructure" OR "urban green space" OR "blue space") AND ("mental well-being" OR "psychological health" OR "community happiness") AND ("urban planning" OR "spatial planning"). The selection process followed rigorous inclusion and exclusion criteria: only peer-reviewed empirical studies in English focusing on urban settings and addressing the mental health impact of BGI were considered. Articles focusing on rural contexts, theoretical-only works, or those unrelated to well-being outcomes were excluded. The final set of articles was thematically categorized under five key dimensions: theoretical foundations, influence of spatial components, scale of interventions, spatial measures, and spatial strategies.

Insights from this review guided the identification of study areas, with selection based on three critical factors: the size and quality of existing BGI, the nature of abutting land use (residential, institutional, or commercial), and the proximity to public transit nodes. These criteria were considered essential for understanding how spatial accessibility and environmental integration influence community-level interaction with BGI Diener, (2009).

The next stage involved primary data collection through a structured questionnaire survey administered to local residents and BGI users. The survey captured participant perceptions, experiences, usage patterns, satisfaction levels, and self-reported emotional responses related to neighborhood green and blue spaces. Based on this, the study developed a set of Community Happiness Indicators, reflecting the multidimensional influence of BGI. These indicators include accessibility, proximity, purpose and frequency of use, community participation in maintaining BGI, and the perceived psychological impact of these spaces.

Subsequently, the study conducted an assessment of community happiness, evaluating how the presence, quality, and spatial configuration of BGI influence the mental well-being of residents across different urban typologies. The final phase involved translating these insights into recommendations and spatial strategies, including design-based suggestions for improving BGI accessibility, enhancing its multifunctionality, and embedding mental health priorities into urban planning frameworks.

The overarching goal of this methodology is to provide an evidence-based approach that supports urban planners and policymakers in shaping more inclusive, psychologically supportive, and sustainable urban environments, where BGI plays a strategic role in fostering community well-being.

To assess the influence of Blue-Green Infrastructure (BGI) on community well-being, a structured methodological framework was developed by analysis responses collected through structures questionnaires across multiple parks in Chennai City. The questionnaire was categorized in four thematic dimensions includes, Demographics (Q1-Q5), Questions based on BGI Accessibility and Usage (Q6-Q12), Mental well-being (Q13-Q16), and Urban Planning Perception based questions on attaining mental well-being (Q17-Q20). This categorization reflects the multidimensional structure of urban happiness as defined in various researches (Florida, 2002; Mouratidis, 2019). The detailed structure of questions includes, frequency of BGI visits (Q7), reason for visiting green spaces (Q8), and proximity to BGI (Q9) exhibited strong positive correlations with mental well-being. In contrast, elements like perceived urban-nature balance (Q18), satisfaction with availability (Q10), and opportunities for social interaction (Q16) showed low or negative correlation, indicating that the presence of BGI alone may not guarantee positive mental health unless supported by equitable access, quality design, and a sense of social inclusivity. Urban planning components such as maintenance quality (Q17) and support for community involvement (Q20) reflected varied impacts, highlighting the role of governance and user-centered planning in shaping perceived well-being.

Each questions; responses were normalized to compute a relative proportion of responses per option. If R_{ijk} represents the number of responses to option k for question i in park j, and T_{ij}, is the total responses for that question, the normalized proportion is computed as:

$$Pijk = \frac{Rijk}{Tij}$$

To evaluate the semantic intensity of responses, a numerical score S_{ik} was assigned to each option using a 5-point Likert scale, where "Strongly Agree" = 5 to a range of "Strongly Disagree" = 1. The weighted score for each option was then calculated as:

$$Wijk = Pijk \cdot Sik$$

Where *Pijk*, is the proportion of respondents selecting an option and *Sik*, assigns the numeric score for that option and *Wijk* is the weighted scores.

The aggregate score for each question was derived by summing all its weighted components:

$$Q_{ij} = \sum_{k} W_{ijk}$$

A critical part of the analysis focused on identifying the influence of each question on mental well = being, measured through the defined questions. Qij is the total score for question i in park j, where i = 13 to 16, questions specifically focused on mental well-being, on two of four thematic questions as a divisions. The average mental well-being score MWB_j for each park was computed as:

$$MWB_j = \frac{1}{4} \sum_{16}^{i=13} Q_{ij}$$

To assess the strength of correlation between the other survey components and mental well-being, a correlation index was defined as:

$$C_{ij} = \frac{Q_{ij}}{MWB_j}$$

Value of $C_{ij} > 1$ indicate that the corresponding question positively contributes to perceived mental well-being, while values <1 may indicate weaker or negative associations. This approach aligns with quantitative behavioral studies that assess subjective well-being (Kahneman and Krueger (2006).

In parallel, a comprehensive Community Happiness Index (CHI) was developed for each park by integrating scores across the major dimensions. Each dimensions was assigned a weight with 30% for BGI Accessibility, 30% for Mental well-being, 30% for components of Urban Planning and the remaining 10% for control variables as Demographic variables. The final CHI was derived as:

$$CHI_j = \sum_{d \in D} wd.\left(\frac{1}{n_d} \sum_{i \in d} Q_{ij}\right).20$$

- d: Dimension (e.g., Mental Well-being, BGI Accessibility, Urban Planning).
- D: The set of all such dimensions.
- w_d: The weight assigned to dimension d (e.g., 0.30 for each core category).
- n_d: Number of questions in dimension d.
- Q_{ij}: The question score from earlier.
- The multiplication by 20 scales the index to a 0–100 range.

Where D denotes the set of indicators, w_d the respective weight for dimension *d*, and n_d the number of questions in that indicators. The multiplication by 20 scales the CHI to a standardized 0–100 range. This weighted aggregation method ensures that all dimensions of urban experience are equitably represented, in line with multi-criteria decision making (MCDM) practice (Saaty, 2008; & Tanguay et al. (2010). This approach not only quantifies the overall level of happiness and satisfaction associated with BGI but also helps identify specific infrastructure and planning components that most significantly impact community mental well-being. (Jennings & Bamkole, 2019; WHO, 2017).

7.4 Result and Discussion

The structured methodological framework was applied to assess the relationship between Blue-Green Infrastructure (BGI) attributes and perceived mental well-being (MWB) across six urban parks in Chennai City. The result indicates significant spatial and perceptual variation in BGI Influences psychological outcomes, shaded by contextual characteristics and levels of user engagement.

Based on the application of the structured methodological framework to various identified parks of Chennai City, the correlation analysis between individual survey questions and mental well-being reveals significant insights into how Blue-Green Infrastructure (BGI) influences community happiness. Figure 7.3 indicates the correlation between the attributes of BGI with respect to mental well-being, reflects a strong positive correlation on frequency of visits to the BGI and the purpose of visit, highlights that BGI plays a vital role on achieving mental health. Study also observed that aesthetic and spatial integration of green elements, even at greater distances, supports mental well-being.

Figure 7.4 evidenced the engagement of community at the Anna Tower Park. However, the people perception indicates the requirement of BGI at the residential community to enhance the mental well-being, accessibility and community bonding, challenges the effectiveness of BGI as a mental health resource.

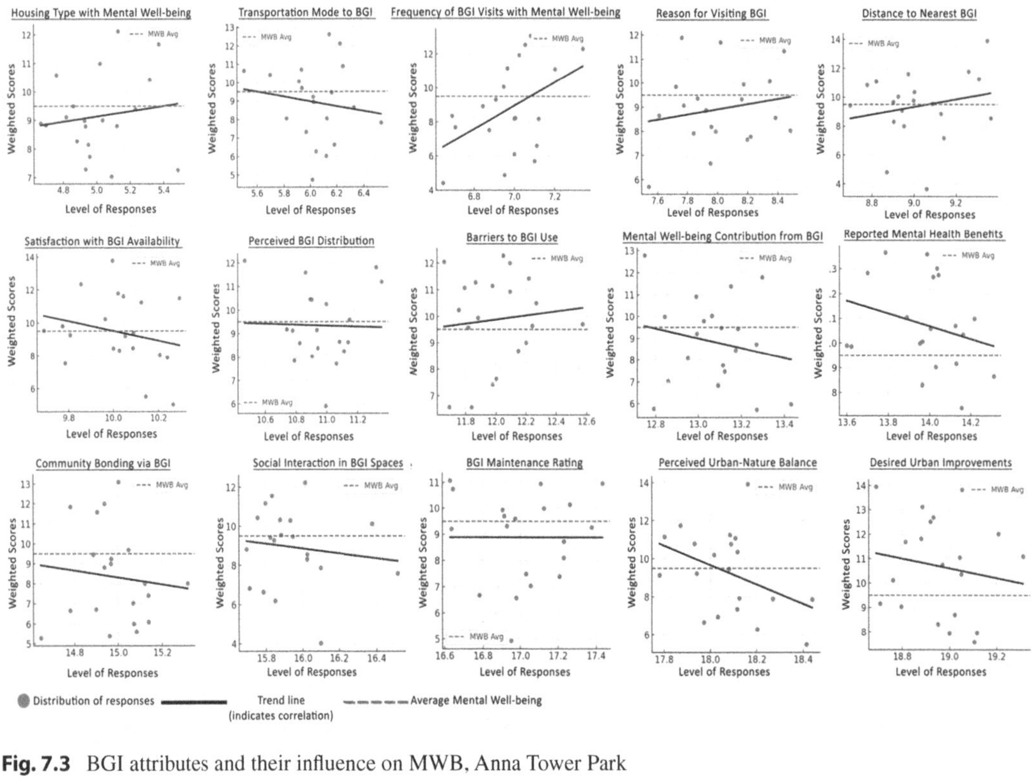

Fig. 7.3 BGI attributes and their influence on MWB, Anna Tower Park

Fig. 7.4 User engagement at Anna Tower Park

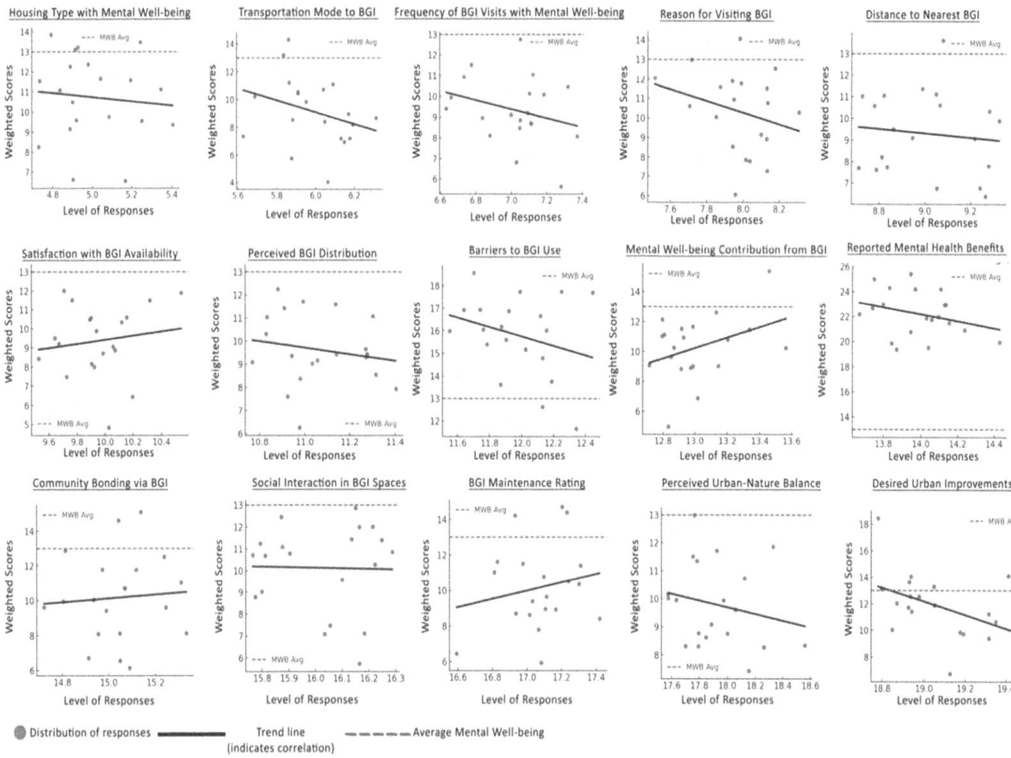

Fig. 7.5 BGI attributes and their influence on MWB, Eco Park, Chetpet

Fig. 7.6 User engagement at Eco Park, Chetpet

Figure 7.5 demonstrates the correlation matrix assessment at the Chetpet, Eco Park, shows the strong positive associations with frequency of BGI Visits (Slope +1.2) and the purpose of visit (slope +0.9), indicates that frequent and purposeful engagement with green infrastructure significantly enhances the individual's mental health. Perceived urban-nature balance with slope +0.7 and distance to nearest BGI imply that visual access and ecological integration of BGI, even when not in immediate proximity, contribute positively to mental health.

Figure 7.6 indicates the open seating zones, shaded pathways, and a small water body with embedded tree canopy encourages the visitors. This spatial layout encourages physical activity and casual recreation, supporting mental rejuvenation. However, despite the park's well maintained appearance and evident use, public perception indicates a gap in residential proximity based BGI access Mouratidis, (2020), suggesting the demand of centralized parks at the community level.

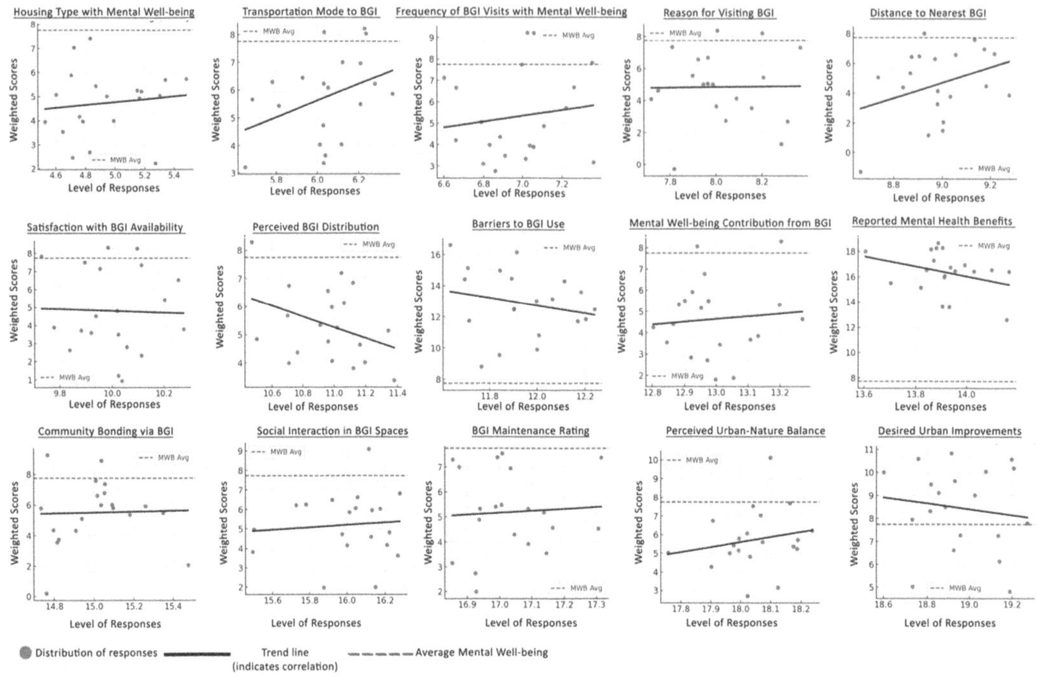

Fig. 7.7 BGI attributes and their influence on MWB, Sathyamurthy Park

Fig. 7.8 User engagement at Sathyamurthy Park

The correlation plots for Sathyamurthy Park reveal moderate positive trends in frequency of BGI visits (slope +0.7), reason for visiting (slope +0.8), and urban-nature balance (+0.6). Indicating beneficial effects of purposeful and regular park use on mental well-being. A notable inverse relationship in Perceived BGI Distribution (−0.9), and MWE benefits (−0.1) suggests that despite park quality, gaps in BGI access across neighbourhoods may dampen broader psychological benefits (Fig. 7.7).

Figure 7.8 highlights shaded walking trails; elderly focused gathering spots, and relaxed seating, signifying diverse user engagement Evans, & Wenner, (2007). However, the relative low MWB average (−7.8) implies that while spatial quality is evident, psychological gains remain limited without BGI access and inclusive urban improvement.

Comparing all the other parks, Semmozhi Poonga, is considered to be the largest BGI in the identified list, as indicated in Fig. 7.9 a high average MWB(16.4), with positive trends for perceived BGI distribution, social interaction with slope (+1.3), and desired urban improvements (slope +1.0). Moderate gains are seen in BGI Maintenance and Urban-Nature Balance with slope (+0.7). Negative trends in accessibility to

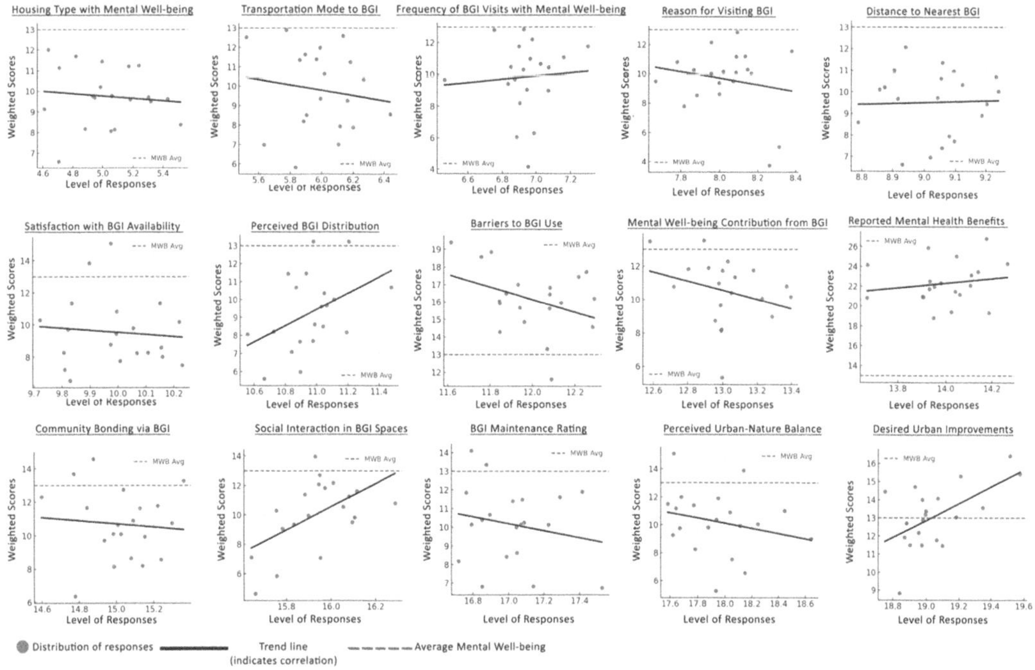

Fig. 7.9 BGI attributes and their influence on MWB, Senganthal Park

Fig. 7.10 User engagement at Senganthal Park

the BGI, mode of transportation, highlights the usage issues of BGI.

Figure 7.10 is the visual evidence of Semmozhi Poonga, show a highly mainted and diverse BGI setting, with shaded pathways and play zones. The park attracts varied users and supports both active and passive use. Despite centralized success, responses indicate a demand for similar residential-level BGI to enhance accessibility and community-level mental health benefits.

As indicated in the Figs. 7.11 and 7.12, Boganvilla Park was identified to experiment the smaller size parks in relevance with the effect of BGI, a like it shows a moderate average MWB, with slope of (11.2), having the positive correlation only because of the users' housing typology and frequency of visit, suggesting that frequent, nearby access contributes to better mental health. Negative trends in satisfaction with BGI availability and requires desired urban improvements, indicates user dissatisfaction despite visible greenery.

Field images highlight reflecting good infrastructure, however survey result suggest a need for more inclusive and decentralized yet continued BGI closer to residential communities to improve daily engagement and MWB.

Reflects a positive correlation as indicated in Fig. 7.13, the responses of Corporation Park at Anna Nagar, residential neighbourhood, displays

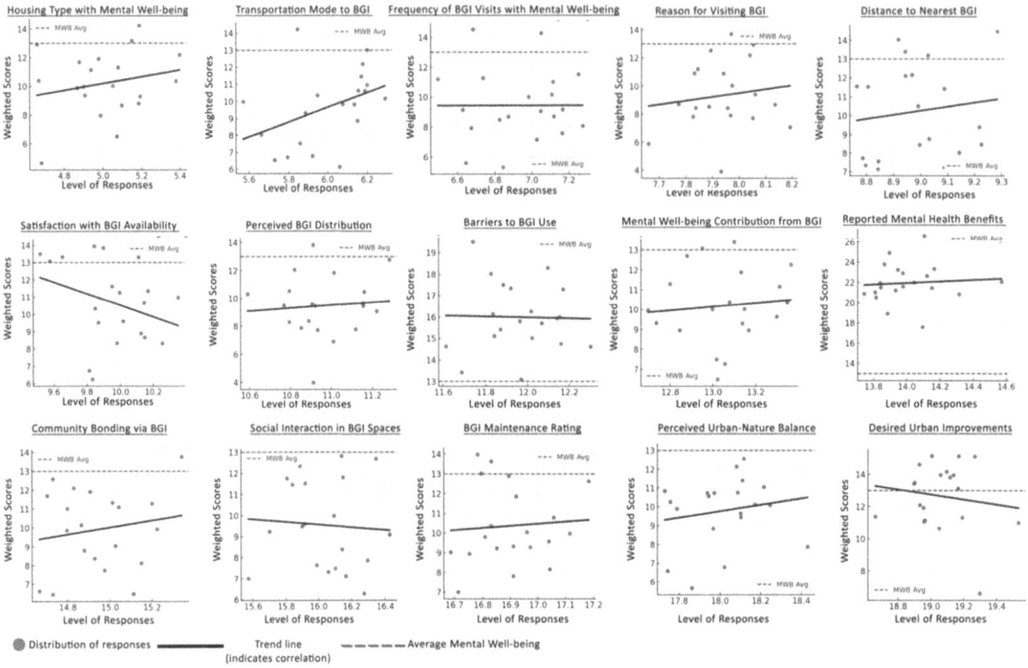

Fig. 7.11 BGI attributes and their influence on MWB, Boganvilla Park

Fig. 7.12 User engagement at Boganvilla Park

a notably high average MWB (34.5) with positive correlation in mental well-being.

As shown in Fig. 7.14 and based on the survey responses, despite the physical quality, there is a need to expand interactive and inclusive elements to support broader community engagement and ensure long-term mental health outcomes.

As indicated in Fig. 7.15. High-performing spaces such as Corporation Park and Semmozhi Poonga demonstrated the most significant positive correlations between BGI attributes and MWB. In Corporation Park, the highest CHI score (MWB = 34.5) was linked to strong social interaction, residential accessibility, and inclusive spatial design. Similarly, Semmozhi Poonga, the largest and most ecologically diverse park in the study, recorded an average MWB score of 16.4, driven by positive user feedback on spatial distribution, perceived ecosystem balance, and visual engagement. These findings reinforce the hypothesis that multifunctional and well-maintained BGI settings—when designed for proximity and inclusivity—offer not only environmental benefits but also foster psychological restoration and community well-being.

Moderately performing spaces such as Eco Park and Sathyamurthy Park revealed more nuanced outcomes. Although they scored positively on indicators such as frequency of use

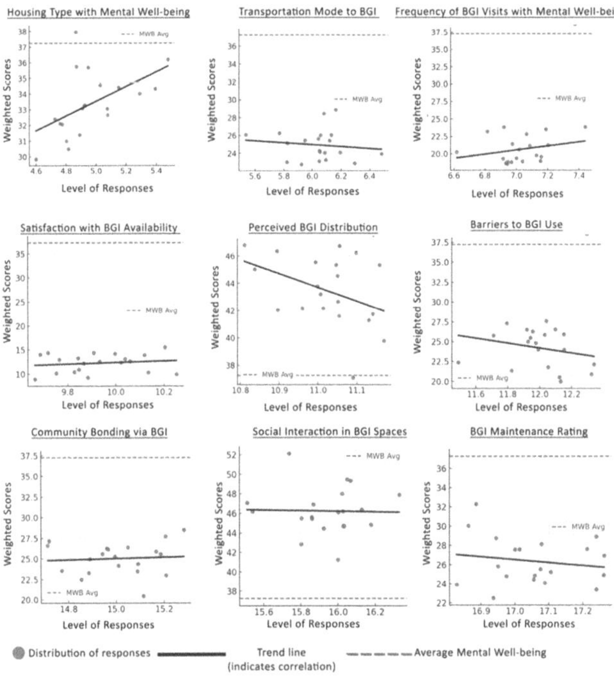

Fig. 7.13 BGI attributes and their influence on MWB, Corporation Park, Anna Nagar

Fig. 7.14 User engagement at Corporation Park, Anna Nagar

and perceived purpose, these benefits were tempered by negative correlations with accessibility and spatial equity. In Sathyamurthy Park, perceived spatial inequity and limited park distribution diminished the broader psychological impact despite its quality amenities Mouratidis, & Poortinga, (2020). The results suggest that even well-designed spaces may fail to deliver mental health benefits if not equitably distributed or adequately integrated into daily community life.

Smaller-scale BGI such as Boganvilla Park reflected mixed outcomes. Though moderate MWB scores were observed, these were primarily attributed to frequent use by nearby residents rather than spatial design or ecological richness. This indicates the vital role of decentralized, community-based BGI in fostering everyday mental well-being, particularly in high-density urban environments where access to larger parks is limited.

The correlation analysis across all parks consistently revealed that frequency of visits, purposeful engagement, and urban-nature balance were the most positively associated indicators of mental well-being. Conversely, attributes such as mode of transportation, transit barriers, and perceived spatial inequity were consistently negative, highlighting the impact of mobility and land use integration on the effectiveness of BGI as a health-promoting infrastructure.

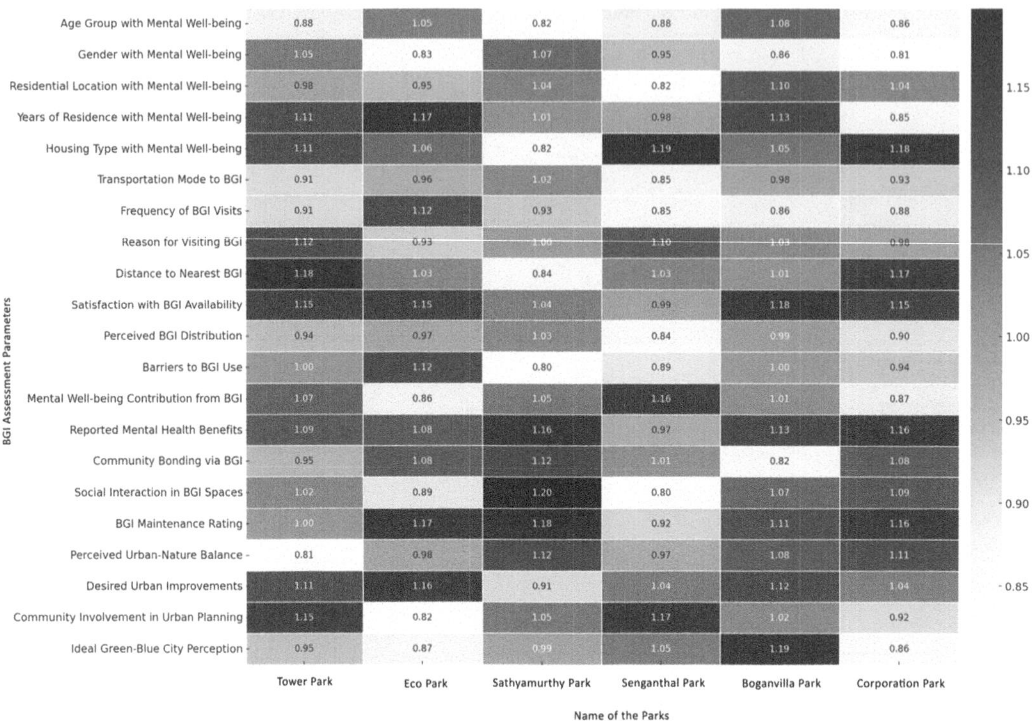

Fig. 7.15 Community happiness index across various parks of Chennai City

Taken together, the findings advocate for BGI as resilient urban waterscapes multifunctional systems that integrate ecological restoration, hydrological performance, and psychological resilience. These systems, when planned with equity and user experience in mind, can serve as both mental health infrastructure and climate adaptation tools. This repositions BGI as a critical element of sustainable water management, where its influence extends beyond stormwater control to include restorative, inclusive, and community-nurturing functions.

7.5 Conclusions, Limitation and Future Research

This study reinforces the growing recognition of Blue-Green Infrastructure (BGI) as a vital component of resilient, health-oriented urban planning. By examining spatial and perceptual responses across diverse neighborhoods in Chennai, the research empirically demonstrates how the quality, proximity, and ecological design of BGI significantly influence mental well-being outcomes. Parks that are accessible, multifunctional, and contextually integrated—such as Corporation Park and Semmozhipoonga, recorded the highest mental health benefits, whereas limited accessibility and uneven spatial distribution diminished the effectiveness of other green spaces despite good design quality. The study introduces the Community Happiness Index (CHI) as a replicable evaluation tool to assess both psychological and spatial equity impacts of BGI, offering a multi-criteria framework that enables urban planners to embed well-being into land use strategies while contributing to climate resilience. This perspective underscores the evolving role of BGI as essential urban infrastructure, one that integrates ecological functionality with human-centered benefits, supporting both sustainable water management and the psychological resilience of urban communities. However, the study acknowledges certain limitations, including the restricted geographic scope to six parks in Chennai, reliance on self-reported data that may be subject to perception bias, limited consideration of seasonal or temporal variations, and underrepresentation of certain demographic groups such as the elderly, children, and differently-abled individuals. Future research should address these limitations by expanding to longitudinal and cross-city comparative studies, integrating sensor-based physiological data, exploring co-benefits of BGI such as air quality and urban cooling, developing participatory co-design models, and linking CHI with hydrological performance indicators to inform dual-purpose BGI design that supports both sustainable water management and community mental well-being.

Funding: The research was not supported by any funding.

Potential conflicts of interest: On behalf of all authors, the corresponding author states that there is no conflict of interest.

Informed Consent: Informed consent was obtained from all individual participants included in the study.

Authors Contribution: All authors have contributed equally to this work.

Data availability statement: The datasets during and/or analyzed during the current study available from the corresponding author on reasonable request.

AI Discloser Statement: The author affirm that no generative artificial intelligence tools, including language models, data generators, were used in the conceptualization, data collection, analysis, interpretation, drafting, or editing of this manuscript. All aspects of the research were conducted by the authors through manual methods and domain expertise.

References

Astell-Burt, T., & Feng, X. (2021). Urban green space, metal health, and longevity. *Nature Sustainability, 4*(6), 534–540.

Bhan, M., et al. (2020). Green spaces and mental health: A review for India. *Indian Journal of Public Health, 64*(3), 238–243.

Bloom, D. E. (2014). Exposure to nature versus relaxation during lunch breaks and recovery from work: Development and design of an intervention study to improve workers' health, Well-being, work performance and creativity. *BMC Public Health, 14*, 488.

Bowler, D. E., et al. (2010). Urban greening to cool towns and cities: A systematic review of the empirical evidence. *Landscape and Urban Planning, 97*(3), 147–155.

Bratman, G. N., et al. (2019). Nature experience reduces rumination and subgenual prefrontal cortex activation. *Proceedings of the National Academy of Science, 112*(28), 8567–8572.

Diener, E. (2009). *Assessing well-being: The collected works of Ed Diener*. Springer.

Diener, E., et al. (2010). National accounts of subjective well-being. *American Psychologist, 65*(8), 852–860.

Elmqvist, T., et al. (2015). Benefits of restoring ecosystem services in urban areas. *Current Opinion in Environmental Sustainability, 14*, 101–108.

Evans, G. W., & Wenner, R. E. (2007). Crowding and personal space invasion on the train: Please don't make me sit in the middle. *Journal of Environmental Psychology, 27*(1), 90–94.

Florida, R. (2002). *The rise of the creative class*. Basic Books.

Gascon, M., et al. (2017). Outdoor blue spaces, human health, and well-being: A systematic review of quantitative studies. *International Journal of Hygiene and Environmental Health, 220*(8), 1207–1221.

Gill, S. E., et al. (2007). Adapting cities for climate change: The role of the green infrastructure. *Built Environment, 33*(1), 115–133.

Grahn, P., & Stigsdotter, U. A. (2010). The relation between perceived sensory dimensions of urban green space and stress restoration. *Landscape and Urban Planning, 94*(3–4), 264–275.

Guizzo, A., et al. (2021). *Sustainable solutions in urban health: Transdisciplinary directions in urban planning for global public health* (pp. 279–296). Springer Nature.

Haddad, N., et al. (2015). Habitat fragmentation and its lasting impact on earth's ecosystem. *Science Advances, 1*(2), e1500052.

Harpham, T. (1994). Urbanization and mental health in developing countries: A research role for social scientists, public health professionals and social psychiatrists. *Social Science & Medicine, 39*(2), 233–245.

Hepat, A., & Khode, D. (2024). A narrative review of urbanization and mental health: An Indian perspective. *National Library of Medicine*. PMCID: PMC10982837.

Hinshaw, S. P. (2000). Stigma and mental disorder: Conceptions of illness, public attitudes, personal disclosure, and social policy. *Development and Psychopathology, 12*(4), 555–598.

Hunter, R. F., et al. (2019). Environmental and social interventions for mental health. *The Lancet Public Health, 4*(11), e567.

Jennings, V., & Bamkole, O. (2019). The relationship between social cohesion and urban green space: An avenue for health promotion. *International Journal*

of Environmental Research and Public Health, 16(3), 452.
Kabisch, N., Qureshi, S., & Haase, D. (2017). Human-environment interactions in urban green spaces—A systematic review of contemporary issues and prospects for future research. *Environmental Impact Assessment Review, 50*, 25–34.
Kahneman, D., & Krueger, A. B. (2006). Developments in the measurement of subjective well-being. *Journal of Economic Perspectives, 20*(1), 3–24.
Kaplan, S. (1995). The restorative benefits of nature: Toward an integrative framework. *Journal of Environmental Psychology, 15*(3), 169–182.
Keyes, C. L. M. (2002). The mental health continuum: From languishing to flourishing in life. *Journal of Health and Social Behavior, 43*(2), 207–222.
Kozaryan, O., & Valente, R. R. (2019). The urban-rural happiness gradient: Do amenities matter? *Cities, 110*, 276–282.
Leyden, K. M. (2003). Social capital and the built environment: The importance of walkable neighborhoods. *American Journal of Public Health, 93*(9), 1546–1551.
Mell, I. C. (2017). Green infrastructure: Reflections on past, present and future praxis. *Landscape Research, 42*(2), 135–145.
Mitchell, R., & Popham, F. (2008). Effect of exposure to natural environment on health inequalities. *The Lancet, 372*, 1655–1660.
Mouratidis, K. (2019). Built environment and social well-being: How does urban form affect social life and personal relationships? *Cities, 74*, 41–50.
Mouratidis, K. (2020). Commute satisfaction, neighborhood satisfaction, and overall happiness. *Travel Behaviour and Society, 21*, 169–179.
Mouratidis, K., & Poortinga, W. (2020). Built environment, urban vitality and social cohesion: Do vibrant neighborhoods foster strong communities? *Landscape and Urban Planning, 204*, 103951.
Nations, U. (2014). *World urbanization prospects: The 2014 revision, highlights*. Department of Economic and Social Affairs. Population Division, United Nations, 32.
Papas, M. A. (2007). The built environment and obesity. *Epidemiologic Reviews, 29*, 129–143.
Saaty, T. L. (2008). Decision making with the analytical hierarchy process. *International Journal of Services Science, 1*(1), 83–98.
Tan, P. Y., Wang, J., & Sia, A. (2013). Perspectives on five years of research on urban greenery and mental health. *Ecopsychology, 13*(4), 239–243.
Tanguay, G. A., et al. (2010). Measuring the sustainability of cities: An analysis of the use of local indicators. *Ecological Indicators, 10*(2), 407–418.
Twohig-Bennett, C., & Jones, A. (2018). The health benefits of the great outdoors: A systematic review and meta-analysis of greenspace exposure and health outcomes. *Environmental Research, 166*, 628–637.
Ulrich, R. S. (1983). Aesthetic and affective response to natural environment. In *Behavior and the natural environment* (pp. 85–125). Springer.
Van de Berg, A. E., et al. (2010). Green space as a buffer between stressful life events and health. *Social Science & Medicine, 70*, 1203–1210.
Van de Ven, F. H. M. (2015). Planning support system for climate adaptation: Composting effective sets of blue-green measures to reduce vulnerability to extreme weather events. *Building and Environment, 83*, 159–167.
Van den Bosch, M., & Ode Sang, A. (2017). Urban natural environments as nature-based solutions for improved public health—A systematic review of reviews. *Environmental Research, 158*, 373–384.
Walsh, C. J., et al. (2005). The urban stream syndrome: Current knowledge and the search for a cure. *Journal pf the North American Benthological Society, 24*(3), 706–723.
White, M. P., et al. (2021). Associations between green/blue spaces and mental health across 18 countries. *Scientific Reports, 11*(1), 8903.
Wilson, E. O. (2009). Biophilia. Harvard University Press. ISBN: 9780674045231
World Health Organisation (WHO). (2004). *Promoting mental health: Concepts, emerging evidence, practice.*
World Health Organisation (WHO). (2014). urban green spaces and health: A review of evidence. WHO Regional Office for Europe.
World Health Organisation (WHO). (2017). Urban green spaces: A brief for action. WHO Regional Office for Europe.
World Mental Health Report: Transforming Mental Health for All. (2022). World Health Organisation.
Xiao, J., et al. (2022). The impact of built environment on mental health: A COVID-19 lockdown perspective. *Health & Place, 77*, 102889.
Zhang, F., & Gao, Y. (2020). Spatial planning, community participation, and subjective well-being: Evidence from urban China. *Sustainable Cities and Society, 63*, 102452.
Zhao, Z., et al. (2022). The impact of urbanization on the delivery of public service-related SDGs in China. *Sustain Cities Society, 80*, 103776.
Zhong, J., & Liu, W. (2022). Role of built environmnets on physical activity and health promotion: A review and policy insights. *Sec. Environmental Health and Exposome, 10*, 950348.

Part III

Smart Technologies and Policy Pathways for Sustainable Water Management

Integrating Geospatial Evidence into Urban Water Management: Lessons from Surface Water Dynamics in Dhaka, Bangladesh

8

Muhtasim Ahmed, Kazi Jihadur Rashid, and Rowshon Ara Toma

Abstract

Surface water has its vital effect—reducing microclimate, sustaining biodiversity and more. In recent decades, the Dhaka Metropolitan Area (DMA) has witnessed a drastic decline in surface water resources due to unplanned urban expansion and weak enforcement of environmental regulations. Despite formal recognition of their importance in urban planning frameworks such as the Dhaka Metropolitan Development Plan (DMDP) and Detailed Area Plan (DAP), surface water bodies continue to be encroached upon at an alarming rate. Analyzing temporal changes in surface water in the DMA using satellite imagery from Landsat-5 and Landsat-8 for the years 2000, 2010, and 2020 and using Modified Normalized Difference Water Index (MNDWI) to classify surface water and non-water areas and to quantify and map spatiotemporal dynamics. The results reveal a substantial decrease in surface water, from 108.43 km^2 in 2000 to just 24 km^2 in 2020, representing a 77.9% decline over two decades. During the same period, non-water surfaces expanded from approximately 225 km^2 to over 309 km^2, indicating a significant conversion of aquatic areas into urban land uses. We discussed the influence of DMDP and DAP on surface water dynamics. This chapter emphasizes the urgent need for integrated spatial data-driven management strategies. By leveraging GIS and remote sensing technologies, urban planners and policymakers can establish a comprehensive digital database for continuous monitoring and evidence-based interventions. Strengthening institutional mechanisms and enforcing urban planning tools such as DAP are critical to curbing further losses and restoring the ecological functions of surface water systems in Dhaka.

Keywords

Environmental planning · Modified normalized difference water index (MNDWI) · Planning policy enforcement · Remote sensing · Surface water dynamics · Urban expansion

M. Ahmed (✉) · R. A. Toma
Remote Sensing Division, Center for Environmental and Geographic Information Services (CEGIS), Dhaka, Bangladesh

K. J. Rashid
Department of Geography, Florida State University (FSU), Tallahassee, FL, USA

U. Chatterjee, A. I. Almulhim (eds.), *Urban Waterscapes*, Sustainable Development Goals Series,
https://doi.org/10.1007/978-3-032-20649-7_8

8.1 Introduction

Surface water is extremely beneficial to the health of the ecosystem and provides vital homes for numerous endangered plants and animal species. It offers a variety of benefits to surrounding communities, including the capacity to protect against the effects of natural disasters, filter water, conserve soil and water, and offer opportunities for recreational activity (Grenier et al., 2007). The world's surface water areas cover at least 7 million square kilometers of land (Melton et al., 2013). However, factors like excessive irrigation, groundwater extraction, and drainage contribute to its deterioration (Koch et al., 2012). As a consequence, mapping and monitoring surface water is crucial for maintaining ecological balance and managing water resources effectively.

Bangladesh is known to be one of the countries where global climate change struck hard and take a high toll. Communities and people who largely rely on natural resources for the support of their livelihoods find it difficult to cope with the uncertainty and size of the ramifications. Climate change may have a substantial impact on the ecological character of surface water and their continued use, and conservation strategies must consider the potentially important role that surface water may play in adapting to and mitigating the effects of climate change. Being the capital city, Dhaka has faced significant environmental challenges, including flooding and water logging. In the past, it was referred to as the "Venice of the East" or "the City of Channels" due to the numerous canals and waterways that facilitated drainage (Mahmud et al., 2011). However, over the years, unplanned and unregulated urbanization across Bangladesh and extensive canal loss have led to the disappearance of surface water and floodplains, exemplifies the consequences of weak planning enforcement. (Mahmud et al., 2011; Rashid et al., 2025). Encroachment and exploitation of riverbanks, poor inter-agency coordination, neglect of system maintenance, and ineffective legislation enforcement have contributed to these issues. As a result, the city experiences severe water logging during the wet season, undermining urban resilience (Kamal & Midorikawa, 2004). A detailed understanding of surface water changes in Dhaka is essential to address these challenges and improve drainage and flood management.

Remote sensing is an efficient tool for studying surface water because it can acquire frequent measurements from locations that are difficult to reach and provide timely information. Satellite remote sensing is particularly effective and cost--efficient for surface water mapping, as it allows for the creation of surface water inventories. These inventories, produced through satellite images, demarcate the area and distribution of surface water over geographical regions. Remote sensing provides updated data, making it a valuable tool for monitoring surface water changes and supporting policy decisions (Ozesmi & Bauer, 2002; Tuli et al., 2024). It can be used to create maps, monitor changes, and quantify surface water dynamics over time. There have been a lot of improvements in determining surface water using remote sensing techniques. Huang et al. (2018) provided a comprehensive review of optical sensors in water surface monitoring and detection. They review several methods such as threshold-based, index-based, and machine learning models for extracting water. Among the index-based techniques, normalized difference vegetation index (NDVI), normalized difference water index (NDWI), modified NDWI, enhanced water index (EWI), and automated water extraction index (AWEI) were used mostly (Bijeesh & Narasimhamurthy, 2020). However, MNDWI is identified as the most robust and easily applicable index separating water from land, especially in urban mix land cover areas (Huang et al., 2018). Moreover, from different sensors, medium--resolution Landsat provide the best trade-offs for multi-decadal analysis and most suitable for spatiotemporal dynamics.

Detailed area plan (DAP) has already been published by the city planning and development authority of Dhaka, RAJUK which underlines precise planning policy for development in metropolitan region of Dhaka. It updates the planning flaws of Dhaka Metropolitan Development plan (DMDP) 1995–2015 and also create more sustainable policy aspect to make the city more livable. It clearly indicates how the surface water protection policy will be implemented throughout the area. In rapid urbanization, encroachment of water areas is a common phenomenon. In background the master plan mentioned that watershed in Dhaka metropolitan area degraded from 29sqkm to 10.28sqkm in between year 1978 to 2009 (RAJUK, 2019), not to mention this resonates the analysis of the chapter. To diminish this unregulated development and lessen the deteriorating effect of urbanization in ecology and biodiversity of Dhaka metropolitan area, this detailed area plan is a guideline for retaining surface water areas for long time. Primarily the plan was to initiate to be implemented from 2016 though it was officially launched in 2022 with some policy revision. Again, Sustainable Development Goals (SDGs) are the cornerstone of the 2030 Agenda for sustainable development, which was endorsed by the United Nations General Assembly in 2015. The 17 basic goals are subdivided into 169 linked targets to guide global development strategy until 2030. These aims and objectives cover the economic, social, and environmental dimensions of sustainable development in an unparalleled manner. Among the 17 goals, goal 06 is about water and sanitation. It stated as—ensure access to water and sanitation for all. (United Nations (UN), 2025) Under that goal 6 targets are given for the countries to achieve. Target 6.6 is to protect and restore water-related ecosystems, including mountains, forests, wetlands, rivers, aquifers, and lakes by 2020. Bangladesh has already achieved great achievement in ensuring that people have access to safe drinking water and sanitation and wants to guarantee that all people have access to safe water. To promote the implementation of SDG6, the government has created an action plan, a national policy on water, water supply, sewerage, environment protection guidelines, and water legislation. The government has created a water management policy that considers all of the country's water-related phenomena and emphasizes the use of surface water to reduce reliance on aquifers.

This chapter will provide a scenario of surface water condition with decadal evidence, and it might be a database for planning policies to manage these watershed areas align with Dhaka metropolitan area detailed area plan (2016–2035). Therefore, objectives of this chapter are to monitor the quantification of surface water changes in the Dhaka Metropolitan Area (DMA) from 2000 to 2020 and management of these in support of Dhaka Structure Plan (2016–2035) for future preservation of ecosystem.

8.2 Justification of the Study

Surface water are vital ecological features that offer numerous economic, ecological, and cultural benefits, such as water quality protection, flood and erosion control, wildlife habitat, and recreational opportunities. In Bangladesh, surface water covers an estimated 7–8 million hectares—around 50% of the country's land area—and include both natural and artificial types(Haroon & Kibria, 2017). However, these ecosystems are under threat due to poor management, urbanization, poverty, and lack of awareness. The decline is further worsened by weakened traditional practices and regulatory oversight. Recent advancements in GIS and remote sensing technologies have given the edge for monitoring and managing surface water. In this chapter, the changes in the surface water of Dhaka city are evaluated using GIS and high-resolution satellite data. Surface water management has become an urgent environmental concern in Bangladesh.

Several research on the land use and land cover change of Dhaka were conducted with a particular focus on urbanization (Dewan & Yamaguchi, 2009; Hossein et al., 2023; Rashid et al., 2023). However, few studies focused on surface water change in the area (Faridatul et al., 2019; Hosen et al., 2025; Hossain et al., 2023; Paul et al., 2019). Paul et al. (2019) using historical CORONA (1967) and Quickbird (2008) satellite data quantifying the loss of 33% of Dhaka's floodplains and surface water bodies and assesses implications for earthquake vulnerability. Faridatul et al. (2019) studied multi-seasonal Landsat imagery (over 27 years) using advanced spatial modeling (GWR) to investigate how seasonal variability, land cover change, and topography influence urban surface water dynamics in Dhaka comparing with Hong Kong. Hossain et al. (2023) focused on surface water decline analyzing Landsat satellite sensor data in GEE platform and linked it to urban sprawl and encroachment. Additionally, Hosen et al. (2025) evaluated urban potential to reduce land surface temperature in Dhaka. Paul et al. (2019) primarily focused on the degradation of surface water and floodplains in and around Dhaka city, linking this environmental loss to increased earthquake vulnerability. Although the study did not provide explicit policy recommendations, it presented a compelling argument for integrating environmental concerns into urban planning. In a more recent contribution, Rahman and Szabó (2021) analyzed changes in land use and land cover and highlighted that wetlands offer the highest ecosystem service value (ESV) among all land use types. Consequently, the degradation of surface water areas results in substantial ecological losses for the city. These studies demonstrate the analytical precision of surface water detection techniques and focus on the environmental impacts of their loss as well as factors that are increasingly making Dhaka less livable.

In light of these findings, this chapter emphasizes the necessity of a planning-level analysis to translate such evidence into actionable policy. The study offers a novel approach by explicitly incorporating the master plan framework, particularly the DMDP and DAP for Dhaka, into its methodology. By aligning empirical surface water analysis with planning instruments, the chapter provides guidance for future urban strategies aimed at preserving existing surface water resources. It acknowledges the inevitable depletion of water bodies due to rapid urbanization but proposes policy-driven pathways to mitigate further ecological damage.

8.3 Materials and Methods

8.3.1 Study Site

The Dhaka Metropolitan Area (also known as DMA) is the focus of this research, and its location coordinates are as follows: longitude 90°20′E to 90°30′E and latitude 23°40′N to 23°55′N (Fig. 8.1). The Dhaka Metropolitan Area (DMA) spans a total area of 306 square kilometers with a population reported to be 15.4 million in 2011 (Ahmed, 2018; Rashid et al., 2023). Dhaka's borders are defined by the many rivers that surround the city. It is in a particularly advantageous position geographically, as it is connected to the rest of the country by a system of rivers. This region took full advantage of the benefits offered by this river system at the beginning of the process of urbanization (Mahmud et al., 2011). The Buriganga River can be found in the south, the Balu River can be found in the east, and the Turag River can be found in the north and west. James Rannell, who surveyed the rivers of east Bengal in 1765, wrote that "the kingdom of Bengal, particularly the eastern part, is naturally the most convenient for trade within itself of any country in the world; for the rivers divided into such a number of branches that the people have the convenience of water carriage to and from any principal place." Rannell made these observations after conducting a survey of the rivers in east Bengal (Ahmed, 1986). Due to its location in the geographic center of what was East Bengal at the time, Dhaka had a significant command over all of these water routes. As can be seen in the Dhaka Metropolitan Area (DMA) is chosen (Fig. 8.1) as the focus of this chapter due to the rapid population increase that the area has experienced over the past 20 years (Rashid et al., 2023). The DMA encompasses both the central city of Dhaka and the areas that surround it. The shifting wetland as a result of the alteration of land cover (LC) is also taken into consideration for the areas surrounding the DMA.

8.3.2 Data Description

We collected Landsat satellite sensors data for this research from Google Earth Engine (GEE) archive. Surface Reflectance Landsat archive was chosen to get pre-processed images. Images from the late post-monsoon season were considered as the surface water features are at their fullest after the monsoon rainfall. To be more specific, November is the most suitable time for peak detection of permanent waterbodies and wetlands with lowest cloud cover. More characteristics of the data can be found below (Table 8.1).

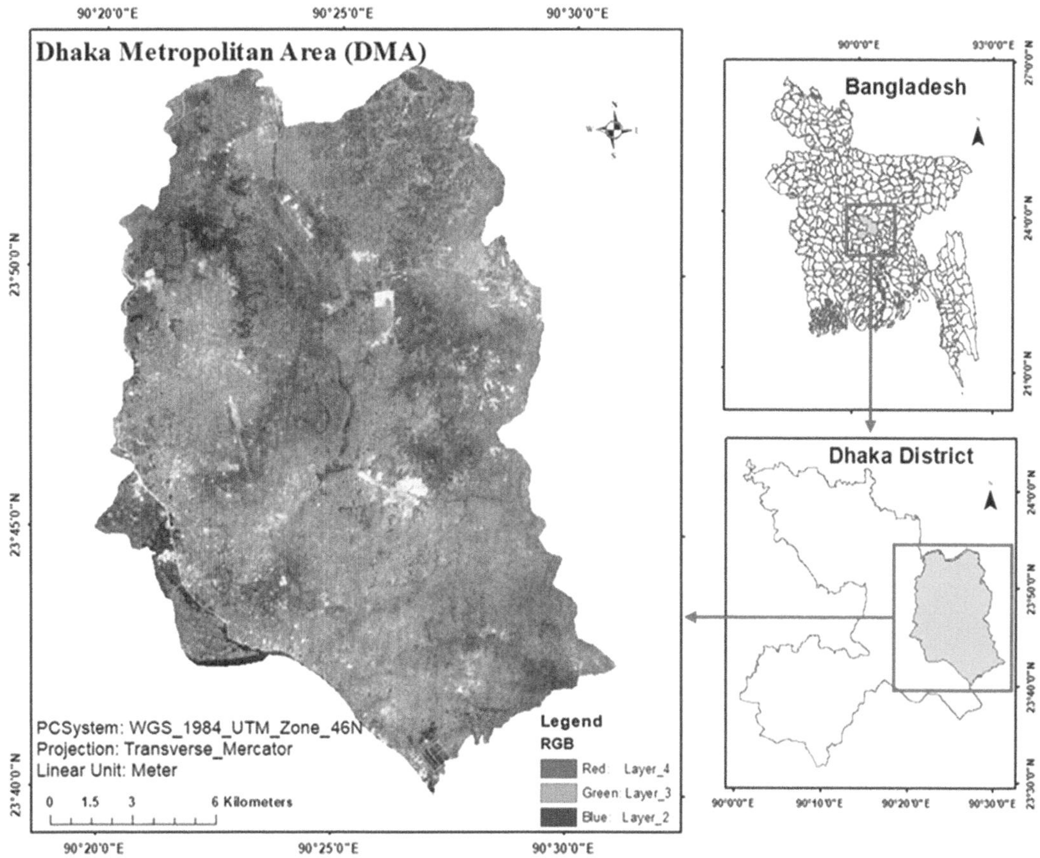

Fig. 8.1 Study area showing administrative boundary

Table 8.1 Characteristics of Landsat satellite images data used for the study

Sl. No	Sensor	Row/Path	Acquisition Month	Resolution
1	Landsat 5	44/137	November, 2000	30 m
2	Landsat 5	44/137	November, 2010	30 m
3	Landsat 8	44/137	November, 2020	30 m

8.3.3 Methods

Initially, we accessed Landsat data in GEE to delineate surface water distribution and extract it to post-process in RStudio. The steps followed in this chapter are shown (Fig. 8.2).

8.3.4 Image Collection and Selection

At first, the tier 1 surface reflectance of Landsat 05 Thematic Mapper and 08 Operational Land Imager was imported from the GEE archive datasets. From these collections, we filtered images based on months and the area of interest in the GEE platform. It is crucial to select the most suitable month for the study area due to the seasonal cloud cover and a mixture of seasonal and permanent surface water, along with growing vegetation, making clear water delineation more challenging. Therefore, we needed a time when the water bodies were adequately filled but not flooded, and when clear sky conditions prevailed. The month of November was selected as mentioned earlier for all the years to minimize sea-

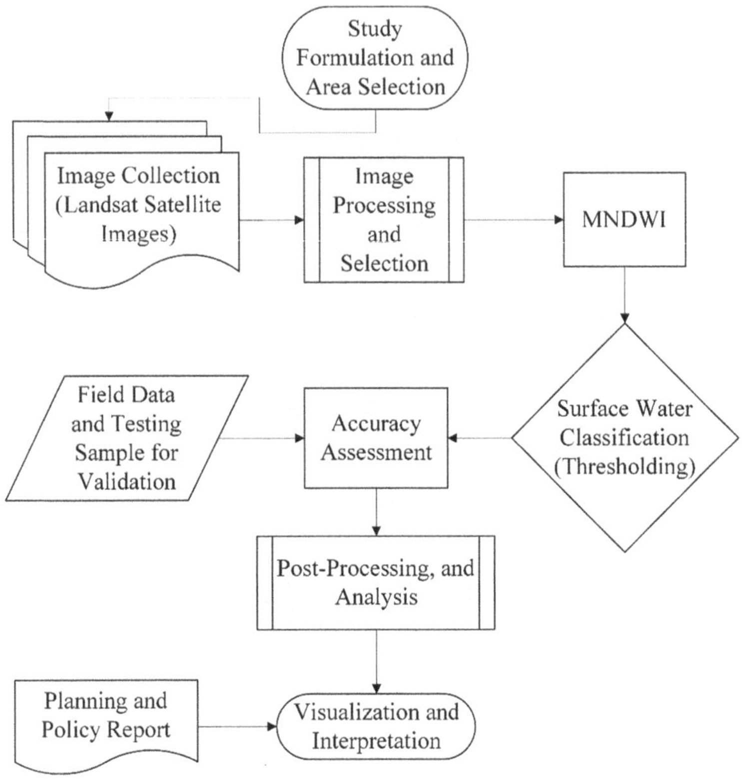

Fig. 8.2 Flowchart of research methodology

sonal variability and consistency across datasets among targeted years. After filtering with dates and DMA boundary, the median function was used to get a single image for each year. Then, it was clipped using the DMA boundary. To facilitate visual inspection of land cover changes, False Color Composites (FCC) were created by selecting the appropriate near infrared (NIR), red, and green bands for each Landsat sensor. These composites were also exported as GeoTIFF files for further analysis.

8.3.5 Surface Water Delineation and Visualization

From the output of the filtering process, images were used for indexing to delineate water in GEE platform. The modified normalized difference water index (MNDWI), an improved technique of the traditional NDWI (Ali et al., 2019), was used for this purpose. The urban area, like Dhaka, poses challenges to detect surface water, especially due to built-up and vegetation noises. This modified index is better suited for urban mixed areas to capture the water surface properly. It enhances the capability of NDWI by replacing near-infrared (NIR) with shortwave infrared (SWIR1) by helping greatly to suppress noise from built-up and vegetation and distinguish open water surface. The MNDWI is calculated using the Green and Shortwave Infrared (SWIR) bands from Landsat sensors (Eq. 8.1).

$$NDWI = \frac{Green - SWIR}{Green + SWIR} \quad (8.1)$$

For Landsat 5, the green band corresponds to Band 2 and the SWIR band to Band 5. For Landsat 8, the green band is Band 3, and the SWIR band is Band 6.

MNDWI values also range between −1 and +1, where positive values typically indicate open water bodies, while negative values are associated with non-water features such as soil, vegetation, and built-up surfaces (Gautam et al., 2015).

We classified the MNDWI into binary water masks (2 = water, 1 = non-water) using the threshold (Eq. 8.2). This threshold has been widely used in prior studies and provides robust water detection in urban and peri-urban areas.

$$W_t(x,y) = \begin{cases} 2 & \text{if } MNDWI_t(x,y) < 0 \\ 1 & \text{Otherwise} \end{cases} \quad (8.2)$$

Where W_t *and* $MNDWI_t$ are the binary water mask and the water index image for t year, respectively.

Post-classification processing was conducted in R using the terra and ggplot2 packages. Raster files were cleaned by setting zero and NA values to missing values to refine the water masks. Water and non-water classes were quantified by calculating the total area (in km^2) and their percentage share within the study region. Raster data were then visualized through thematic maps, illustrating water and non-water classes for each year. Additionally, the distribution of MNDWI values was examined using density plots. These plots allowed comparison of the frequency distribution of MNDWI values across the three study years, providing insights into the variation and potential shifts in surface water extents over time.

8.4 Results and Discussion

8.4.1 Surface Water Distribution and Decadal Change

Classified maps of surface water and non-water areas for 2000, 2010, and 2020 are displayed (Fig. 8.3). A visible and consistent decline in water extent is evident across the three-time steps. In 2000, large contiguous areas of water were concentrated primarily in the eastern, north-western, and south-western zones, with significant coverage throughout the city's periphery. By 2010, much of this water had diminished in all the mentioned zones where most of the uprising housing project underwent in this timeframe. By 2020, the water bodies became highly fragmented and sparse, persisting mostly in the northwestern and eastern pockets due to continuing the development process. This spatial trend reflects severe urban encroachment into wetlands and water bodies, often driven by land reclamation, infrastructure development, and population pressures. These findings visually confirm the ongoing reduction in natural hydrological surfaces across the metropolitan region.

We can see that the surface water constituted approximately 32.52% of the study area, with non-water areas covering 67.48%, indicating a

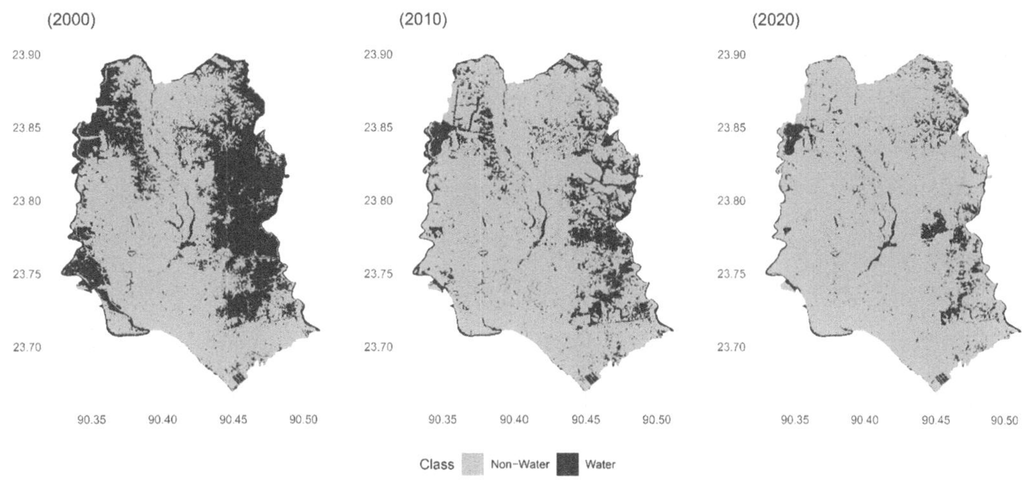

Fig. 8.3 Surface water features including rivers, lakes, ponds, and wetlands in the year 2000, 2010, 2020

relatively balanced distribution in the year 2000 (Table 8.2). However, by 2010, a marked shift had occurred: water bodies accounted for only 15.10%, while non-water surfaces expanded to 84.90%. This significant reduction—over 17% points—suggests accelerated loss of surface water within just a decade. The trend continued into 2020, where surface water area fell further to 7.26%, leaving 92.74% of the region covered by non-water classes. In absolute terms, surface water area declined from 108.43 km^2 in 2000 to 50.34 km^2 in 2010, and then to 24.19 km^2 in 2020, indicating a cumulative loss of more than 84 km^2 over two decades. Conversely, non-water areas mostly including urban surface increased steadily from 224.98 km^2 in 2000, to 283.07 km^2 in 2010, and reached 309.22 km^2 by 2020. The largest drop in water extent occurred between 2000 and 2010, during which approximately 58 km^2 of surface water was lost, likely due to urban land conversion and infrastructure development. This degradation trend highlights the escalating encroachment upon aquatic ecosystems, driven by rapid urbanization and unchecked land-use change.

Table 8.2 shown classification statistics from 2000 to 2010

Year	Class	Area (km^2)	Percentage (%)
2000	Non-water	224.98	67.48
2000	Water	108.43	32.52
2010	Non-water	283.07	84.90
2010	Water	50.34	15.10
2020	Non-water	309.22	92.74
2020	Water	24.19	7.26

The density plots of MNDWI values for each year showing the distribution difference at decadal interval (Fig. 8.4). These plots help to further understand the statistical behavior of pixel values across the landscape. The peaks of between 0 and 0.4 represent surface water whereas the top skewed peaks below 0 are non-water surfaces especially including urban pixels. In 2000, the density curve shows a broader spread and higher peaks in the positive MNDWI range, indicating a substantial number of water pixels. There are two peaks in that year, one being skewed between 0.15 and 0.25 representing deep waterbodies and another being flattened represents mostly shallow seasonal surface water. By 2010, the density shifted slightly leftward with fewer pixels exceeding the threshold of 0, suggesting reduced surface water. Following the trend, thc distribution became sharply centered

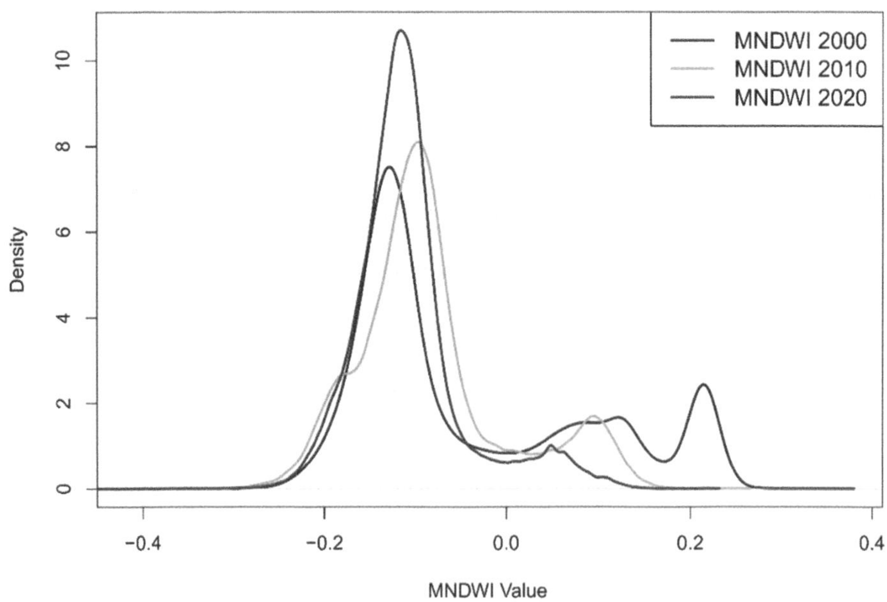

Fig. 8.4 Decadal change detection of surface water using density plot

around −0.1, with minimal density above 0, reflecting a significant decrease in water-related pixel values in 2020. However, the clear difference between shallow and deep water diminished in 2010 and 2020. The possible reason could be white pollution reducing the depth of surface water bodies as well as increased water pollution affecting pixel reflectance. The progressive leftward shift of the curves illustrates the spatial contraction of water bodies and intensification of non-water land cover over time. This aligns with both the map and tabular results and reinforces the conclusion of widespread water loss in DMA. This rapid conversion is largely attributable to urban sprawl, particularly the encroachment of built-up areas into low-lying floodplains, wetlands, and khals, driven by population growth and the increasing demand for housing and infrastructure.

Historical analyses of Dhaka's urban expansion reveal that prior to the late 1990s, the city remained relatively compact. However, the post-2000 period witnessed accelerated land conversion, especially in water-rich zones, aligning with the findings of this chapter. This pattern is further supported by earlier research. For example, Faridatul et al. (2019) reported a marked decline in surface water from 38.47% to 24.91% between 2006 and 2016, alongside a sharp increase in impervious surfaces from 12.81% to 22.99%, an indication of the replacement of waterbodies with built infrastructure. Similarly, a land use/land cover (LULC) change analysis by Rahman and Szabó (2021) observed that while waterbodies dominated the landscape in 1990 (85.85 km^2), they shrank to just 34.94 km^2 by 2020, whereas the built-up area became the most prominent class. Though their study incorporated multiple LULC categories, the overarching trend—diminishing aquatic spaces amidst expanding urban footprints—converges with the current analysis.

A comparable decline is also evident in a recent assessment using the Water Ratio Index (WRI) by Hossain et al. (2023). Their study reported that surface water increased slightly from 26.23 km^2 in 1990 to 32.99 km^2 in 2000, but then followed a steep downward trajectory, dropping to 28.67 km^2 in 2010, and reaching only 15.82 km^2 by 2021. Despite employing a different methodological approach, their results corroborate the consistent reduction in surface water observed in this chapter. These independent yet aligned findings highlight a clear and alarming trajectory—one where urban growth persistently comes at the cost of natural hydrological features. Taken together, the convergence of evidence across multiple studies and classification techniques emphasizes that the loss of surface water in Dhaka is neither an anomaly nor limited to a single methodology. Instead, it represents a widespread and ongoing environmental transformation, closely tied to urbanization dynamics that prioritize land development over ecological sustainability. The sharp reduction in surface water has serious ecological and urban implications. Water bodies play a crucial role in flood regulation, urban cooling, groundwater recharge, and biodiversity. Their disappearance increases vulnerability to climate extremes and undermines sustainable urban development. Immediate policy interventions are required to protect remaining water bodies and restore degraded wetlands to ensure ecological resilience in the face of rapid urban growth.

8.4.2 Surface Water Management in Planning and Policy

There have been both long-term and short-term plans and policies developed to make the Dhaka Metropolitan Area (DMA) a sustainable urban region. The previous master plan, the Dhaka Metropolitan Development Plan (DMDP), was formulated in 1995 (gazetted in 1997) and remained in effect until 2015. It placed strong emphasis on surface water retention and the protection of rivers and floodplain areas by prohibiting encroachment and unauthorized development through landfilling. However, in contrast to these policies, surface water bodies were increasingly filled in, which facilitated rapid urbanization. Private land developers converted wetlands into urban land despite DMDP's explicit restrictions against permanent development in such ecologically sensitive areas. Although the plan aimed to make Dhaka a livable city, rather than simply accommodating the influx of people into the metropolitan area, it gradually lost focus on preserving surface water. As a result, more than 50% of surface water bodies have been lost over recent decades. Roy (2007) reported that flood retention ponds, mandated by the DMDP to comprise 12.5% of land for development projects over 2.02 hectares, were often ignored or filled in for profit. The JICA (2000) study on groundwater noted that unregulated development in wetlands and lowlands has reduced Dhaka's natural recharge capacity, intensified water stress and increasing flood vulnerability. The report echoed the DMDP's concerns about land elevation changes, typically 3 to 7+ meters through sand filling, which disrupts natural hydrology and reduces surface retention. RAJUK's delay in finalizing the Detailed Area Plan (DAP) and the overlapping responsibilities of 42 agencies have severely undermined enforcement of the DMDP (gazetted in 1997), enabling unauthorized development due to weak zoning control and political interference (Roy, 2007). To amend the shortcomings of the previous master plan and better accommodate population growth and housing demand, a new master plan—DAP (2016–2035)—was formulated with updated policies and management strategies. This plan was designed to be effective for 19 years, with a provision for revisions every 5 years based on emerging needs. Importantly, it also includes measures for the protection of hydrologically sensitive areas. However, after being gazetted in 2022, the DAP was quickly revised—extending its effective period from 2022 to 2035—under pressure from land developers. This indicates the possibility that DAP, like its predecessor DMDP, may fail to achieve its intended goals. While plans like the DMDP and DAP outline comprehensive strategies for sustainable urban development, their effectiveness is repeatedly hindered by poor governance, inadequate funding, lack of inter-agency coordination, widespread illicit activities (such as land grabbing), and difficulty in enforcing policies that often face resistance and political disregard.

The surface water bodies covering approximately 7% of the total land area is crucial in maintaining ecological balance, regulating urban flooding, and enhancing the quality of urban life (Tuli et al., 2024). The Dhaka Structure Plan (2016–2035) proposes a range of policies aimed at conserving and restoring these natural assets to create a more functional, resilient, and livable city. However, whether these strategies will prove effective remains a critical concern. One of the major proposals is the creation of Special Conservation Zones to protect rivers, canals (*khals*), and wetlands from encroachment, introducing pre-emption rights that allow the government to purchase land within flood flow and retention zones to prevent unregulated private development. It emphasizes the management and preservation of these water bodies as essential for urban flood resilience and calls for their integration into recreational spaces through community participation, including a proposal to establish 55 water parks. The plan also recommends recognizing canals, lakes, and river corridors as ecological and urban amenities, with regulated development along these corridors and the formation of a city-wide "blue network" to ensure surface water sustainability. While the Natural Wetland Conservation Act (2000) prohibits the filling of water bodies, it does not apply to privately owned ones, weakening its effectiveness.

Focusing on the SDGs, to achieve goal 6 one of the targets is by 2020, protect and restore water-related ecosystems, including mountains, forests, wetlands, rivers, aquifers and lakes. Regarding this target Bangladesh did not create a success story as per this chapter analyzed already. As it has been seen that for recent years government put more emphasize on safe drinking water and proper sanitation (SDG Cell Bangladesh Bureau of Statistics, 2025). It also aligns its national priority targets (NPTs) on these issues. As a result, less concern to protect and restore the surface water areas reflected on the data shown (Table 8.2). The extent of surface water in the Dhaka Metropolitan Area has been steadily diminishing, with its total area nearly halved in each successive decade. This trend directly contradicts the objectives of Sustainable Development Goal (SDG) 6.6, which aims to protect and restore water-related ecosystems.

Although planning policies exist for the retention of surface water in the Dhaka Metropolitan Area, significant degradation has occurred over each decade. While the DAP includes mechanisms to protect surface water areas, it lacks provisions for expanding or designating new areas as reservoirs. Its focus is limited to protecting existing features such as lakes, rivers, and canals, many of which continue to face encroachment and shrinkage due to political influence and local muscle power, undermining the implementation of DAP policies. Alarmingly, between 2000 and 2020, surface water area in Dhaka declined from 108.43 km^2 to 24.19 km^2, a staggering 77.9% reduction. This decline is largely attributable to weak governance and unchecked urban sprawl. Zhang et al. (2017) examined stormwater management in Berlin and Singapore, highlighting that although centralized systems incur significant initial costs, they offer long-term economic and environmental benefits. Strategies involving greater retention, natural infiltration, and decentralization have improved the health of urban water bodies in Berlin. In contrast, Dhaka's failure to implement water-sensitive urban design has contributed to the continued fragmentation and degradation of its water bodies. While the Dhaka Structure Plan (2016–2035) outlines goals for water conservation and protected zones, slow and inequitable implementation severely undermines its effectiveness. This situation underscores the need for Dhaka to adopt more efficient, decentralized approaches to urban water management. As described by Leigh and Lee (2019), decentralized water infrastructure offers advantages in resilience and sustainability. However, implementation at scale is often hampered by institutional inertia and rigid governance systems, factors that have contributed to Dhaka's sharp decline in surface water over the past two decades. Despite these challenges, Leigh and Lee argue that centralized systems, when tailored to specific land uses and socio-economic conditions, particularly in low-income or rapidly urbanizing areas, can provide flexible, localized solutions. Such systems could help restore Dhaka's water balance through community-managed initiatives like rainwater harvesting and permeable surface installations, guided by spatial data and targeted interventions. This comparison highlights that the absence of effective hydraulic infrastructure is intrinsically linked to the lack of a robust planning governance system. Leigh and Lee propose a two-dimensional regulatory and planning framework that combines local autonomy with institutional oversight to enhance resilience. The City Water Resilience Framework (CWRF) offers a comprehensive governance-based model for urban water resilience and has been successfully implemented in cities like Miami and Cape Town (Saikia et al., 2022). CWRF integrates governance, finance, infrastructure, and social inclusivity to address structural vulnerabilities and enable adaptive management. In contrast, zoning restrictions within the DAP lack integrated monitoring and feedback systems that are central to CWRF. The 77.9% loss of surface water in Dhaka indicates the governance failures and policy enforcement delays that CWRF aims to address. Integrating CWRF elements in Dhaka, such as stakeholder-led planning, resilience scorecards, and iterative policy development, could facilitate collaborative governance and resource restoration, offering a path toward reclaiming lost water assets and reforming outdated planning frameworks.

These studies are global approaches towards surface water management through policies and planning process. Comparing with management policies in Dhaka metropolitan area would give a better understanding that other implementations, programs, approaches can be added as there is provision for revising DAP every 5 years. Environmental sustainability and resilience are core guiding principles throughout the Dhaka Structure Plan. Surface water is not merely seen as passive environmental features but is positioned as active infrastructure for flood management, biodiversity conservation, and urban cooling. The plan emphasizes integrated planning approaches, community involvement, and institutional coordination to achieve long-term conservation outcomes. Urban surface water has been shown to reduce surrounding land surface temperatures and UHI intensity significantly, reinforcing the Structure Plan's vision of water bodies as climate-resilient infrastructure (Hosen et al., 2025; Rashid et al., 2025). Without proactive preservation, such thermal benefits will be lost amid Dhaka's intensifying urban heat. Unlike terrain-constrained cities like Hong Kong, Dhaka's flat landscape has facilitated widespread urban encroachment, emphasizing the necessity for stringent land-use regulation and pre-emptive government acquisition of water-sensitive zones (Faridatul et al., 2019). While the Structure Plan presents a comprehensive framework for wetland and surface water protection, significant challenges remain in translating policy into practice. Institutional weaknesses, competing land pressures, and enforcement limitations threaten the effectiveness of these initiatives. Therefore, robust governance mechanisms, strong political will, and public awareness will be crucial for realizing the plan's vision for sustainable surface water management in Dhaka.

8.5 Conclusion

Surface water ecosystems in Bangladesh have experienced significant degradation in recent decades, despite being among the most productive and ecologically diverse components of the landscape. The findings of the chapter point to a significant environmental change in the urban environment of Dhaka: uncontrolled urban development and a lack of strict enforcement of rules have resulted in a 77.9% decrease in surface water area over the past 20 years. Although the analysis is based on the local context of Dhaka, the trends it uncovers speak to a larger global issue that rapidly urbanizing cities in the global south are dealing with the increasing threat that unsustainable development is bringing to surface water ecosystems. This chapter not only tracks hydrological change but also provides a replicable method for long-term surface water monitoring in areas with limited data, utilizing a multi-temporal remote sensing technique that incorporates MNDWI and Landsat data. This technical contribution enhances global research on urban sustainability by offering a geographical framework for assessing how land use affects water systems and promoting planning responses grounded in evidence. Moreover, in comparison to worldwide case studies like resilience governance framework in Cape Town or decentralized infrastructure in Berlin, this research emphasizes the need for integrated, flexible, and participatory water management approaches. In addition, it would be a remarkable measure if the policy makers of Bangladesh put their concern more on protect and restore surface water in capital urban domain. To step ahead in restore the historical waterscapes and safeguard existing water areas by law, rule or guidelines can change the urban dynamics as well as the ecology in Dhaka Metropolitan area. Globally significant, resilient water governance is highlighted by the Dhaka case, which underscores the necessity for flexible planning tools that find a balance between ecological protection and growth. In this way, the chapter contributes to both academic knowledge and discussion about practical policy. The lessons learned from Dhaka provide timely guidance on protecting vital urban water resources in the face of increasing environmental stress as cities around the world struggle with the conflicting demands of population growth and climate change. Urban expansion, particularly in Dhaka, is inevitable; however, it must be guided by comprehensive hydrological assessments rather than unregulated, demand-driven development. Water management should be central to any future urban planning efforts in the region, given the unique environmental context. The destruction of surface water habitats and the widespread infilling of ponds and reservoirs are significantly undermining the capacity to sustain these ecosystems. Protecting surface water areas requires a holistic approach that includes preserving the surrounding environments. While the conservation of government-owned khas ponds is important, it is not sufficient. Privately owned or man-made ponds and reservoirs can also contribute to ecological balance if managed appropriately. To ensure long-term sustainability and a livable environment, it is essential to implement robust policies and legal frameworks focused on surface water protection. Public awareness and community engagement must also be prioritized to foster a collective sense of responsibility toward water resource conservation. Only through coordinated efforts can surface water ecosystems be preserved for future generations.

References

Ahmed, S.U. (1986). Dacca: A Study in Urban History and Development (1st ed.). Routledge. https://doi.org/10.4324/9781351186759

Ahmed, S. U. (2018). *Dacca: A study in urban history and development*. Routledge.

Ali, M. I., Dirawan, G. D., Hasim, A. H., & Abidin, M. R. (2019). Detection of changes in surface water bodies urban area with NDWI and MNDWI. *Methods, 9*(3), 946.

Bijeesh, T. V., & Narasimhamurthy, K. N. (2020). Surface water detection and delineation using remote sensing images: A review of methods and algorithms. *Sustainable Water Resources Management, 6*(1), 68. https://doi.org/10.1007/s40899-020-00425-4

Dewan, A. M., & Yamaguchi, Y. (2009). Land use and land cover change in greater Dhaka, Bangladesh: Using remote sensing to promote sustainable urbanization. *Applied Geography, 29*(3), 390–401. https://doi.org/10.1016/j.apgeog.2008.12.005

Faridatul, M. I., Wu, B., & Zhu, X. (2019). Assessing long-term urban surface water changes using multi-year satellite images: A tale of two cities, Dhaka and Hong Kong. *Journal of Environmental Management, 243*, 287–298. https://doi.org/10.1016/j.jenvman.2019.05.019

Gautam, V. K., Gaurav, P. K., Murugan, P., & Annadurai, M. (2015). Assessment of surface water Dynamicsin Bangalore using WRI, NDWI, MNDWI, supervised classification and K-T transformation. *Aquatic Procedia, 4*, 739–746. https://doi.org/10.1016/j.aqpro.2015.02.095

Grenier, M., Demers, A. M., Labrecque, S., Benoit, M., Fournier, R. A., & Drolet, B. (2007). An object-based method to map wetland using RADARSAT-1 and Landsat ETM images: Test case on two sites in Quebec, Canada. *Canadian Journal of Remote Sensing, 33*(sup1), S28–S45.

Haroon, A. K. Y., & Kibria, G. (2017). Wetlands: Biodiversity and livelihood values and significance with special context to Bangladesh. In *Wetland science: Perspectives from South Asia* (pp. 317–346). Springer India. https://doi.org/10.1007/978-81-322-3715-0_17

Hosen, M. I., Hasan, M. M., Talha, M., Akter, M. M., & Nasher, N. R. (2025). Exploring the cooling benefits of urban lakes: A multi-year analysis of Dhaka, Bangladesh. *HydroResearch, 8*, 361–373. https://doi.org/10.1016/j.hydres.2025.01.001

Hossain, M. J., Mahmud, M. M., & Islam, S. T. (2023). Monitoring spatiotemporal changes of urban surface water based on satellite imagery and Google earth engine platform in Dhaka City from 1990 to 2021. *Bulletin of the National Research Centre, 47*, 150. https://doi.org/10.1186/s42269-023-01127-5

Hossein, M. K., Ahmed, A., Uddin, M. J., Saifullah, A. S. M., Babu, M. A. H., & Sarker, T. (2023). Prediction of land use and land cover changes (LULC) and urban growth analysis in Dhaka metropolitan area using CA-Markov model and geospatial techniques. *Arabian Journal of Geosciences, 16*, 468. https://doi.org/10.1007/s12517-023-11570-3

Huang, C., Chen, Y., Zhang, S., & Wu, J. (2018). Detecting, extracting, and monitoring surface water from space using optical sensors: A review. *Reviews of Geophysics, 56*(2), 333–360. https://doi.org/10.1029/2018RG000598

JICA & BWDB. (2000). *The study on the groundwater development of Dhaka metropolitan area in the people's republic of Bangladesh: Final report (Vol. 1)*. Japan International Cooperation Agency (JICA) and Bangladesh Water Development Board (BWDB). https://openjicareport.jica.go.jp/pdf/11996774_06.pdf

Kamal, A. M., & Midorikawa, S. (2004). GIS-based geomorphological mapping using remote sensing data and supplementary geoinformation: A case study of the Dhaka city area, Bangladesh. *International Journal of Applied Earth Observation and Geoinformation, 6*(2), 111–125.

Koch, M., Schmid, T., Reyes, M., & Gumuzzio, J. (2012). Evaluating full polarimetric C-and L-band data for mapping wetland conditions in a semi-arid environment in Central Spain. *IEEE Journal of Selected Topics in Applied Earth Observations and Remote Sensing, 5*(3), 1033–1044.

Leigh, N. G., & Lee, H. (2019). Sustainable and resilient urban water systems: The role of decentralization and planning. *Sustainability (Switzerland), 11*(3), 918. https://doi.org/10.3390/su11030918

Mahmud, M. S., Masrur, A., Ishtiaque, A., Haider, F., & Habiba, U. (2011). Remote sensing & GIS based Spatio-temporal change analysis of wetland in Dhaka city, Bangladesh. *Journal of Water Resource and Protection, 3*(11), 781.

Melton, J. R., Wania, R., Hodson, E. L., Poulter, B., Ringeval, B., Spahni, R., et al. (2013). Present state of global wetland extent and wetland methane modelling: Conclusions from a model inter-comparison project (WETCHIMP). *Biogeosciences, 10*(2), 753–788.

Ozesmi, S. L., & Bauer, M. E. (2002). Satellite remote sensing of wetlands. *Wetlands Ecology and Management, 10*(5), 381–402.

Paul, S. S., Akhter, S. H., Hasan, K., & Rahman, M. Z. (2019). Geospatial analysis of the depletion of surface water body and floodplains in Dhaka City (1967 to 2008) and its implications for earthquake vulnerability. *SN Applied Sciences, 1*, 565. https://doi.org/10.1007/s42452-019-0582-5

Rahman, M. M., & Szabó, G. (2021). Impact of land use and land cover changes on urban ecosystem service value in Dhaka, Bangladesh. *Land, 10*(8), 793. https://doi.org/10.3390/land10080793

Rajdhani Unnayan Kartripakkha (RAJUK). (2019). *Dhaka structure plan 2016–2035.*

Rashid, K. J., Akter, T., Imrul Kayes, A. S. M., & Yachin Islam, M. (2023). Exploring the Spatio-temporal patterns and driving forces of urban growth in Dhaka megacity from 1990 to 2020. In U. Chatterjee, N. Bandyopadhyay, M. D. Setiawati, & S. Sarkar (Eds.), *Urban commons, future smart cities and sustainability*. Springer Geography. Springer. https://doi.org/10.1007/978-3-031-24767-5_17

Rashid, K. J., Das Tuli, R., Liu, W., & Mesev, V. (2025). Quantifying the drivers of the spatial distribution of urban surfaces in Bangladesh: A multi-method geospatial analysis. *Remote Sensing, 17*(12), 2050. https://doi.org/10.3390/rs17122050

Roy, S. (2007). *Private residential land development activities in Dhaka and their compatibility with the DMDP Structure Plan 1995–2015* (Master's thesis, Bangladesh University of Engineering and Technology). Department of Urban and Regional Planning, BUET.

Saikia, P., Beane, G., Garriga, R. G., Avello, P., Ellis, L., Fisher, S., Leten, J., Ruiz-Apilánez, I., Shouler, M., Ward, R., & Jiménez, A. (2022). City water resilience framework: A governance based planning tool to enhance urban water resilience. *Sustainable Cities and Society, 77*, 103497. https://doi.org/10.1016/J.SCS.2021.103497

SDG Cell Bangladesh Bureau of Statistics. (2025). *SDG tracker Bangladesh's development mirror*. https://sdg.gov.bd/

Tuli, R. D., Rashid, K. J., & Akter, T. (2024). Impact analysis of the 2022 flood event in Sylhet and Sunamganj using Google earth engine. In *Application of geospatial technology and Modeling for River Basin management (chapter 3)*. Elsevier. https://doi.org/10.1016/B978-0-443-23890-1.00003-7

United Nations (UN). (2025). *Sustainable development goals*. https://www.un.org/sustainabledevelopment/water-and-sanitation/

Zhang, D., Gersberg, R. M., Ng, W. J., & Tan, S. K. (2017). Conventional and decentralized urban stormwater management: A comparison through case studies of Singapore and Berlin. *Germany. Urban Water Journal, 14*(2), 113–124. https://doi.org/10.1080/1573062X.2015.1076488

9 Preliminary Field Observation on Impact of Urban Growth on Ground Water in Agra City by Using Geo-Informatics Applications

S. Kumar and D. S. Parihar

Abstract

Urbanization, or the fast expansion of urban areas, presents considerable problems to the sustainable management of water resources in cities worldwide. Agra, a city rich in history and culture, is facing these issues in the midst of rapid urbanization. The objectives of present study are the complex relationship of urban growth and water resources in Agra by using geospatial technologies to find out the geographical developments, trends, and effects of urban expansion on groundwater levels and urban water balance. In 1901, city had total 188,022 populations, which continue growth and in the census 2011 population is 1,760,285 where growth in numbers is 1,572,263. Based on AMC survey in 2001, there were 59,569 households, including residential as well as industrial buildings, whereas in the year 2023 (after 22 year) there conducted a waste management survey, reports are 326,284 households which showed rapid growth of houses total 266,715 (12123.41 houses per year) and rapidly growths in AMC ward from 40 in 1981 to 100 wards in 2024. Study investigates its relation to past urban growth patterns, and essential factors which influence water dynamics. For the completion of this study, we comprehensively outlined the procedures and techniques adopted and describe the complexities of the data gathering procedure, the geospatial analysis tools utilized. Through extensive research, the study indicates spatial trends, identify areas of increased groundwater depletion, and reveal locations on the verge of water crisis.

Keywords

Impacts of urbanization · Concrete forest · Preliminary field observation · Geo-informatics · GPS · GWD · SDG-6.4

Abbreviations

AMC	Agra Municipal Corporation
GEP	Google Earth Pro
GIS	Geographic Information System
GPS	Global Positioning System
GWD	Ground Water Decrease

S. Kumar
Ericsson India Global Private Limited, Noida, India
e-mail: sunilkumar2002091@dei.ac.in

D. S. Parihar (✉)
Department of Geography, Kumaun University, Nainital, India
e-mail: devendraparihar@kunainital.ac.in

U. Chatterjee, A. I. Almulhim (eds.), *Urban Waterscapes*, Sustainable Development Goals Series,
https://doi.org/10.1007/978-3-032-20649-7_9

Highlight The effect of urban growth on groundwater is a worldwide issue because of rising population demands, greater infrastructure, and fewer natural areas for water replenishment. Quick urban development results in excessive groundwater extraction and less water filtration, which harms both the water supply and quality. Tackling this challenge is vital for fostering sustainable city growth and ensuring water security around the globe. This chapter studied the impact of urban growth on groundwater availability and the rise in pollution levels caused by heavy industrial wastewater. It draws attention to the urgent necessity of implementing sustainable groundwater use and oversight in urban areas. In India, the number of ***"zero-day"*** cities—where water resources are nearly depleted—is rapidly increasing due to the lack of sustainable planning and implementation. Based on primary field observations, this chapter presents the ground truth of impacts of urban growth on groundwater and offers recommendations and sustainable approaches to water resource management in urban settings.

9.1 Introduction

Urbanization can be understood as the continued population growth, emigration, or shifting of the movement of people from rural to urban regions (Liang & Wang, 2020). The major driving factors for rural to urban transition are industrialization, employment, better standard of living, and social security benefits (Aithal & Ramachandra, 2016). In the twenty-first century, urbanization blooms as the biggest problem for many developing and developed countries all over the world (Jaeger et al., 2010). Groundwater is the most crucial and abundant water source and comprises 26% of the Earth's unfrozen freshwater located beneath the Earth surface (Arulbalaji et al., 2019). Urbanization, in general, has four immediate re percussions on the hydrological cycle: flooding (e.g. as a result of increased soil sealing), water shortage (e.g. due to rising consumption), alteration in the river and groundwater regimes and water pollution (Yang et al., 2016). Urban areas serve as significant sources of both nonpoint and point contaminants, impacting groundwater quality through various means (Wakode et al., 2018). Point sources, such as leaking underground storage facilities and accidental spills, contribute to groundwater pollution Jha et al., 2008. Rapid expansion of urban regions manifests in two primary effects on groundwater resources (Baier et al., 2014), the reduction of natural aquifer recharge due to ground sealing with concrete, and contamination from drainage leakage, industrial waste, and effluents (Abanyie et al., 2023). Understanding the water balance of urban aquifers is complex, as urbanization introduces additional recharge sources and numerous abstraction points scattered across the urban landscape (Dutta et al., 2024).

Hydro-geologists now widely acknowledge that water supply and storm drainage infrastructure contribute significantly to groundwater recharge through leaks (Lerner, 2002). Despite the land sealing effect caused by paving and construction, which hinders natural recharge from precipitation (Wakode et al., 2018), the substantial volume of the circulation of water within underground water-wastewater infrastructure, often under high pressure (Wartalska et al., 2024), along with losses from pipe leakage or soak-ways draining built-up areas, findings show a rise in the total volume of recharge in the conurbation (Mao et al., 2024).

In such environments, new sources of contamination may significantly increase total infiltration compared to pre-urban conditions (Morris et al., 2003). Urban areas often import large quantities of water, leading to high flow pressure within the pipe network used for water transportation and distribution (Beker & Kansal, 2024). Any leakages in these networks can contribute to substantial recharge in the urban water as well as urban ground water (Kiflay et al., 2025; Van Broekhoven et al., 2024). Although leakages from sewer networks are less significant in the urban recharge process compared to water supply mains, they can still serve as major sources of contaminants (Ducci et al., 2023; Serafeim et al., 2024). Similarly, Li et al., 2019 studied on the urban water balance dynamics in

Beijing, China, utilizing advanced geospatial methods to assess water demand and supply patterns. Soltani et al., 2018 investigated the impacts of urban expansion on groundwater resources in Tehran, Iran, employing geospatial analysis to evaluate changes in groundwater levels and quality. Additionally, Tziritis et al., 2020 explored the implications of urbanization on urban water resources management in Athens, Greece, utilizing geospatial technology to develop water management strategies. Gupta & Srivastava, 2016 examined the effects of urbanization on groundwater resources in Delhi, India, employing geospatial techniques to assess groundwater depletion trends.

These research endeavours encompass a diverse array of studies focusing on various facets of urbanization's influence on groundwater dynamics and water balance Gupta et al., 2017. For instance, Gupta and Srivastava (2016) evaluated groundwater potential zones in tonified conditions, offering insights into the spatial distribution of groundwater resources. Jha et al., 2019 examined the effect of urban growth on groundwater quality, shedding light on the contamination risks associated with quick built-up area growth. Tiwari et al., 2020 investigated the effects of land-use changes and urban expansion on groundwater quality and quantity using geospatial techniques. Furthermore, Singh et al., 2021 demonstrated the utility of remote sensing and GIS in assessing urbanization's impact on water bodies and hydrological processes. Sharma et al., 2018 focused on assessing effect of urban growth on groundwater recharge rates and quality in the Agra district, shedding light on the specific factors influencing groundwater availability and contamination. Yadav et al., 2020 investigated the spatial patterns of urban expansion and land-use changes in the Agra region, highlighting their implications for groundwater resources and urban water balance. Sharma et al., 2024 utilized geospatial techniques to model and analyze the hydrological processes and water balance dynamics in Agra, providing insights into the local distribution of water resources and their susceptibility to urban growth pressures. Groundwater plays an important role in the economic and social health of the developing world's city population (Yar, 2020).

This chapter explores how much water scarcity and groundwater depletion occur in Agra's urban areas by using Geo-Informatics application and GPS technology for field observations. It examines the effects of quick urban growth and the spread of concrete buildings on how groundwater is recharged. The research investigates how changing natural areas into crowded constructions influences the groundwater situation in the city, emphasizing how urban development leads to lower groundwater levels in Agra (Fig. 9.1).

9.2 Study Area

The Agra district occupies the western most part of the Uttar Pradesh. The district is bounded by the Rajasthan and Madhya Pradesh states in the West and South and by Mathura, Etah, Firozabad, Mainpuri and Etawah district from north to east respectively (Singh, 2013). It encompasses an area of 4027 km^2 and Agra lies between latitude "26^{0}44′10 to 27^{0}24'30" North and longitude "77^{0}30′15 to 78^{0}51'30" East (Fig. 9.2). The city enjoys excellent accessibility, facilitated by a well-connected network of national and state highways, including the Yamuna expressway, linking it efficiently to metropolitan centres, i.e., Delhi and Jaipur. With around 1.6 million residents, Agra ranks as the fourth largest city in Uttar Pradesh and is the 23rd largest across India. According to the Census of India, 2011 Agra's city population stands at 1,585,704, while its metropolitan population is reported at 1,760,285. The sex ratio in Agra is 875 females/1000 males, with a child sex ratio of 857 girls per 1000 boys. The city boasts an average literacy rate of 73.11%, with male and female literacy rates at 77.81% and 67.74%, respectively. Hinduism prevails as the dominant religion in Agra, with 80.68% of the population adhering to it, followed by Islam at 15.37%. Minor religious affiliations include Jainism, Sikhism, Christianity, and Buddhism, constituting 1.04%, 0.62%, 0.42% and 0.19% of the population, respectively. Approximately 1.66% of the populace declared having no religious affiliation.

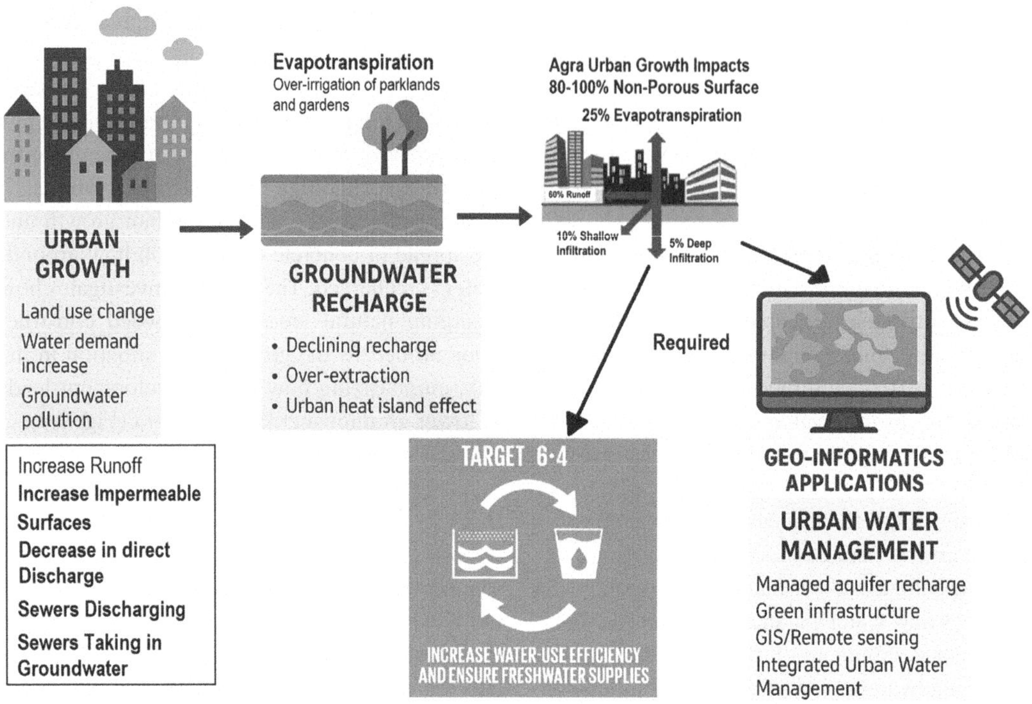

Fig. 9.1 Graphical abstract of the study. (after author)

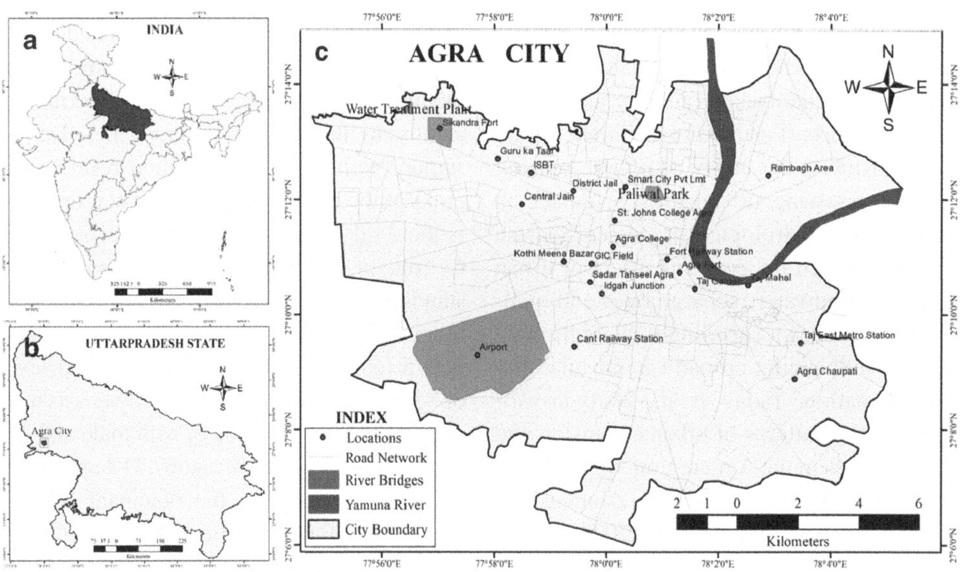

Fig. 9.2 Location map of the study area: (A) India, (B) Uttar Pradesh and (C) Agra city region. Source: Field surveying and GPS, April 2024

9.3 Material and Methods

This chapter focuses on illuminating the methodology used throughout the study, outlining the characteristics of the study location, and explaining the data collection process. Methodologies cover a wide range of techniques and approaches used to effectively full fill research objectives (Fig. 9.4C).

9.3.1 Instruments and Software Uses

Instrument refers to any tool or device that is used to collect measure or analyze data related to the earth's surface and its features. The GPS instruments were used in the field for collecting the geographical coordinate of the point where the tube well, hand pump, construction field, rivers, lakes, wells, etc. are located in Agra city. In this chapter, we use different software to help us understand and analyze the data we collect. These applications are indispensable to preparing maps of the study area location, sample sites, and ground water status in the study area. Software is ArcGIS-10.8.2 (*Licensed*) used for final mapping, QGIS-2.18 used for data conversion, and GEP used for Geo-tagging and data verification. During the field surveys, the mobile app GPS Map Camera (version 1.4.25) was used to take photos that included GPS coordinates.

9.3.2 Site Selection Method and Sample Size Determination

Choosing a site is an essential part of learning about city planning, managing the environment, and developing infrastructure. This process started by setting clear goals and creating criteria for selection which followed by gathering information, analysing it, evaluating the results, and making informed decisions. Important elements that affect site selection include the land's shape, weather conditions, and the presence of natural resources, as well as social factors such as the number of people living in an area, how land is used, and how easy it is to reach. In cities, environmental issues—like fast urban growth, protecting habitats, ensuring good water quality, reducing surface water pollution and avoiding the depletion of groundwater—significantly impact the choices made during site selection. Other important factors to think about are the existing infrastructure, legal requirements, and what the community desires. For this research, we identified 197 sites for collecting groundwater data (see Fig. 9.3).

The size of this sample was calculated using the Cochran formula, created by William G. Cochran, a well-known statistician recognized for his work in sampling theory (refer to Fig. 9.4B). As per the census of 2011, the population of Agra city is 1,760,285. The last census was conducted in 2011. Let us assume the reliable population of Agra city is an unknown variable; the William Cochran formula is used, furthermore, add chances of error are 5%. In Agra city, there are a total of 100 wards, and for sampling taken the number of wards is 20, which are considered highly affected areas. Collecting the 23 responses from each ward, these are purposive samples (Figs. 9.4A and B).

The formula is expressed as:

$$n = \frac{Z^2 \times P \times (1 - P)}{E^2}$$

Where:

- n = required sample size
- Z = Z-score corresponding to the desired level of confidence (e.g., 1.96 for 95% confidence level)
- P = estimated proportion of the population with the attribute of interest (if unknown, use 0.5 for maximum variability)
- *EE* = desired margin of error (expressed as a proportion)

9.3.3 Field Work, Data Collection and Processing

Carefully organized fieldwork allowed researchers to gather extensive data directly

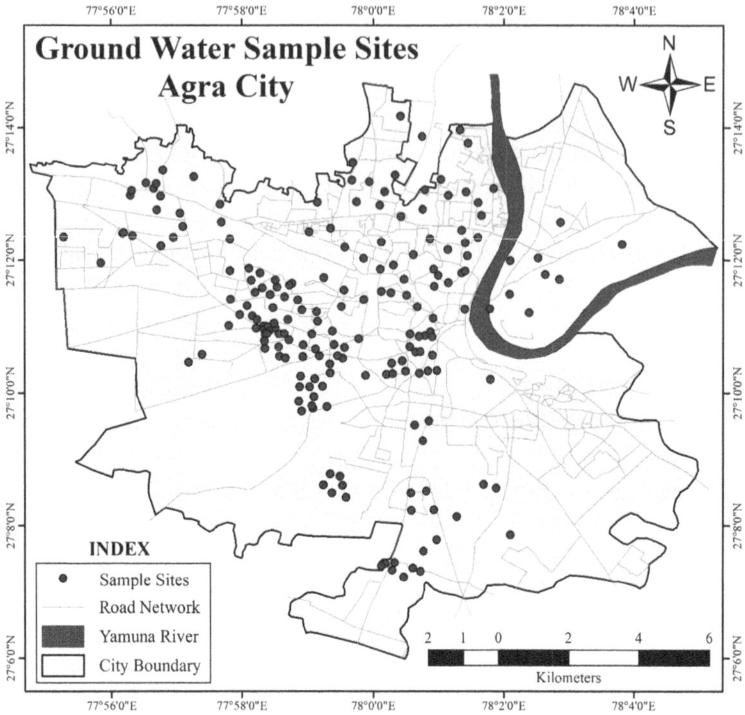

Fig. 9.3 Ground water sample sites at Agra city of different wards. Source: Field surveying and GPS, April 2024

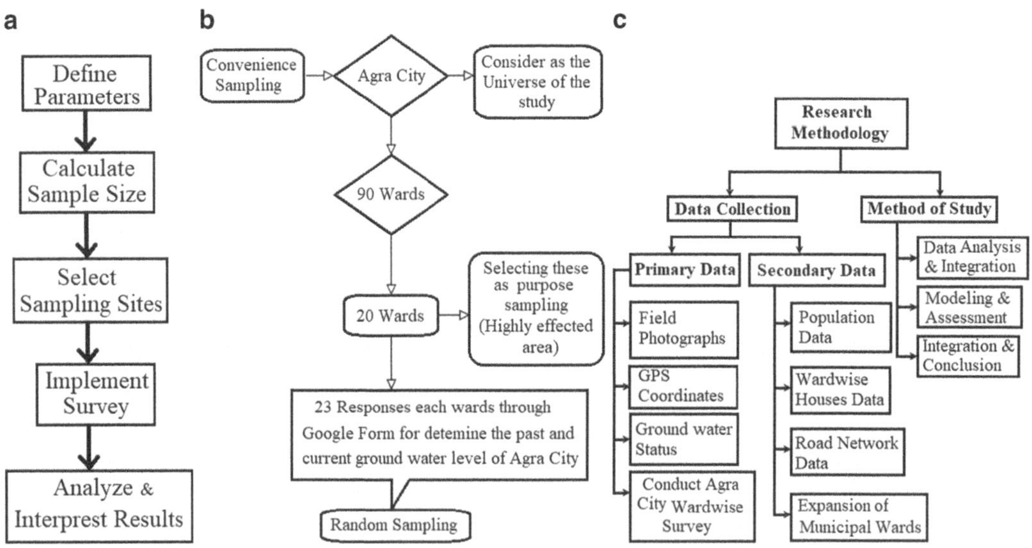

Fig. 9.4 Methodology flow chart: (A) sampling methodology, (B) sample size determination and (C) overall research methodology

from the selected area, helping to understand groundwater behaviours and the urban water balance. A vital part of this work included collecting historical (before 2000) and current (2024) information about groundwater levels from various sources such as tube wells and hand pumps throughout Agra. With the help of GPS devices, the exact locations of each site were marked to ensure precise mapping. Along with physical data, surveys using questionnaires were carried out to capture the views of residents. These surveys focused on discovering how water is used, opinions on its quality, and observations on how urban development has impacted groundwater levels. Such community input greatly enhanced the comprehension of urban water dynamics. Working together with local groups, such as the Agra Water Department, was essential for gaining access to additional datasets. By combining geospatial technology with data collected from the field and involving the community and institutional partners, this phase of research provided a complete picture of how urban growth influences groundwater resources and the overall urban water balance. This stage also included sorting, processing, and analysing the gathered data to uncover valuable insights. The workflow for processing the data can be seen in Fig. 9.4.

9.4 Results and Discussion

9.4.1 Field Observation Results on Ground Water

The growth of urban settlements, as seen in Agra, has substantial consequences for water supplies in the surrounding area using bore wells, wells, hand pumps, tube wells etc. As urbanization grows, water sources are under increasing pressure from growing populations, concrete forest and metaled road network, industrial operations, and municipal needs. This has a significant impact on both surface water bodies and groundwater reserves. In Agra, growing urbanization has resulted in greater extraction of water from surrounding rivers, such as the Yamuna, for housing, municipal and industrial use. This has resulted in lower flow rates and poorer water quality, harming both aquatic ecosystems and downstream communities. Furthermore, the expansion of urban development into floodplains and wetlands has reduced these areas' natural ability to absorb and regulate water, increasing the risk of floods during monsoon seasons. Groundwater depletion is becoming a major concern in Agra as a result of excessive extraction for use in daily life. As cities grow, there is a higher dependency on groundwater for agriculture and drinking water supplies, resulting in diminishing water levels.

This has significant long-term ramifications for the region's water security and sustainability, i.e., urbanization contributes to water damage by allowing untreated sewage, industrial effluents and solid waste to enter water bodies. The expansion of paved road networks with the growth of Agra city, consisting of asphalt, concrete and other impermeable surfaces, has a significant negative impact on groundwater recharge (Fig. 9.5). Instead of soaking into the ground, rainwater is rapidly diverted into drains and storm water systems, leading to increased surface runoff, threat of urban flooding and reduced groundwater recharge. Surface runoff water carries so many types of water contaminants, including heavy metals, viruses and organic contaminants degrade water quality, endangering human health and aquatic ecosystems. Water quality degradation limits the availability of clean water for drinking, sanitation, and agricultural reasons, worsening water scarcity difficulties in peri-urban areas.

9.4.2 Ground Water Status in Agra City (Before 2000 and in 2024)

Groundwater depletion is becoming a major concern in Agra as a result of excessive extraction. As cities grow, there is a higher dependency on groundwater for agriculture and drinking water supplies and resulting in diminishing water levels. This has significant long-term ramifications for the region's water security and sustainability, i.e., urbanization contributes to water damage by allowing untreated sewage, industrial effluents, and solid waste to enter water bodies.

Results of community-based survey using GPS coordinates (197 point in Table 9.1 and Fig. 9.6) and questionnaire on ground water extraction sources depth (hand pump, wells, tube wells etc.) before 2000 and in April 2024. Ground

Fig. 9.5 Status of tube wells (TW) and hand pumps (HP) sample sites in Agra city: (A) TW site-1, (B) TW site-2, (C) non-running HP-1, (D) running HP, (E) TW site-3 and (F) TW site-4, (G) non-running HP-2 and (H) non-running HP-3. Source: Field surveying and GPS, April 2024

Table 9.1 Results on ground water depth

S. N.	GW Depth (feet)	Before 2000		2024	
		No.	Percentage	No.	Percentage
1.	<50	2	1.02	0	0
2.	51–100	48	24.36	17	8.63
3.	101–150	83	42.13	58	29.44
4.	151–200	38	19.29	65	32.99
5.	201–250	16	8.12	22	11.17
6.	251–300	8	4.06	27	13.71
7.	301–350	2	1.02	3	1.52
8.	351–400	0	0	4	2.03
9.	>400	0	0	1	0.51
Total		197	100	197	100

Source: Field surveying and GPS, April 2024

water depth statuses were before 2000 (Fig. 9.6A) under 50 m (2 sites), 100 m (48 sites), 150 m (max, 83 sites) and over 300 m (2 sites). In 2024 (Fig. 9.6B), there are maximum sites exist under 151–200 m (65) but no sites under 50 m, only 17 sites found between 51–100 m depth and 8 sites found over 300 m depth (1 site holds more than 400 m depth).

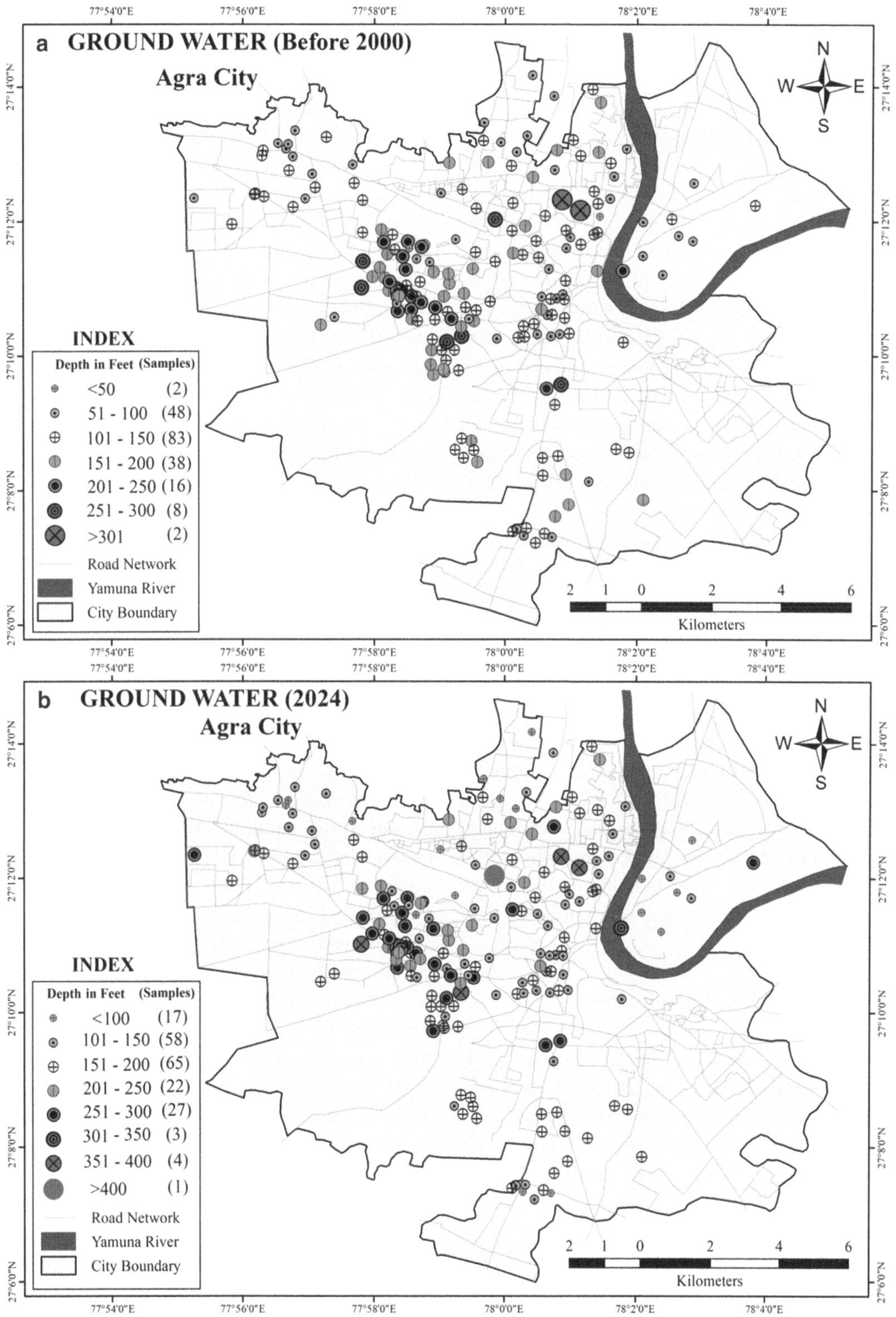

Fig. 9.6 Ground water level (in feet) at different sites on Agra city: (A) before year of 2000 and (B) in year 2024. Source: Field surveying and GPS, April 2024

9.4.3 Community Perception on Urbanization and Ground Water Decreasing

The rise in urban expansions and the resulting decrease in groundwater levels are a significant concern, leading to varied community perceptions. Some community people may perceive the water scarcity as a direct consequence of increased population and infrastructure, while some others may view it as a natural cycle or even acknowledge the role of overuse and pollution in the decline. It is essential to consider these varying opinions to create successful water management plans. Many residents, particularly in areas experiencing water shortages, are not aware of the decreasing groundwater levels and their implications for their daily lives. Urbanization can also lead to groundwater contamination from industrial waste, sewage, house hold, agriculture and other pollutants, raising concerns about water quality and health risks. Perceptions of the causes of groundwater depletion vary. Some may attribute it to urbanization and its associated infrastructure, while others may focus on factors like overuse for agriculture, industrial activities, or even the impact of climate change. Community members may have different ideas about how to address the issue. Some may support government interventions and stricter regulations, while others may advocate for individual behaviour changes or community-based solutions.

9.4.4 Growth in Population

Agra ranks as the fourth-largest city in Uttar Pradesh and is listed among the 23 most populous cities in India. As per the 2011 Census, the population exceeded 1.5 million, residing in 100 municipal wards, with males accounting for 53% and females 47%. The Agra cantonment Board, which includes eight wards, had a population of 53,053 (54.90%) male and 45.10% female. By 2021, municipal records indicated that the city's population had risen to about 1.9 million. Over the past century, there has been notable population growth, evident in various census reports. In 1901, Agra had a population of 188,022, which slightly fell to 185,449 by 1911, showing a small decadal decline of −1.36%. The 1921 Census reflected a minimal growth to 185,532, resulting in an exceedingly low decadal growth rate of 0.04%. A larger increase was recorded in 1931, with the population reaching 205,487, which meant a growth of 10.75%. This upward trend persisted in later decades, as the population grew to 257,368 in 1941 and then to 333,530 in 1951, showcasing growth rates of 25.24% and 29.59%, respectively. The population boom in the 1950s was mainly due to an influx of refugees and regional migration, as Agra started providing more economic opportunities. In 1961, changes in census definitions and city boundary adjustments resulted in a recorded population of 462,020, indicating a decadal growth of 38.52%, which was the highest at that time. However, in the following decades, the growth rate began to decline.

By 1971, the population reached 591,917 with a 28.11% growth rate, which fell to 17.27% in 1981. During the period from 1981 to 1991, resurgence happened, with the population increasing to 891,790 and a growth rate of 28.46%. This surge aligned with India's economic reforms and the increasing acknowledgment of Agra's cultural heritage, especially after the Taj Mahal gained UNESCO World Heritage status. This period saw a significant rise in domestic and foreign tourists, contributing to the city's expansion. By 2001, Agra's population surpassed 1 million, reaching 1,275,134. The years from 1991 to 2001 showcased the highest urban demographic growth rate recorded at 42.98%, fuelled by rapid urbanization and greater integration into the tourism economy. By 2011, the population grew to 1,585,704, though the decadal growth rate fell to 24.35%, suggesting a slowdown. Over the same timeframe, the Agra Cantonment Board's population grew steadily, rising from 49,755 in 1991 to 50,968 in 2001, and then to 53,053 in 2011. By 2022, Agra's urban population was estimated at around 1.9 million, with an expected growth rate of about 19.95% between 2011 and 2021. The decadal growth rates and population trends in Agra from 1911 to

Table 9.2 Population grow trajectory as per census data from 1911–2021

S. N.	Census year	Total population	Per census	
			Growth rate	Growth in percent
1	1911	185,449	0	0
2	1921	185,532	+83	+4.47
3	1931	229,764	+44,232	+23.84
4	1941	284,149	+54,385	+23.67
5	1951	375,665	+91,516	+32.20
6	1961	508,680	+133,015	+35.40
7	1971	634,622	+125,942	+24.76
8	1981	747,318	+112,696	+17.75
9	1991	948,063	+200,745	+26.86
10	2001	1,567,709	+819,649	+65.35
11	2011	1,760,285	+192,576	+12.28

Source: Census Bureau, AMC Records, 2024

2011 are summarized in Table 9.2 and Figs. 9.7A, B and C.

9.4.5 Growth in No. of Houses and Municipal Wards

A survey by the AMC in 2001 identified 59,569 households, which included both residential and industrial spaces. Fast forward to 2023, and a waste management assessment revealed that this number skyrocketed to 326,284 households, marking an astonishing gain of 266,715 households. This translates to an average of about 12,123 new households annually. The notable growth across all municipal wards emphasizes the swift development of residential infrastructure in Agra. Several reasons likely fuelled this remarkable increase, such as rising population, urban development projects, enhanced economic opportunities, and changing real estate trends. Figure 9.7D illustrates the trend of household growth. The municipal framework of Agra has transformed, showcasing the city's on-going urban change. Back in 1981, the city was organized into 40 wards. This figure climbed to 67 by 1991, reflecting a significant jump in both population and urban expansion during that time. In 2001, further adjustments resulted in 80 wards to better handle the growing population. The trend persisted in 2011 with 90 wards, and projections indicate an increase to 100 wards by 2024, according to AMC data.

This continuous growth signifies not only population demands but also the necessity for improved urban management. The establishment of new wards typically coincides with the development of vital public infrastructure, such as transportation networks, utility services, and community amenities, to cater to the increasing population's needs. The rise in municipal wards illustrates both the physical and demographic evolution of Agra while highlighting the need for adaptive administrative structures to oversee a more complex urban environment. This growth pattern emphasizes the critical role of thoughtful urban planning and governance in ensuring fair access to services, sustainable city development, and an enhanced quality of life for residents. In summary, from 1911 to 2011, Agra experienced steady population growth, as detailed in Table 9.2 and Fig. 9.7A, supported by census information and demographic studies that paint a detailed picture of the city's evolving population dynamics.

9.4.6 Impacts of Urban Growth in Natural Resources

The environment, vegetation, water resources, and agriculture are being significantly impacted by rapid urbanization and population increase, leading to the on-going growth of metropolitan areas (see Fig. 9.8). As cities expand, they create major changes in land use, social interactions, infrastructure development, and the consumption of resources. To effectively create policies, encourage sustainable growth, and develop resilient urban systems, it is crucial to comprehend the intricate effects of urban growth. The swift development of urban regions has resulted in considerable environmental harm, producing challenges like insufficient housing, poorer water quality, heightened air pollution, increased noise and dust, urban heat islands, and improper waste management, including solid and hazardous materials. A key issue is the rising pressure on the existing water infrastructure—encompassing supply networks, treatment facilities, and distribution sys-

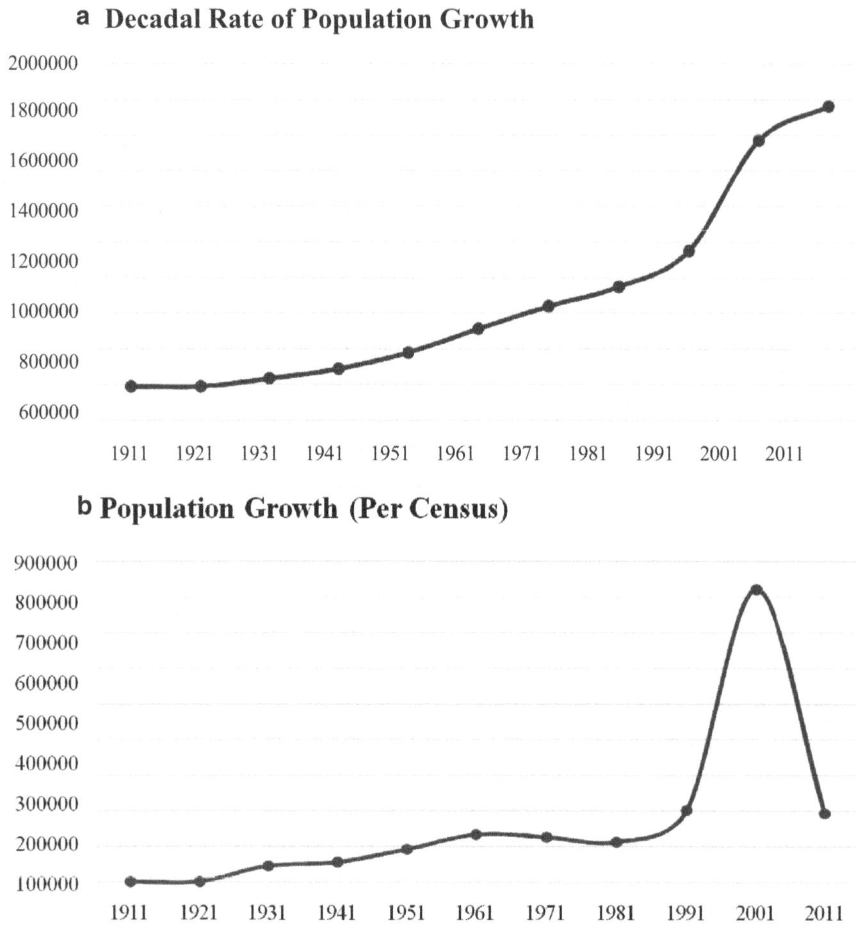

Fig. 9.7 Population, wards and houses growths in Agra city: (A) decadal rate of population; (B) per census population growth rate graph in percent; (C) per census population growth and (D) number of houses growth rate. Source: Census Bureau, AMC Records, 2024

tems—driven by the growing population. This increasing demand highlights the critical necessity to enhance and expand water infrastructure to satisfy the need for more freshwater.

As Agra's population increases, so does its need for water. Analysis of water usage patterns combined with population forecasts indicates a clear connection between growing numbers and rising water requirements. This situation demands immediate and sustainable water management solutions. If proactive measures are not taken, Agra could experience severe water shortages, particularly during peak demand times or seasonal changes. Additionally, urban growth replaces natural vegetation and permeable land with surfaces that do not absorb water, such as concrete, asphalt, and brick. Consequently, rainwater is more likely to run off to drainage systems and rivers rather than infiltrate the ground, which elevates the risk of flooding and diminishes groundwater recharge. Over the years, water quality has worsened due to increased sedimentation and pollution from surface runoff, a problem exacerbated by uncontrolled urban development. These insights highlight the need to incorporate population trends into urban planning to tackle Agra's increasing water issues. It is vital to embrace comprehensive planning strategies that account for population predictions, anticipated water needs, and sustainable development principles to secure the long-term resilience and sustainability of the city's water supply.

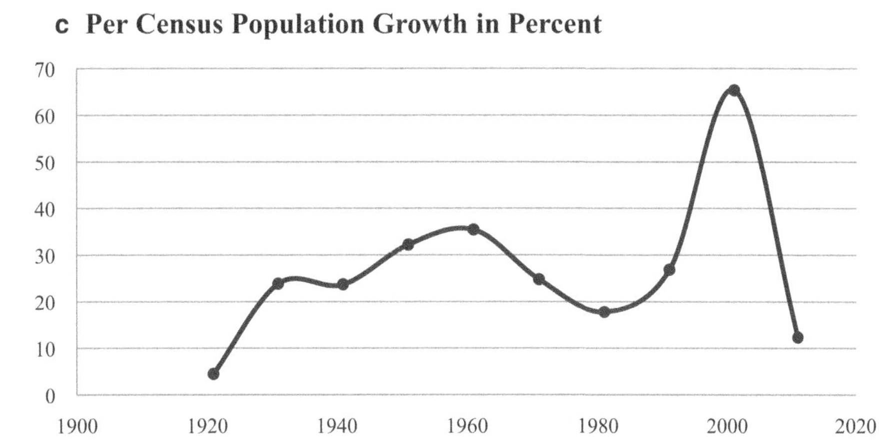

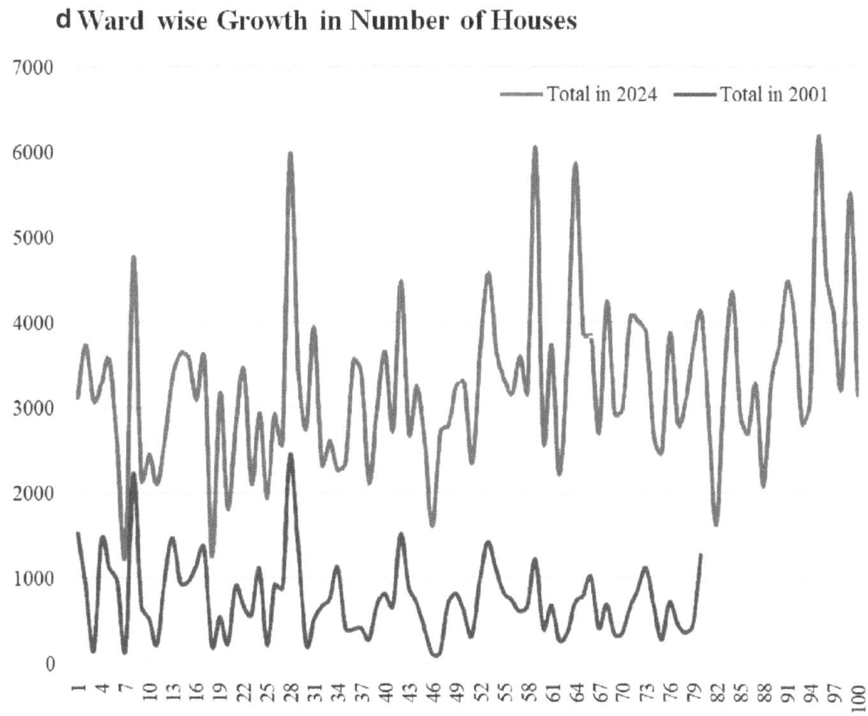

Fig. 9.7 (continued)

9.4.7 Increased Urban Area Relation to Ground Water

The study of the relationship between urban expansion and groundwater dynamics in Agra city reveals a complex interaction of factors that influence the city's hydrological landscape. Agra city's urban footprint has expanded significantly in recent years, as per census data of 2001 in Agra city, there were 67 wards and today, as per Nagar Nigam Agra, there are 100 wards, this increment only owing to causes such as population increase, industrialization, and infrastructure development. Rainwater cannot soak into the ground; thus it collects as runoff, potentially causing localized flooding, erosion, and surface water contamination. Converting natural landscapes to metropolitan contexts alters groundwa-

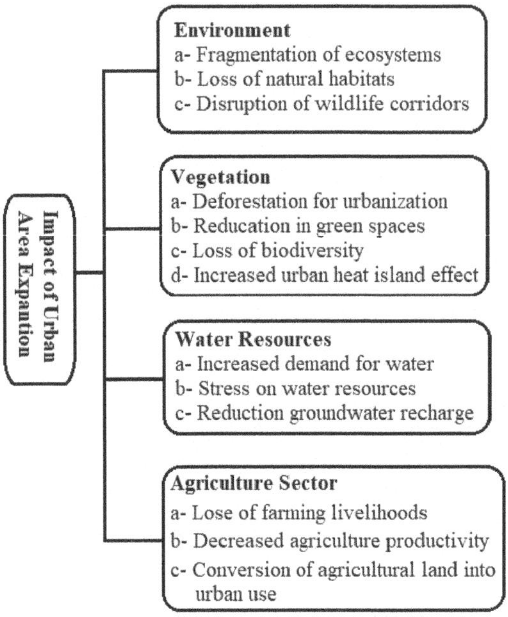

Fig. 9.8 Impact of urban area expansion. Source: Authors

ter recharge processes, altering aquifer replenishment. Studies using hydrological modelling and groundwater monitoring data show that recharge rates are decreasing in urbanized regions as infiltration opportunities decline and surface runoff increases. Urban development frequently entails excavation, land leveling, and construction activities that alter subsurface geological properties. These changes can impact groundwater flow patterns, aquifer recharge zones, and the overall hydro-geological regime in the area. GIS-based groundwater recharge analysis and evidence models reveal subsurface changes related to urbanization Industrial discharge, poor waste disposal and the use of agro-chemicals all contribute to groundwater contamination, raising concerns about water quality and human health.

9.4.8 Increase Concrete Forest Relation to Ground Water Recharge

The swift growth of Agra city areas has resulted in the creation of *"Concrete Jungles"* consisting of dense groups of buildings, roads, and other surfaces that do not absorb water. This shift disrupted the natural water cycle, particularly hindering the recharge of groundwater. In Agra city, where urban growth has been picking up speed for a long time, replacing open spaces and green areas with concrete and asphalt restricts the ability of rainwater to soak into the soil and renew underground water reserves. Under normal circumstances, rainwater seeps slowly into the earth, helping to refill groundwater supplies. Yet, in cities where hard surfaces prevail, a large portion of rainfall turns into surface runoff. This runoff swiftly flows into drainage systems and rivers, failing to benefit the groundwater supply. Consequently, water that could sustain a long-lasting water source often contributes to urban flooding or evaporates unused. This decrease in the ability to recharge is especially worrisome in regions where groundwater serves as a critical source of fresh water for homes, farming, and industry. For example, in Agra, the rising population has increased the need for groundwater, while the city's expanding concrete structures diminish natural recharge areas. The mismatch between groundwater extraction and its replenishment can cause water table declines, empty wells, and persistent water shortages. To alleviate these consequences, urban development must embrace sustainable practices, including using permeable pavements, creating green spaces, collecting rainwater, and safeguarding natural recharge areas. Addressing or reducing the harmful effects of concrete jungles is vital for re-establishing groundwater equilibrium and securing water availability in fast-growing cities.

9.5 Conclusion and Recommendations

9.5.1 Recommendation and Policy Perspectives

The research emphasizes how vital it is to include groundwater issues in urban planning and development. By using green infrastructure, efficient storm water management, and sustainable land-use practices, cities can reduce the negative effects that urban growth has on groundwater while ensuring long-term sustainability and resilience. One major suggestion is to establish systems for managing reclaimed water to boost groundwater recharge within city limits. Reclaimed water, which comes from treated wastewater generated by households, industries, or municipalities, can be safely used for purposes that do not require drinking water, like watering landscapes, supporting industrial activities, and aiding ecological restoration. Through careful management of the treatment, use, and disposal of this reclaimed water, urban areas can maximize their benefits for groundwater replenishment and enhancement. Additionally, the study presents further specific suggestions aimed at improving groundwater sustainability:

9.5.1.1 Water Control Measures and Increase Groundwater Table

Water is essential for life, and managing this crucial resource properly is important for keeping the environment healthy, ensuring crops grow well, and supporting people's health. In various areas, particularly in India's Himalayan and dry regions, groundwater levels are dropping because of uncontrolled removal and inadequate water management. It is important to put water management strategies in place to help replenish groundwater and bring back the natural flow of water. In cities, the management of reclaimed water plays a key role in tackling wastewater issues and aiding groundwater restoration and longevity. By utilizing this reclaimed water for recharging groundwater, urban areas can reduce water shortages and improve their overall water stability. Proper management requires implementing safety measures that optimize benefits while reducing risks to both public health and the environment. Such measures involve thorough treatment methods, dependable distribution networks, various reuse opportunities, and robust regulations. Collectively, these initiatives foster the safe, effective, and sustainable use of reclaimed water, enhancing urban water management practices.

Effective water management strategies are crucial for restoring groundwater levels and ensuring sustainable water use. These measures connect traditional methods with contemporary scientific approaches, particularly in delicate ecological areas such as the Indian Himalayas. By implementing well-designed water control plans, we can raise the groundwater level, which benefits both human necessities and the well-being of entire ecosystems.

9.5.1.2 Groundwater Recharge and Surplus Supply Control Measures

This recommendation advocates for the combined use of groundwater recharge initiatives and measures to manage surplus water supplies in urban areas, like Agra. To effectively and environmentally manage groundwater resources within cities, a comprehensive strategy that integrates both groundwater replenishment and control of excess water supply is essential. By enhancing groundwater recharge techniques and optimizing surplus water management, cities can foster resilience and sustainability. Groundwater recharge initiatives focus on restoring aquifers through either natural or artificial means, while surplus management strategies are designed to mitigate excess water to avoid flooding and ensure efficient water usage. It is beneficial to encourage the collection and storage of rainwater from rooftops, roads, and other surfaces for future irrigation, landscaping, or groundwater replenishment. Additionally, promoting the use of permeable materials for roads, sidewalks, and parking areas helps rainwater seep into the ground instead of increasing surface runoff. Constructing infiltration basins or trenches at key sites will aid in capturing and retaining excess storm-water, allowing it to seep into the ground over time. Controlling groundwater recharge and surplus supply is essential rather than optional. We need a coordinated approach that involves integrated watershed management, raising public awareness, encouraging community involvement, and putting strong policies in place to create effective recharge systems. This approach will help us secure water resources for future generations, support the lives of people in rural areas, and address the increasing issues related to climate change and water shortages.

9.5.2 Conclusion

Research indicates that rapid urban expansion and shifts in land use are notably hindering the replenishment of groundwater. These difficulties increase the likelihood of groundwater shortages and pollution, highlighting the critical importance of comprehensive water management solutions. Such solutions should prioritize boosting recharge rates, controlling demand, and matching land use with water sustainability objectives. The research also highlights the significance of making decisions based on data in the management of groundwater. Policies, resource distribution, and adaptable strategies should all be informed by scientific data to guarantee the enduring sustainability of groundwater supplies. The findings suggest that a cooperative effort is needed, bringing together government bodies, research organizations, community members, and civil society to develop effective long-term solutions, i.e., master plan for artificial recharge to ground water by Central Ground water Board (CGWB), Jal Shakti Abhiyan (JSA), Mission Amrit Sarovar, National Aquifer Mapping (NAQUIM), community involvement by MGNREGA Initiatives.

Urban growth often reduces groundwater recharge, but if urban water management integrates recharge enhancement, efficient water use, and sustainable withdrawal strategies, it makes a direct contribution to SDG 6.4 by ensuring long-term water security in Agra city. Expansion of impervious surfaces (roads, concrete, rooftops) reduces natural infiltration and groundwater recharge. Population and industrial growth increase dependence on groundwater. Leaks, waste, and untreated sewage threaten aquifers, reducing usable reserves. Due to declining recharge groundwater table is falling, and over-extraction impacts aquifer stress and subsidence. Urban heat island effects altered evapotranspiration, changing the hydrological balance. Managed Aquifer Recharge for using stormwater harvesting, recharge wells, and percolation tanks. Increase Green Infrastructure, i.e., wetlands, Parks, and permeable pavements for infiltration. Use in practice GIS/Remote Sensing applications for monitoring urban sprawl, mapping recharge zones, and modelling aquifer stress. Support and planning for Integrated Urban Water Management (IUWM), i.e., Linking surface water, groundwater, and wastewater in planning. Target 6.4.1 (Water-use efficiency)—Agra city areas that protect and enhance groundwater recharge can reduce dependence on imported water and improve recycling efficiency. Smart metering, leakage reduction, and water-efficient appliances complement recharge-based supply security. Target 6.4.2 (Sustainable withdrawals and stress)—Balanced Agra city growth with recharge zones ensures withdrawals remain within renewable limits, and by reducing overexploitation, Agra cities move towards sustainable water stress levels.

Between 1901 and 2011, Agra experienced considerable urban growth, with an increasing population, broader built environments, and modified land use patterns. This urban development has stressed natural resources, especially groundwater, due to heightened demand and fewer chances for natural replenishment. To counteract the adverse impacts of hard surfaces such as paved roads, the study recommends incorporating permeable pavements, green infrastructure, and rainwater collection systems into urban design and planning. The spread of *"concrete jungles"* threatens both groundwater availability and ecological balance. Therefore, sustainable urban planning that includes permeable areas and efficient water conservation methods is crucial for ensuring the future water supply of Agra. Finally, the study calls on policymakers, urban planners, and community leaders to make groundwater recharge a priority and to embrace comprehensive and sustainable water management practices. By utilizing these insights and promoting collaborative, knowledge-driven decision-making, Agra can enhance its resilience and safeguard water security for both present and future populations.

Acknowledgments Authors are thankful to the Head, Department of Geography, Kumaun University, Nainital for providing GIS lab for completing this research work and also thankful to Dr. Naveen Ram, Assistant Professor, Dept. of Economics, Uttarakhand Higher Education for helping in applying statistical methodology in this research. We are grateful to the local communities, Jal Sansthan Department, Ground Water Department of the Agra city for providing us with relevant data and information in the study area.

Funding: The research was not supported by any funding.

Potential conflicts of interest: On behalf of both authors, the corresponding author states that there is no conflict of interest.

Informed Consent: Informed consent was obtained from all individual participants included in the study.

Author Contribution: All authors have contributed equally to this work.

Data Availability Statement: The datasets analyzed during the current study are available from the corresponding author on reasonable request.

AI Discloser Statement: I hereby declare that no artificial intelligence (AI) tools were used and all work presented herein is entirely my own.

References

Aithal, B. H., & Ramachandra, T. V. (2016). Urbanisation and its implications on land use/land cover dynamics of Jaipur city using satellite data. *Journal of Environmental Science, Computer Science and Engineering and Technology, 5*(2), 142–153.

Arulbalaji, P., Rajkumar, P., & Srinivasamoorthy, K. (2019). Assessment of groundwater quality for drinking and irrigation purpose in Kodaikanal taluk, Tamil Nadu, India. *Groundwater for Sustainable Development, 9*, 100223.

Baier, K., Schmitz, S., Azzam, R., & Strohschon, B. (2014). Groundwater vulnerability to urbanization—a case study. *Environmental Earth Sciences, 71*(3), 1279–1291.

Beker, B. A., & Kansal, M. L. (2024). Complexities of the urban drinking water systems in Ethiopia and possible interventions for sustainability. *Environment, Development and Sustainability, 26*, 4629–4659. https://doi.org/10.1007/s10668-022-02901-7

Ducci, L., Rizzo, P., Pinardi, R., Solfrini, A., Maggiali, A., Pizzati, M., Balsamo, F., & Celico, F. (2023). What is the impact of leaky sewers on groundwater contamination in urban semi-confined aquifers? A test study related to fecal matter and personal care products (PCPs). *Hydrology, 10*(1), 3. https://doi.org/10.3390/hydrology10010003

Dutta, J., Choudhury, R., & Nath, B. (2024). Quantification of urban groundwater recharge: A case study of rapidly urbanizing Guwahati city, India. *Urban Science, 8*(4), 187. https://doi.org/10.3390/urbansci8040187

Gupta, A., Gajbhiye, S., & Paliwal, S. (2017). Hydrological modelling of a river basin for sustainable water resources management using remote sensing and GIS: A case study of Agra district, Uttar Pradesh, India. *Applied Water Science, 6*(1), 29–42.

Gupta, M., & Srivastava, P. K. (2016). Evaluating groundwater potential zones in urban environment: A case study of Delhi. *Journal of the Geological Society of India, 90*(1), 81–90.

Jaeger, J. A., Bertiller, R., Schwick, C., & Kienast, F. (2010). Urbanization and its effects on habitat features in a mountain landscape. *Landscape Ecology, 25*(4), 601–614.

Jha, M. K., Kumar, D., & Bansod, B. K. S. (2019). Impact of urbanization on groundwater quality in Dhanbad district, Jharkhand, India. *Groundwater for Sustainable Development, 8*, 258–271.

Jha, M. K., Singh, V. P., & Vatsa, K. K. (2008). Groundwater contamination studies: Some important aspects and problems. *Water, Air, and Soil Pollution, 193*(1–4), 129–146.

Kiflay, E., Schirmer, M., Foppen, J. W., & Moeck, C. (2025). Impact of urbanization on groundwater recharge: Altered recharge rates and water cycle dynamics for Arusha, Tanzania. *Hydrogeology Journal, 33*, 33–47. https://doi.org/10.1007/s10040-024-02870-3

Li, J., Gao, X., & Li, H. (2019). Temporal-spatial patterns and driving forces analysis of urban water balance: A case study of Beijing. *Water, 11*(8), 1623.

Liang, S., & Wang, J. (2020). *Urbanization: monitoring and impact assessment, Chapter-23.* Advanced Remote Sensing (2nd ed.), Academic Press, 833–870. https://doi.org/10.1016/B978-0-12-815826-5.00023-4

Mao, J., Wang, Y., Zhang, H., & Jing, X. (2024). Study on the influence of urban water supply pipeline leakage on the scouring failure law of cohesive soil subgrade. *Water, 16*(1), 93. https://doi.org/10.3390/w16010093

Morris, B. L., Lawrence, A. R., Chilton, P. J., Adams, B., Calow, R. C., & Klinck, B. A. (2003). *Groundwater and its susceptibility to degradation: A global assessment of the problem and options for management.* United Nations Environment Programme.

Sharma, M., Rawat, S., Kumar, D., Awasthi, A., Sarkar, A., Sidola, A., Choudhury, T., & Kotecha, K. (2024). The state of the Yamuna River: A detailed review of water quality assessment across the entire course in India. *Applied Water Science, 14*, 175. https://doi.org/10.1007/s13201-024-02227-x

Sharma, P., Rai, S. P., & Pandey, A. (2018). Impact of urbanization on groundwater recharge and quality:

A case study of Agra district, Uttar Pradesh, India. *Environmental Earth Sciences, 77*(23), 815.

Singh, A., Shukla, A., & Sinha, R. (2021). Assessment of urbanization impact on water bodies using geospatial techniques: A case study of Lucknow city, Uttar Pradesh, India. *Journal of Geo-visualization and Spatial Analysis, 5*(1), 3.

Singh, A. V. (2013). *District ground water brochure Agra district, Uttar Pradesh*. https://www.cgwb.gov.in/old_website/District_Profile/UP/Agra.pdf

Soltani, F., Mardani, A., Zakikhani, P., & Mohammadi, M. (2018). The impact of urban expansion on groundwater resources using geospatial technology: A case study of Tehran, Iran. *Geocarto International, 33*(5), 499–510.

Tiwari, A. K., Kumar, M., Keshari, A. K., & Kumar, M. (2020). Evaluation of land—use change and urbanization impact on groundwater quality in Varanasi city using geospatial techniques. *Environmental Monitoring and Assessment, 192*(1), 13.

Tziritis, E., Giannikos, P., & Katsanou, K. (2020). An approach to improve urban water resources management: A case study in Athens. *Water Resources Management, 34*(7), 2343–2362.

Van Broekhoven, F. J. G., Griffioen, J., Dekker, S. C., Sharma, M. K., Bhagwat, A., & Schot, P. P. (2024). Linking recharge water sources to groundwater composition in the Hindon sub-basin of the Ganges River, India. *Science of the Total Environment, 954*, 176399. https://doi.org/10.1016/j.scitotenv.2024.176399

Wartalska, K., Grzegorzek, M., Bełcik, M., Wdowikowski, M., Kolanek, A., Niemierka, E., Jadwiszczak, P., & Kazmierczak, B. (2024). The potential of rain water harvesting systems in europe–current state of art and future perspectives. *Water Resources Management, 38*, 4657–4683. https://doi.org/10.1007/s11269-024-03882-0

Yadav, R. K., Sarkar, S., & Sankar, M. K. (2020). Dynamics of urban growth and land use pattern in Agra region: A GIS based analysis. *Journal of the Indian Society of Remote Sensing, 48*(2), 287–300.

Yang, L., Smith, J., Baeck, M. L., Smith, B., Tian, F., & Niyogi, D. (2016). Structure and evolution of flash flood producing storms in a small urban watershed. *Journal Geophysical Research: Atmospheres, 121*, 3139–3152. https://doi.org/10.1002/2015JD024478

Yar, P. (2020). Urban development and its impact on the depletion of groundwater aquifers in Mardan City, Pakistan. *Groundwater for Sustainable Development, 11*, 100426. https://doi.org/10.1016/j.gsd.2020.100426

10 Resilience and Spatial Justice in Urban Waterscapes: A GIS-Based Analysis of Sylhet City, Bangladesh

Shataparna Das, Subrata Das, Farha Moon, and Mohammad Rakibul Islam Anik

Abstract

Urban waterscapes provide significant social, economic, and environmental advantages; however, inequities in their distribution and accessibility compromise social equity and resilience within rapidly urbanising areas. This study utilises Geographically Weighted Regression (GWR) in Geographic Information Systems (GIS) to analyse spatial disparities in the accessibility of waterbodies throughout Sylhet, Bangladesh, a monsoon city undergoing severe loss of blue spaces. Examining the influence of residential density, waterbody stability, and total waterbody area, our findings identify four distinct pathways to resilience shaped by spatial heterogeneity: (1) density-driven revitalisation within urban centres, (2) hydrological fragmentation occurring in transformed floodplains, (3) isolation-preserved functionality evident in outlying zones, and (4) topography-imposed efficiency limitations in hilly regions. Notably, regions characterised by elevated population densities experience a pronounced acceleration in blue space loss. In contrast, peripheral zones maintain waterbodies that remain inaccessible due to shortcomings in planning, despite showing higher stability metrics. Case studies of successful urban revitalisation initiatives illustrate how context-responsive interventions can transform urban pressures into opportunities for stewardship. We advocate for the implementation of terrain-adaptive design principles to guide equitable governance of waterscapes, arguing that achieving spatial justice in urban resilience requires embracing shaped by spatial heterogeneity, rather than ignoring it.

Keywords

Waterscape · Resilience · GWR · Waterbody conservation · Spatial heterogeneity

10.1 Introduction

10.1.1 Urban Waterscape

Urban waterscapes, including ponds, lakes, rivers, and wetlands, play a critical role in enhancing the overall livability, promoting sustainability, and fostering resilience within urban areas. These distinct blue spaces are instrumental in

S. Das · S. Das (✉) · F. Moon
Department of Architecture, Shahjalal University of Science and Technology, Sylhet, Bangladesh
e-mail: shataparna-arc@sust.edu; ar.subrata-arc@sust.edu; farhamoon-arc@sust.edu

M. R. I. Anik
Department of Architecture, Rajshahi University of Engineering & Technology, Rajshahi, Bangladesh
e-mail: rakibul.islam@arch.ruet.ac.bd

U. Chatterjee, A. I. Almulhim (eds.), *Urban Waterscapes*, Sustainable Development Goals Series,
https://doi.org/10.1007/978-3-032-20649-7_10

providing essential ecosystem services such as water storage, groundwater recharge, and flood regulation, which help biodiversity and environmental stability (Ribbe et al., 2024; Upadhyay, 2023). They also mitigate urban heat island effects through the Water Cold Island (WCI) phenomenon by regulating local temperature and humidity and maintaining vital microclimates (Ampatzidis & Kershaw, 2020; Ma et al., 2024; Manteghi et al., 2015; Sun & Chen, 2012; Upadhyay, 2023). Beyond their ecological value, urban waterbodies also offer recreational and cultural benefits that significantly enrich the physical and mental well-being of urban residents (Alikhani et al., 2022; Das et al., 2022; Fletcher et al., 2013; Khalaji et al., 2024). However, these benefits are not distributed evenly. In many rapidly urbanising cities, access to waterbodies is increasingly shaped by socio-spatial disparities, resulting in uneven water resilience and environmental justice.

10.1.2 Waterbody Access per Capita and Spatial Justice

Access to waterbodies per capita is an essential yet often overlooked dimension of public health, environmental justice, and urban resilience. However, the spatial distribution and accessibility reflect deep-rooted inequalities embedded within urban environments. In this context, (Nelson et al., 2023) argued that urban disparities are expressed through spatial dimensions, where historically underprivileged groups often experience significant constraints in accessing vital environmental amenities. (Sarkar et al., 2024), further highlight how urban morphology and imbalanced growth patterns can worsen socio-economic inequalities, particularly in areas undergoing rapid urbanisation. Global research shows that marginalised groups often face challenges accessing urban blue spaces, limiting their ecological and social benefits. In Helsinki (Viinikka et al., 2018), it was found that affluent and well-educated people lived closer to water bodies. In contrast, communities characterised by low income and immigrant status were disproportionately and systematically excluded from such vital resources. Similarly, (Marçal et al., 2024) assessed water security inequalities in Brazil, particularly within the context of the Global South, revealing an uneven spatial clustering that highlighted the disadvantages faced by residents in peripheral neighbourhoods, thereby emphasising the urgent need for equitable resource distribution. While over 50% of the global population lives within 3 kilometres of a freshwater source (Kummu et al., 2011), this proximity often fails to translate into meaningful, equitable access. In Misiones, Argentina, for instance, large segments of the population still lack access to improved water sources, highlighting how affordability and infrastructure gaps worsen spatial injustice (Antonio & González, 2020). Moreover, these challenges are particularly acute in South Asian cities, including those in Bangladesh, where the pressures of rapid urbanisation, land commercialisation, and weak planning frameworks have intensified blue space degradation and exclusion (Ashik et al., 2024; Rahman & Zhang, 2018; Reyad & Ullah, 2022). As a result, waterbodies that once served as vital ecological and communal resources are increasingly being lost, fragmented, or becoming inaccessible, further entrenching spatial and social inequities. This decline not only affects social justice but also undermines water system resilience, as marginalised populations are more vulnerable to disruptions in water access during crises (Han et al., 2024).

10.1.3 Urban Waterscapes in Bangladesh: Challenges and Inequities

In Bangladesh, urban waterscapes are increasingly shaped by a complex interplay of demographic pressure, unregulated land use change, weak planning capacity, and socio-spatial inequality. Despite their ecological, cultural, and public utility, urban waterbodies across the country are under significant threat. Research indicates that residents of high-density slums in

Dhaka face severely limited access to both green and blue spaces, despite an overwhelming demand for these essential amenities (Rahman & Zhang, 2018). Sylhet, a rapidly developing northeastern city, presents a compelling case of ecological loss and spatial exclusion. Studies show that between 1988 and 2006, the total area of waterbodies in Sylhet district fell dramatically from 81,535.2 hectares to just 28,435.6 hectares, a decline primarily attributed to unregulated urbanisation and aggressive land acquisition by private real estate developers (Gupta et al., 2018; Haque et al., 2008). As a result, water bodies that were once integral to the ecological and cultural fabric of the city are increasingly subject to encroachment or made inaccessible to marginalised communities, leading to significant social and environmental ramifications (Das et al., 2022). This clear spatial inequity not only results in a marked decline in environmental quality, but it also worsens existing social disparities, as access to these urban blue spaces is frequently stratified along critical lines such as income levels, population density, and geographical location (Ashik et al., 2024; Viinikka et al., 2018). While some waterbodies remain in Sylhet's peripheral areas, they are frequently disconnected from residential zones and public use due to inadequate infrastructure and weak integration into the broader urban system (Das et al., 2022; Reyad & Ullah, 2022).

The degradation and loss of traditional water bodies, as observed in Sylhet, significantly harm water resilience by disrupting essential ecosystem functions and compromising long-term urban water security (Falkenmark et al., 2019). It also raises serious concerns regarding equity and environmental justice. Recent hydrological modelling by (Dutt et al., 2024) further emphasises the consequences of these land use changes, showing increased surface runoff, diminished groundwater recharge, and heightened flood risk across urban Sylhet. These findings highlight the urgent need for spatially informed planning interventions that prioritise both environmental sustainability and social inclusion.

10.1.4 Geospatial Approaches to Urban Waterscape Analysis

Geospatial analysis has become an essential tool for assessing urban inequality, particularly in access to infrastructure such as waterbodies. By mapping spatial disparities, Geographic Information Systems (GIS) enable planners to identify underserved areas and evaluate the equity of urban development. In rapidly growing cities, especially in the Global South, these methods support data-driven, context-sensitive planning. For example, (Reyad & Ullah, 2022) used GIS overlay analysis to expose Dhaka-centric development patterns and the disproportionate concentration of infrastructure in affluent neighbourhoods. Similarly, (Mashrur et al., 2015) developed a spatial equity index based on accessibility to public facilities, offering a foundation for evaluating fairness in urban planning. Furthermore, the application of network-based analytical techniques, as exemplified in the work of (Kurniawan & Roychansyah, 2023), offers a robust methodological framework for scrutinising the nuances of walking access to urban waterscapes, particularly within densely populated cities such as Sylhet, thereby enhancing our understanding of spatial dynamics and equity in urban settings. Among advanced tools, Geographically Weighted Regression (GWR) stands out for its ability to capture spatial heterogeneity. Unlike global models, GWR reveals how relationships between variables—such as population density and water access—vary across neighbourhoods. In a city like Sylhet, where land use, elevation, and waterbody distribution differ widely across Strategic Planning Zones (SPZs), GWR provides a nuanced understanding of where and why access gaps emerge, and how resilience varies spatially.

10.1.5 Towards Resilient Waterscape

Addressing the spatial inequities in urban waterscapes requires policy interventions that are firmly grounded in robust geospatial evidence and data-driven insights. As urban pressures intensify, especially in rapidly growing secondary cities like Sylhet, equitable access to waterbodies per capita must be treated as a core component of urban resilience, not just a matter of ecological preservation. (Wu & Liu, 2022) have proposed an innovative framework that intricately links urban design principles with overarching equity goals, thereby emphasising the vital need for inclusive planning practices that account for the diverse needs of all community members. In a similar context, (Kurniawan & Roychansyah, 2023) advocate for comprehensive reforms in the allocation of public spaces, specifically aimed at benefiting the most vulnerable and marginalised groups within society. In the specific case of Sylhet, the adoption of such progressive approaches could serve as a guiding framework for the protection of water bodies that hold significant cultural value, while simultaneously enhancing accessibility for communities that have been historically underserved and marginalised. These strategies are essential for fostering resilient urban waterscapes—systems capable of absorbing shocks, adapting to change, and maintaining core water functions in the face of ongoing socio-environmental pressures (Falkenmark et al., 2019; Unnikrishnan et al., 2017). Ultimately, resilient urban waterscapes depend not only on ecological restoration but also on spatial justice. Policies that integrate equity with hydrological function—guided by tools like GWR—can help cities transition from fragmented and exclusionary water systems toward inclusive, adaptive, and sustainable blue infrastructure.

10.1.6 Research Aim and Objectives

Despite these emerging frameworks, there remains a notable gap in localised empirical research that evaluates waterbody accessibility and social equity in secondary Bangladeshi cities. While policy discourse increasingly recognises the ecological value of blue infrastructure, local empirical research in secondary cities like Sylhet remains limited, especially on questions of spatial justice and per capita accessibility. Existing urban master plans tend to focus on conservation rather than equitable distribution, overlooking the lived experiences and spatial needs of vulnerable populations (Reyad & Ullah, 2022).

This chapter addresses that gap by conducting a comprehensive examination of the spatial inequities related to urban waterscape accessibility in Sylhet, Bangladesh. It focuses on the complex interplay between the distribution of waterbodies, population density, and the principles of social equity. Leveraging the capabilities of Geographic Information Systems (GIS), the study analyses how these disparities are manifested across Sylhet's Strategic Planning Zones (SPZs) and explores their broader implications for environmental justice, urban resilience, and community well-being. Drawing upon a rich tapestry of global and regional scholarship, including geospatial analyses that illuminate urban inequalities (Kurniawan & Roychansyah, 2023; Mashrur et al., 2015) and pertinent case studies derived from other South Asian cities (Bhor & Mayavel, 2024; Marçal et al., 2024), we aim to identify critical gaps in the accessibility of waterscapes and subsequently propose informed policy measures that can effectively address these inequities.

The novelty of this study lies in its geospatial analysis of waterbody distribution to population density and social equity. By shifting the focus from ecological conservation to questions of access, resilience, and fairness, this research contributes to a more holistic understanding of inclusive and resilient urban waterscape development. It underscores the necessity of integrated strategies that address both environmental degradation and socio-spatial disparities, advancing sustainable water management in line with the core theme of this chapter.

The key objectives of this study are to:

1. Analyse the impact of various urban factors on the water access per capita.
2. Evaluate the relationship between urban waterscape management and spatial justice.
3. Propose evidence-based policy recommendations to promote equitable and resilient waterscape integration in future urban development.

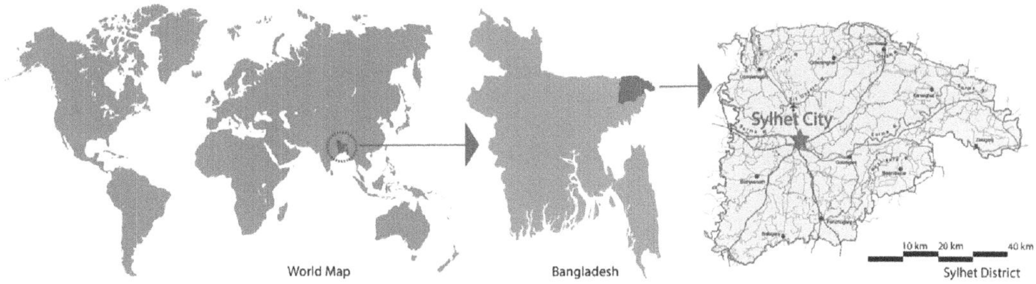

Fig. 10.1 Location map of Sylhet City

10.2 Study Area

Situated in the northeastern region of Bangladesh along the Surma-Kushiyara river system, Sylhet City (24.8917°N, 91.8833°E) presents a compelling narrative of urban growth and hydrological change (Fig. 10.1). The city's unique geographical setting, positioned between the floodplains of the Surma River and the foothills of the Meghalaya plateau, has historically supported an intricate network of waterbodies that have shaped its cultural and ecological identity. Over time, natural and human-made water features, including ponds, dighis (communal water reservoirs), khals (canals), and haors (seasonal wetlands) have served multiple roles—from providing essential water resources to supporting livelihoods, regulating microclimates, and maintaining spiritual and community practices.

The contemporary urban landscape of Sylhet is organised into Strategic Planning Zones (SPZs), a framework that reveals striking spatial patterns in waterbody preservation and access. SPZ 1–6 are part of the future expansion area of the city, and SPZ 7–12, confined within the red boundary, indicate the existing Sylhet City Corporation limit (Fig. 10.2). The city's historic core zones have experienced the most dramatic hydrological transformations, where significant portions of waterbodies have been lost to relentless urban expansion driven by population growth and land pressure (Gupta et al., 2018; Haque et al., 2008). In these areas, unregulated infill construction and weak regulatory enforcement have converted ponds into garbage-filled ditches or built environments, stripping them of their historical and environmental functions (Dewan & Yamaguchi, 2009). This stands in stark contrast to peripheral areas where waterbodies have persisted, though often in compromised states. The eastern expansion areas present a more nuanced scenario, where water management reflects socioeconomic disparities through selective preservation and aesthetic repurposing of water features.

Sylhet's tropical monsoon climate, with seasonal rainfall exceeding 4000 mm annually, makes its waterbodies critical for flood mitigation, stormwater retention, and groundwater recharge. However, urban development has progressively compromised these functions. The widespread filling of ponds and khals in the city's centre has significantly reduced Sylhet's ability to manage flood events, leading to measurable increases in flood duration and intensity. The resulting rise in surface runoff and decline in aquifer recharge have increased both ecological risk and infrastructural vulnerability (Ameen et al., 2015). Meanwhile, surviving waterbodies across the city face growing pressures from pollution, encroachment, and competing land uses. The northern zones, with their distinctive tea garden landscapes, present unique water management challenges where commercial agricultural interests sometimes conflict with community water needs. This complex urban waterscape forms the foundation for the current study, which examines how variations in waterbody accessibility are shaped by localised physical and socio-spatial factors across Sylhet's 6 Strategic Planning Zones. In particular, the GWR model

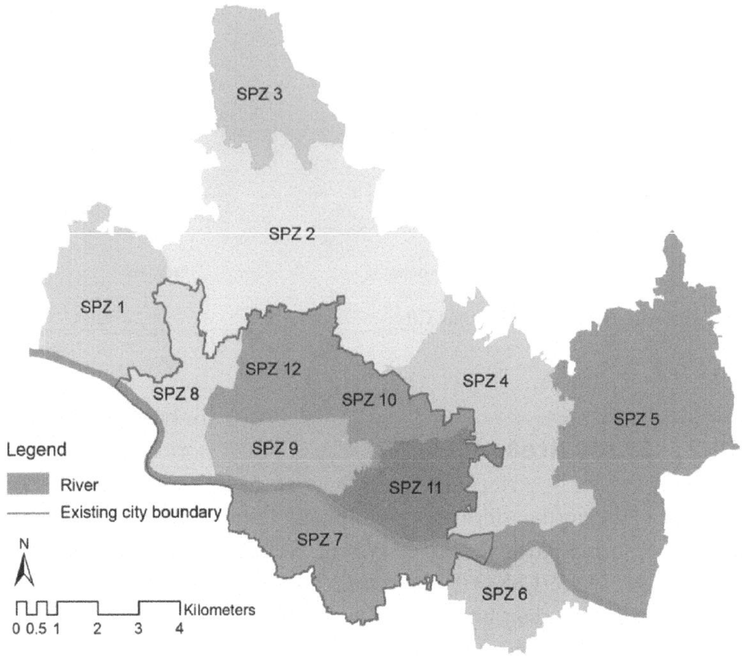

Fig. 10.2 Location of Sylhet City in Sylhet district (left) and SPZs boundaries of the city

uses three core explanatory variables: population density, waterbody area, and waterbody stability. These variables were selected to reflect the interaction between demographic pressure, the physical presence of waterbodies, and their relative persistence over time. The following sections will explore how Sylhet's distinctive combination of geographical factors, urban growth patterns, and planning decisions has shaped its contemporary waterscapes, with special focus on the equity dimensions of waterbody access and preservation across the city's strategically planned zones.

10.3 Materials and Methods

10.3.1 Conceptual Framework and Analytical Approach

This study employs Geographically Weighted Regression (GWR) as a spatial decision-making tool to analyse inequities in urban waterscape accessibility. GWR can address spatial non-stationarity, where relationships between variables shift across geography due to contextual factors like topography, infrastructure, or socio-economic gradients (Xu & Ouyang, 2018). GWR's utility in capturing place-specific dynamics is exemplified in studies of urbanisation impacts, such as a spatiotemporal analysis of socio-economic drivers on landscape patterns in Shenzhen, China (Liu et al., 2020). This approach is particularly suited for assessing waterscape resilience pathways in Sylhet, where accessibility dynamics vary significantly across Strategic Planning Zones (SPZs). The methodological flowchart integrates detailed analysis and comparative study through GIS and real-world case studies to generate policy recommendations (Fig. 10.3).

10.3.2 Spatial and Demographic Data on the Study Area

Primary data collected from Sylhet City Corporation, including the demographic data (Table 10.1) and GIS shapefiles to support urban analysis. Waterbody count, total area, building density, and residential density were acquired from GIS through the spatial join tool.

Figure 10.4 illustrates the distribution of waterbodies throughout Sylhet city and total area of waterbodies in each SPZ. The central and

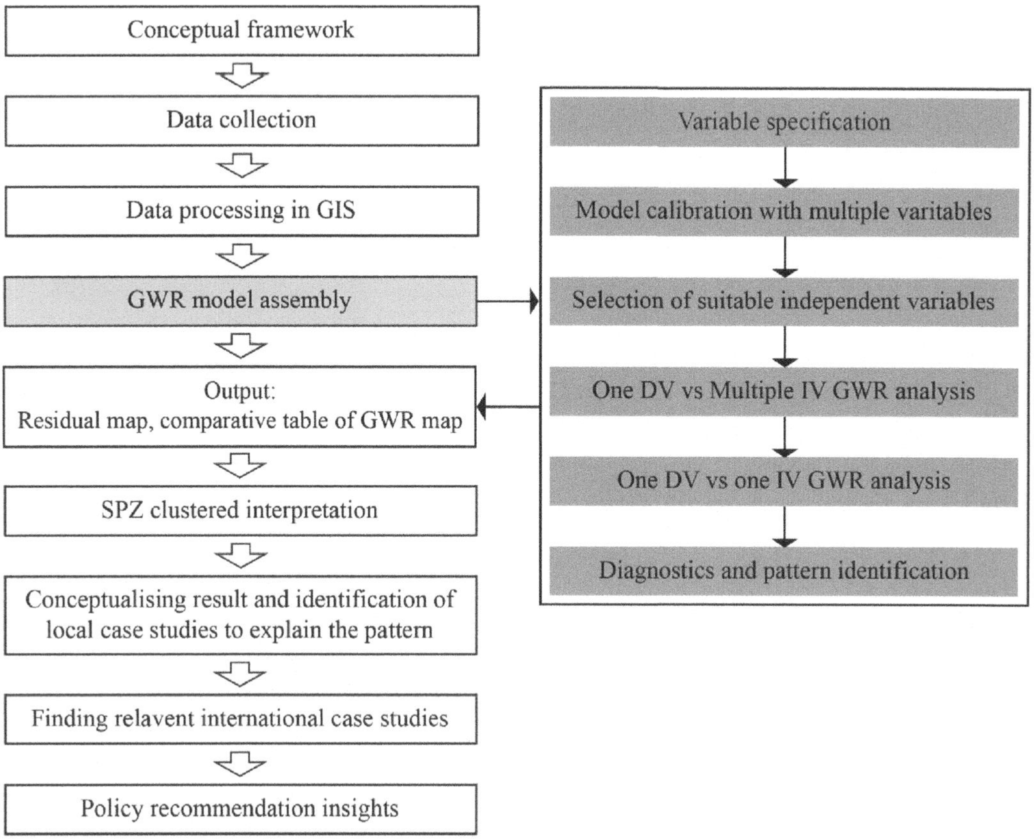

Fig. 10.3 Methodological flowchart

Table 10.1 Waterbody distribution and Socio-Spatial Metrics by SPZ

SPZ	Waterbody %	Population	Total density (persons/acre)	Residential density (persons/acre)	Residential land (%)
7	13.8	63,155	37.5	83.3	45.0%
8	12.9	97,080	84.8	123.2	68.8%
9	6.3	98,776	104.9	188.6	55.6%
10	1.8	103,254	105.7	145.9	55.5%
11	8.9	76,943	70.1	135.3	51.8%
12	2.1	94,140	104.2	144.9	72.0%

northern zones (SPZ 9, 10, 12) have comparatively low concentrations of water bodies. River separated SPZ 7 has the highest amount of waterbodies, whereas the other two zones (SPZ 8, 11) share a large waterfront and have plenty of waterbodies as they are developed on low-lying flat land. Especially the southeastern part of the city was previously a part of '*Rupali Haor*', a wetland ecosystem, which was landfilled to develop the high-density housing area.

Figure 10.5 illustrates the building footprints of Sylhet city, where some large voids in SPZ 7, 8, and 11 indicates the presence of large open spaces and wetlands in this zone. Interestingly, there are many other pockets in the city core area, some of them are created from small hillocks, and some others are urban ponds. The residential population density map at the right side reflects the variation of residential land available at different SPZs and volume of residential buildings

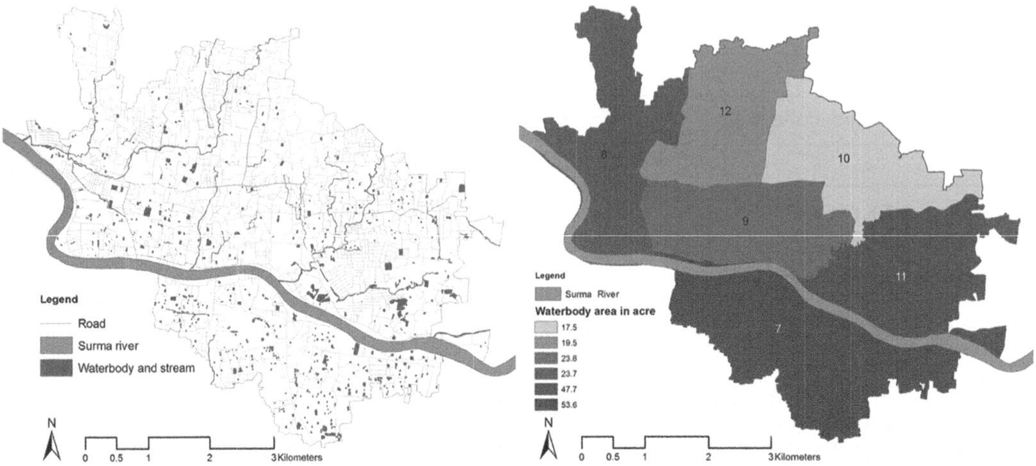

Fig. 10.4 Distribution of waterbodies (left) and total waterbody area in different SPZs

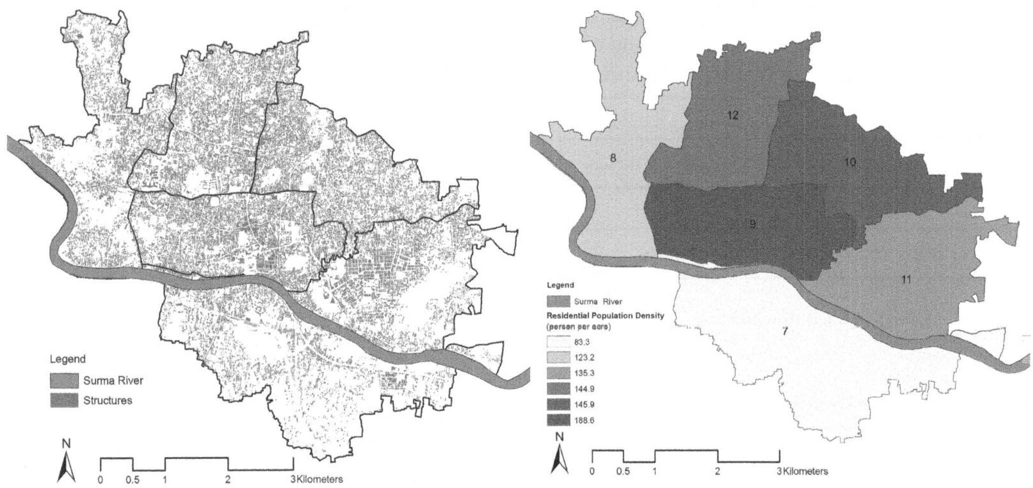

Fig. 10.5 Building footprint (left) and residential density (right) in different SPZs

on them. Total floor area of building calculated to generate this map, so buildings heights played in crucial role to define the density. Comparatively high land elevation of land at SPZ 10, 12, and 9 attracted higher development density compared to the peripheral parts.

Most of the cities in Bangladesh developed in flat terrain, in contrast, Sylhet is popular for its combination of hilly terrain in the north and flat lowland (haor) in the south (Fig. 10.6). The topography of the city limited the number of waterbodies in the northern part; moreover, in SPZ 10 and 12, a high percentage of waterbodies have been encroached on in the last few decades. Historical changes of waterbodies compared with satellite data from 2001 to 2025 and the oldest evidence collected from the georeferenced shapefile of land survey data from 1956 (Fig. 10.6). The southern SPZs are developed from lowlands and the demand of retention pond is higher. The comparative scenario of topography, development density, volume of waterbody, and stability of waterbodies in last seven decades explains the spatial growth pattern. These topographic factors and the distribution of water bodies and building structures provides a crucial basemap for the geostatistical analysis.

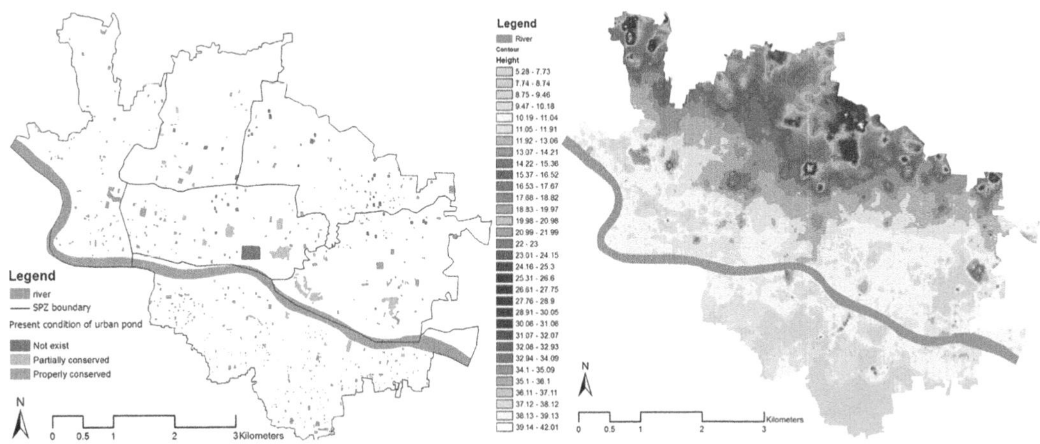

Fig. 10.6 Encroachment of waterbodies in different SPZs over last 70 years (left) and topographic map of Sylhet City (right)

10.3.3 Variable Specification and Selection

From GWR model calibration with these multiple variables, three potential independent variables: residential density, waterbody stability, and total waterbody area, were selected for final analysis. The effect of the three finally selected independent variables on water access per capita was examined in four combinations to compare the global model performance.

Dependent variable:

Water access per capita (m^2/person) = Accessible waterbody/population

Independent variables:

Residential population density (person/acre) = Residential population per SPZ/ Residential Landuse area of SPZ

Stability of waterbody (percentage) = (Total area of waterbody—area of waterbodies faced encroachment) / total area of waterbody

Total area of waterbody per SPZ (acre) = Total area calculated with clip and dissolve tool

Correlation with each individual independent variable and the combined GWR model, compared with localized R^2 values and standard residuals, reveals the spatial distribution pattern and interaction between the variables. Spatial distribution pattern and interaction between variables reveal zone-based characteristics that require further investigation through selective case studies across different SPZs. Some international best practices were studied to formulate the future recommendation pathway, which is aligned with the spatial phenomena identified from GWR analysis.

Table 10.2 Comparative model performance indicators

Model	Global R^2	AICc	Effective no.	Bandwidth
Combined (3 IVs)	0.954	−15,357	4.01	50,905 m
Stability only	0.425	−42,289	4.63	2134 m
Density only	0.947	−60,079	4.73	1997 m
Area only	0.820	−6.74	2.01	50,905 m

10.4 Result

10.4.1 Global Model Performance

Geographically Weighted Regression revealed extreme spatial non-stationarity across Sylhet's Strategic Planning Zones (SPZs). The combined model achieved exceptional performance (R^2 = 0.954) despite Sylhet's complex topography. Bandwidth distances (1997 m–50,905 m) reflected the city's unique monsoon-driven hydrological regimes (Table 10.2).

Global R^2 values confirm density as the strongest predictor (94.7% variance explained), while

Table 10.3 Summary table of GWR 1 (Water access per capita vs waterbody stability, area, and residential population density)

SPZ	Stability coeff	Density coeff	Area coeff (×10^{-6})	Local R^2	StdResid	Interpretation
7	+0.000523	−0.000705	+0.8457	0.954	+1.36	Isolation enhances stability benefits despite landfilling
8	+0.000520	−0.000704	+0.8467	0.954	−0.98	Growth misalignment undermines preservation advantages
9	+0.000521	−0.000705	+0.8461	0.954	+1.41	Density-driven revitalization outweighs space constraints
10	+0.000522	−0.000705	+0.8455	0.954	+0.84	Balanced but topography-limited utilization
11	+0.000523	−0.000705	+0.8452	0.954	−1.38	Landfilling severely disrupts hydrological function despite preservation
12	+0.000520	−0.000704	+0.8462	0.954	−0.36	Slope constraints moderate otherwise strong coefficients

stability alone shows limited explanatory power (42.5%).

10.4.2 Spatial Patterns of Predictor Effectiveness

The combined GWR analysis with waterbody stability, area, and residential population density reveals a distinct spatial pattern. Table 10.3 illustrates how waterbody stability, residential density, and waterbody area affect water access per capita across different SPZs. This table presents statistical model results for different planning zones (SPZ 7–12), examining the relationship between stability and key factors such as density and area. The model itself exhibits remarkable consistency across all zones: the Stability Coefficient remains very stable around +0.000522, the Density Coefficient is consistently negative at approximately −0.000705, and the Area Coefficient is consistently positive around +0.846 × 10^{-6}. The high Local R^2 value of 0.954 in every zone indicates the model explains over 95% of the variation in stability within each specific area, signifying an extremely strong fit.

Despite the remarkably consistent core relationships across all zones, the actual performance of individual zones relative to the model's prediction varies considerably. This variation is revealed by the standardised residuals and explained by unique local conditions (Fig. 10.7)

- **SPZ 7 & 9:** These zones exhibit better-than-predicted stability (positive residuals). This suggests that local factors not captured by the core density/area model are providing significant stability advantages. In SPZ 7, this appears to be related to its separation from urban pressure by the river, though it includes landfilling. In SPZ 9, the benefits associated with increased density seem to be effectively counteracting limitations posed by available space.
- **SPZ 8 & 11:** These zones show worse-than-predicted stability (negative residuals). This indicates the presence of local challenges or inefficiencies undermining the potential benefits suggested by the core model coefficients. For SPZ 8, development patterns are conflicting with conservation efforts, reducing expected gains. In SPZ 11, despite preserva-

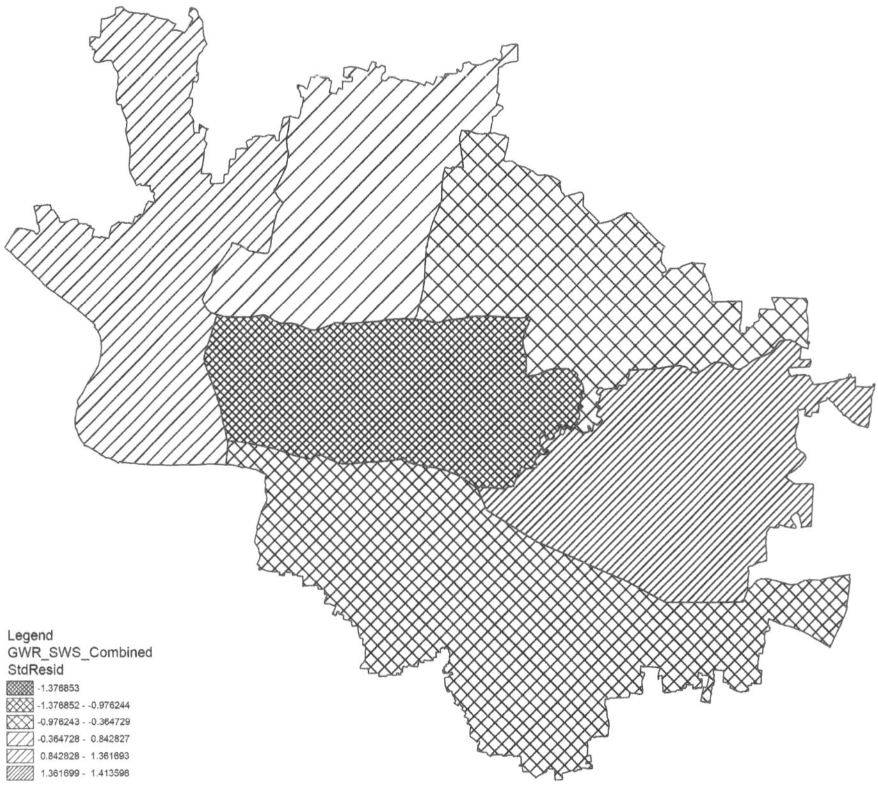

Fig. 10.7 GWR 1 map (Water access per capita vs waterbody stability, area, and residential population density)

tion efforts, significant disruption to natural water systems (likely linked to landfilling) is causing stability to fall short of predictions.

- **SPZ 10 & 12:** These zones demonstrate performance much closer to the model's prediction (smaller residuals). This reflects a situation where the core model relationships hold well, but local physical constraints are limiting the magnitude of the potential stability benefits. SPZ 10 achieves balance but is restricted by its terrain. SPZ 12 has inherently strong potential according to the model, but its steep slopes are acting as a moderating factor, preventing it from reaching the highest possible stability levels predicted solely by density and area.

Waterbody stability demonstrated paradoxical relationships across zones. In SPZ 7 (river-separated), high stability (+0.00342 coefficient) generated exceptional water access (+3.64 residual), likely due to isolation preserving functionality. Conversely, SPZ 9 (urban core) showed moderately high stability coefficients (+0.00220) but negative residuals (−0.74), indicating revitalisation efforts were undermined by density pressures. Most critically, SPZ 11 (watershed development) revealed a planning-implementation gap—high stability coefficients (+0.00305) coexisted with severe accessibility deficits (+1.90 residual), suggesting landfilling severed hydrological connectivity despite formal preservation (Table 10.4).

High waterbody stability increased per-capita access most effectively in isolated zones (SPZ 7), while showing negative returns in high-density cores (SPZ 9) (Fig. 10.8).

Residential population density exhibited equally divergent impacts. SPZ 9 defied conventional expectations—its high density (−0.001584 coefficient) correlated with 224% better access than predicted, demonstrating how revitalisation

Table 10.4 Summary table of GWR 2 (Water access per capita vs waterbody stability)

SPZ	Stability Coeff	Local R^2	Std Resid	Interpretation
7	0.00342	0.804	+3.64	Isolation enhances function
8	0.00097	0.650	−0.53	Growth misalignment
9	0.00220	0.425	−0.74	Density pressure reduces benefit
11	0.00305	0.720	+1.90	Landfilling disconnects function
10	0.00213	0.349	−0.32	Balanced but topography-limited
12	0.00093	0.357	−0.52	Topography limits access

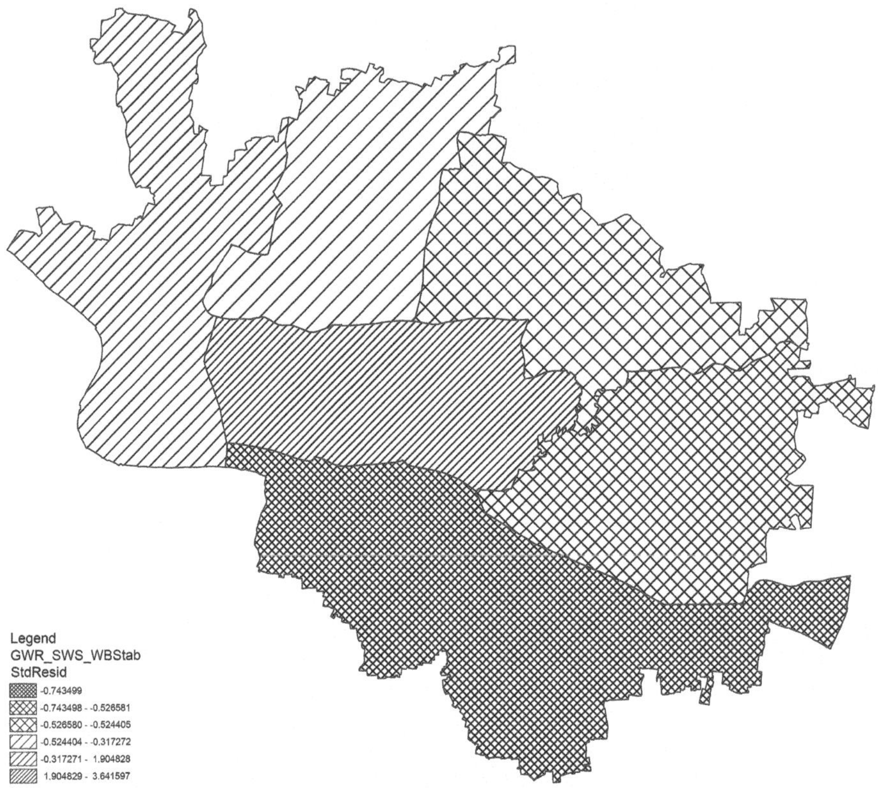

Fig. 10.8 GWR 2 map (Water access per capita vs waterbody stability)

converts density from liability to asset. This contrasted sharply with SPZ 10, where density generated 122% worse access despite similar coefficients, revealing how topographic limitations undermine potential benefits.

Density showed a 224% access premium in revitalised zones (SPZ 9) versus a 122% deficit in topography-limited zones (SPZ 10) (Fig. 10.9).

Waterbody area utilisation efficiency varied significantly across zones, with topography as the critical factor (Fig. 10.10). Flat zones (SPZ 7, 8) leveraged area effectively (+1.80 and +0.70 residuals), while hilly terrain in SPZ 11 caused severe underperformance (−1.79 residual) due to hydrological disconnection of large waterbodies with minimal access (Table 10.5). SPZ 12's near-neutral residual (−0.05), indicating this area neither helps or hinders water access (Table 10.6).

Waterbody area utilisation efficiency varied from 156% (SPZ 7) to 45% (SPZ 10), with topography as the key moderator.

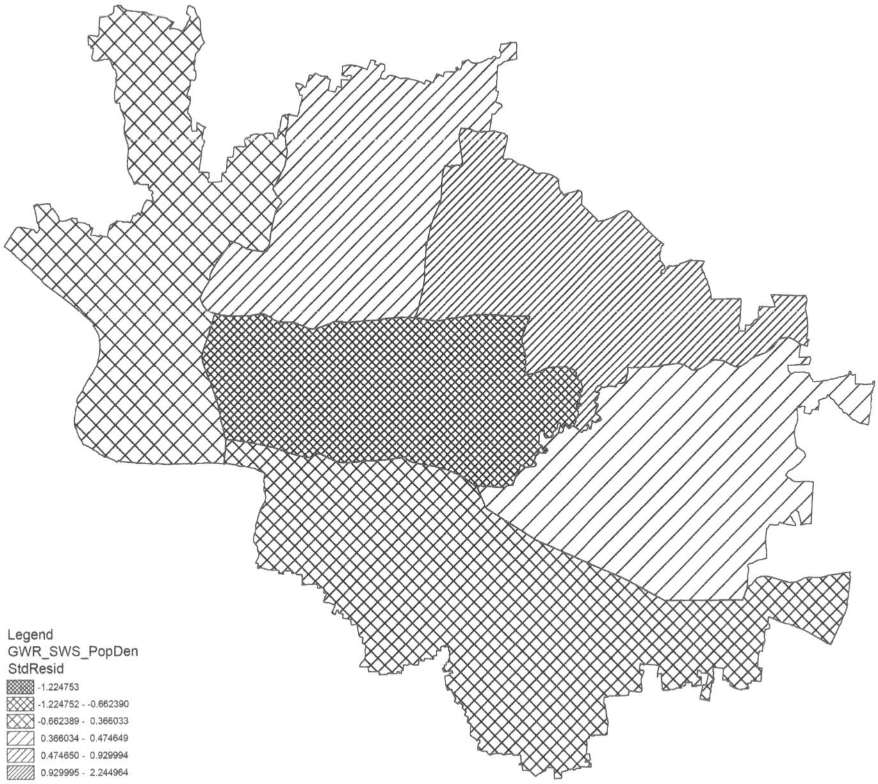

Fig. 10.9 GWR 3 map (Water access per capita vs residential population density)

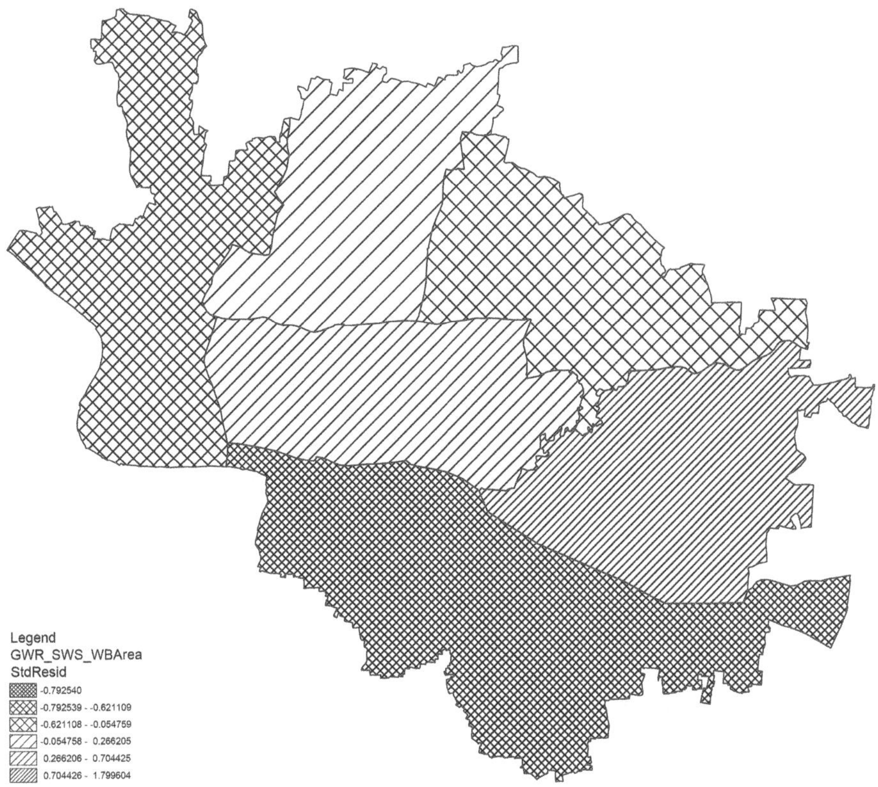

Fig. 10.10 GWR 4 map (Water access per capita vs waterbody area)

Table 10.5 Summary table of GWR 3 (Water access per capita vs residential population density)

SPZ	Stability Coeff	Local R^2	Std Resid	Interpretation
7	−0.00207	0.936	+0.93	Isolation yields positive density effect
8	−0.00086	0.606	+0.47	Preservation balances pressure
9	−0.00158	0.682	+2.24	Revitalization converts density to asset
10	−0.00161	0.535	−1.22	Topography limits utilization
11	−0.00228	0.848	+0.36	Landfilling offsets gains
12	−0.00075	0.371	−0.66	Space constraints

Table 10.6 Summary table of GWR 4 (Water access per capita vs waterbody area)

SPZ	Stability Coeff	Local R^2	Std Resid	Interpretation
7	1.1527	0.820	+1.80	Isolation preserves function
8	1.1526	0.820	+0.70	Preservation maintains access
9	1.1525	0.820	−0.62	Revitalization compensates
10	1.1520	0.820	+0.26	Topography allows moderate efficiency
11	1.1522	0.820	−1.79	Landfilling disrupts access
12	1.1521	0.820	−0.05	Near neutral efficiency

10.4.3 Local Model Performance Clusters

The core urban area in SPZ 9 shows density-driven success with +2.24 residual value, where river separated SPZ 7 took advantage of waterbody stability (+3.64) due to low density pressure and total volume of waterbodies (+1.80) in that zone. The latest urban residential extension of SPZ 11 displays some conflicting indicators when stability is quite high (+1.90), density, and waterbody area also indicate almost neutral residual value (+0.36 and −0.05 respectively), but the combined model performs badly (−1.38) due to landfilling and fragmented nature conservation in this area (Table 10.7).

Table 10.7 Quad-model residual comparison

SPZ	Combined model	Waterbody stability only	Residential density only	Waterbody area only
7	▲ +1.36	▲ +3.64	▲ +0.93	▲ +1.80
8	▼ −0.98	▼ −0.53	△ +0.47	▲ +0.70
9	▲ +1.41	▼ −0.74	▲ +2.24	△ +0.62
10	▲ +0.84	△ −0.32	▼ −1.22	△ +0.26
11	▼ −1.38	▲ +1.90	△ +0.36	▼ −1.79
12	▼ −0.36	▼ −0.52	▼ −0.66	△ −0.05

10.4.4 Four Distinct Resilience Pathways Emerged from Residual Profiling

1. Revitalisation Success (SPZ 9): Dominated by strong density-driven performance (+2.24 residual), compensating for stability/area limitations. This zone demonstrated how strategic investment converts urban pressure into water access benefits.
2. Isolation Advantage (SPZ 7): Showed exceptional stability and area utilisation (+3.64 and +1.80 residuals), with density also performing better than predicted (+0.93), indicating river separation creates accidental water sanctuaries while hindering integrated planning.
3. Planning-Implementation Gap (SPZ 11): Exhibited the most contradictory signals, high stability residuals (+1.90) clashed with the worst combined performance (−1.38), and severe area underperformance (−1.79 residual) diagnostic of landfilling-induced hydrological fragmentation.
4. Transitional Zones (SPZ 8, 10, 12): Displayed intermediate patterns. SPZ 10 achieved balanced but suboptimal outcomes through organic growth, SPZ 12 leveraged compactness for moderate efficiency, while SPZ 8 suffered growth misalignment despite area preservation (+0.70).

10.4.5 Case Studies from the Study Area to Understand the Paradox and Heterogeneity

To contextualise the spatial patterns observed in the GWR analysis in a better way, a set of case studies from key Strategic Planning Zones (SPZs) was selected. These examples help ground the statistical findings in real-world urban conditions and illustrate the underlying drivers of access inequities and resilience differences across Sylhet's waterscape. Dhopa Dighi and the Kean Bridge riverfront development exemplify how strategic interventions in high-density urban cores like SPZ 9 can transform degraded or threatened waterbodies into accessible and vibrant public spaces. Dhopa Dighi, which is one of the largest ponds in Sylhet city, has been revitalised into a leisure activity place for the city people. This pond faced severe encroachment in the last few decades but has now regained its original size through restoration, complemented with urban design initiatives. Figure 10.11 illustrates the chronological change of Dhopa Dighi, alongside the current visuals of the site post-revival work.

Likewise, the riverfront near Kean Bridge, Sylhet—a historically significant site featuring the most iconic bridge of the colonial era and a clock tower, has been reimagined through landscape, integrating a public place along with the north bank of the river. Figure 10.12 below shows

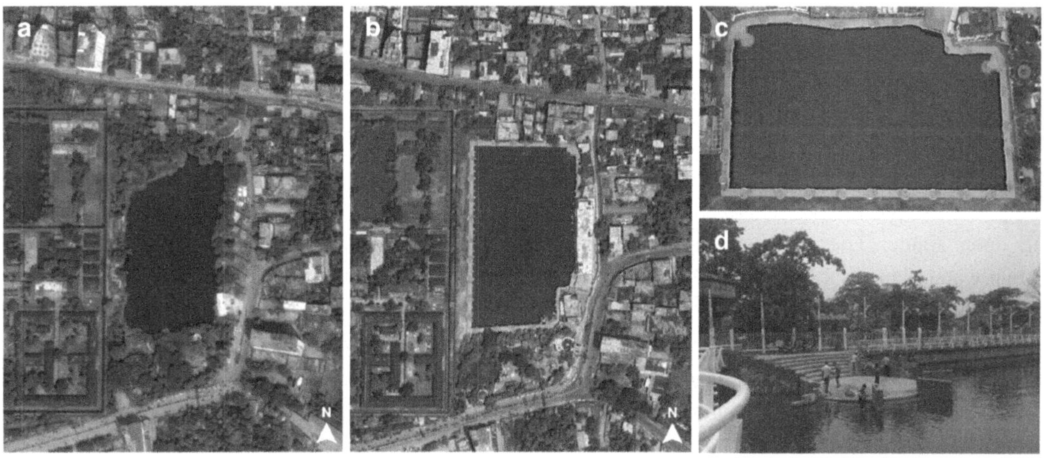

Fig. 10.11 a. Satellite image of Dhopa Dighi in the year 2015 b. Satellite image of Dhopa Dighi in the year 2025 c. Top view of the Dighi after revitalisation d. View of the newly constructed ghat (Source: Google Earth, Author)

Fig. 10.12 a. Kean Bridge and Ali Amzad Clock Tower b. Landscape on the northern bank of the Surma River (Source: Author, Google photos)

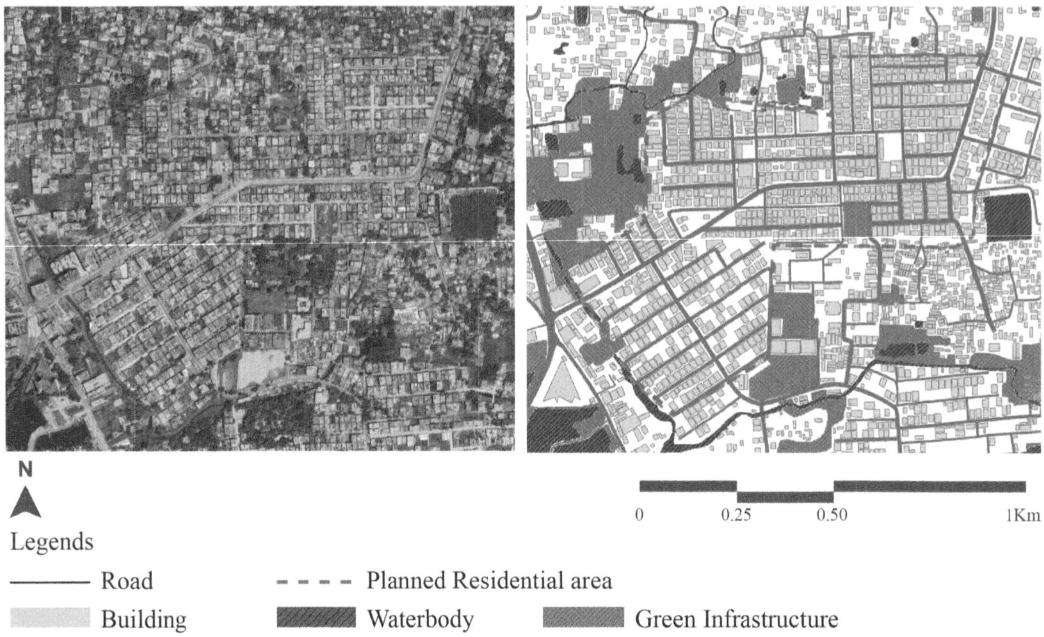

Fig. 10.13 Satellite image (Source: Google Earth) and Map of planned residential area at Upashahar, Sylhet (Source: Author)

the present condition of the site as an active public open space. These are examples of active urban waterscape that help to explain the unusually high residuals observed in SPZ 9, where strategic planning has created a positive impact despite the loss of other waterbodies and high population density. These examples of active urban waterscapes help explain the unusually high positive residuals observed in SPZ 9, where strategic revitalisation has generated public value despite the broader loss of waterbodies and high population density.

In contrast, two case studies from SPZ 11 explain why it performs low despite of having better stability, comparatively lower density, and larger waterbody area. The first case study from this SPZ is the Uposhohor Housing Estate, which is developed by landfilling a haor, representing a planning implementation gap. As illustrated in Fig. 10.13 the planned residential blocks have some open spaces but do not have any water bodies.

The other example shows that the development of the M. A. Muhit sports complex on a former lake, further fragmented the hydrological system, replacing ecological functionality with ornamental infrastructure as shown in Fig. 10.14 These cases reflect the high stability coefficients but negative access outcomes observed in GWR residuals, highlighting how preservation without connectivity undermines resilience.

SPZ 7, located on the periphery and separated by natural river barriers, contains the highest number of intact waterbodies in Sylhet. However, most of these remain inaccessible due to restricted institutional control, limited public amenities, and inadequate physical connectivity. While this spatial isolation has contributed to the preservation of water quality and structural integrity—as reflected in high waterbody stability scores—the lack of public access renders these resources functionally underutilised.

Figure 10.15 illustrates the inconsistent temporal changes in waterbody conditions within a marked area of SPZ 7 between 2001 to 2025. Between 2001 and 2008, two major waterbodies (number ii, iii) were filled. By 2015, the waterbody ii was restored, while the waterbody iii

Fig. 10.14 a. Satellite image of the site during 2008; b. Satellite image of M. A. Muhit sports complex during 2015; c. The image of the complex; d. Image of the complex compound affected by flood during heavy monsoon (Source: Google earth, images)

Fig. 10.15 a. Satellite image in 2001, b. Satellite image in 2008, c. Satellite image in 2015, d. Satellite image in 2025 (Sources: Google earth)

remained unchanged. In the same period, a portion of another adjacent waterbody (number i) was filled and built upon. At present, the largest waterbody (number ii) is once again undergoing partial infill, and the yellow-marked waterbody has been converted into a public park. These fluctuations indicate the absence of a coherent planning strategy and highlight the vulnerability of even ecologically stable zones to sporadic land use change.

These case studies collectively reinforce the key findings from the GWR analysis: that spatial patterns of resilience and inequity are not uniform, and that context-sensitive design and governance are essential to bridging the gap between preservation and accessibility.

10.5 Discussion

Urban waterbodies are vital to sustainable city development, particularly in rapidly urbanizing regions like Sylhet, where topographical variation and land pressures create uneven accessibility. This study, using Geographically Weighted Regression (GWR), reveals how spatial disparities in waterbody access are influenced by multiple factors: conservation status, density, revitalization, and planning. Key findings are elaborated below with specific reference to local projects and international parallels.

10.5.1 Importance of Waterbody Conservation in High-Density Urban Areas

The results of the GWR analysis reveal a compelling insight: conservation alone is insufficient, particularly in high-density urban areas like SPZ 9, where waterbodies face intense land pressure. Nonetheless, SPZ 9 achieved the highest positive residuals (+2.24) among all zones against the residential population density and +1.41 for the combined model, despite having the most negative density coefficient (−0.00158). This implies that access to waterbodies in the area is significantly higher than what would be expected based on population density alone, clearly indicating that factors beyond land availability, such as revitalization and integrated conservation efforts, have played a defining role. Furthermore, while stability coefficients in SPZ 9 are modestly positive (+0.000521), the local R^2 of 0.682 for density shows that revitalisation interventions have dramatically improved how well density explains access outcomes. In other words, the built environment of SPZ 9 has been modified in a way that transforms population pressure into a facilitating factor for waterbody access, not an obstacle.

This contradiction is particularly noteworthy when juxtaposed against SPZ 11 (Surma Floodplain), which has larger waterbody areas and lower density but performs significantly worse. Despite a higher stability coefficient (+0.00305) and similar area coefficient (1.1522×10^{-6}), SPZ 11 has a negative residual (−1.38), revealing lower-than-expected waterbody access. This distinct difference illustrates that active waterbody conservation, especially when paired with public space design, can overcome the limitations typically associated with dense urban cores.

The case of Dhopa Dighi is indicative of this finding. Once encroached and degraded, it has been restored through embankment protection, walking paths, and landscape enhancements. Similarly, the Kean Bridge riverfront integrates public access, heritage value, and landscape architecture into one cohesive design, reinforcing the idea that accessibility and conservation can coexist, even thrive, in high-density conditions. International parallels strengthen this observation. The Dharavi Waterfront Parks Project in Mumbai, India, implemented in one of the most densely populated informal settlements in the world, transformed a polluted nallah (drain) into a linear park system through community-led design, stormwater management, and edge landscaping. The project not only improved local microclimates and reduced flooding but also created usable, dignified green space for a population long deprived of public amenities. The revitalisation strategies applied in the urban waterscapes of the Ruhr Area in Germany also exemplify how restored waterscapes have become symbols of regional advancement and identity serving as sites for economic development, tourism, and leisure after industrial decline (Huning & Frank, 2011). Likewise, the Medellín River Parks Project in Colombia converted canalised, flood-prone riverbanks into accessible green corridors, incorporating flood basins, pedestrian pathways, and native vegetation. Like Sylhet's Kean Bridge riverfront, it illustrates how water-sensitive design can simultaneously restore ecological function and enhance urban liveability. Another notable case is Delhi's Rajokri Lake, where a formerly polluted pond was revitalised using nature-based treatment SWAB (Scientific Wetland with Active Bio-Digestion) and upgraded with community amenities such as walkways and an amphitheatre. These interventions improved water quality and public use, turning a degraded site into a resilient, multifunctional space (Srivastava et al., 2021). These cases illustrate how context-sensitive, community-responsive conservation can succeed in even the most spatially constrained environments.

Hence, these successes fundamentally challenge the presumption that density inevitably degrades blue spaces; rather, they reveal that well-designed hydrological integration converts population concentration into a funding and stewardship mechanism. For rapidly urbanising Bangladeshi cities, this underscores the critical need to mandate water-sensitive design through developer impact fees (minimum 5% of project costs) and community co-management frameworks under the Sylhet Master Plan 2040.

10.5.2 Positive Impact of Revitalisation Projects on Waterbody Access per Capita

Waterbody revitalisation plays a transformative role in enhancing water access, particularly in built-up urban areas where spatial constraints limit traditional conservation strategies. In SPZ 9 of Sylhet, revitalised areas recorded significant improvements in waterbody access, despite high density and limited surface water area.

International case studies further validate these findings. In Singapore, the Active, Beautiful, Clean (ABC) Waters Programme exemplifies a strategic approach to waterbody revitalisation in dense urban environments. This initiative transforms utilitarian canals and drains into aesthetically pleasing, multifunctional spaces that support recreation, environmental education, and ecological resilience. Features such as lookout decks, pedestrian bridges, and landscaped trails have increased public engagement, encouraging walking, jogging, and cycling, and contributing to community well-being (Neo et al., 2022). The ABC Waters model demonstrates how revitalised waterscapes can simultaneously serve as hydrological infrastructure and accessible urban amenities. The Palli Pukur Project in Kolkata is another relevant example. It exemplifies how small-scale, community-driven revitalisation of a single urban pond using native vegetation, low-cost seating, and improved access can enhance the usability, aesthetics, and ecological value of a neighbourhood resource. Despite the limited scale, the intervention improved thermal comfort and created a communal gathering space, highlighting the scalability and replicability of such micro-interventions. These examples affirm that revitalisation not only enhances per capita access to waterbodies but also redefines their role within the urban fabric. By embedding waterbodies into daily life, cities can elevate them from passive ecological features to active civic spaces, yielding both social and environmental returns.

10.5.3 Adverse Effects of Lacking Water-Sensitive Development

While the GWR analysis shows that conservation and revitalisation efforts can significantly improve waterbody access in certain zones (e.g., SPZ 9), it also exposes the serious limitations of planning practices that fail to incorporate water-sensitive urban development. This failure is particularly evident in zones where waterbodies are present in adequate quantity or stability but remain functionally disconnected due to poor spatial design, lack of access, or landfilling. SPZ 11 in the Surma Floodplain is the most illustrative case of this paradox.

In SPZ 11, the model recorded a stability coefficient of +0.00305 and a waterbody area coefficient of 1.1522×10^{-6} both among the highest in the dataset. Moreover, the zone's population density is moderate, indicating that physical and spatial conditions should, in theory, support good waterbody access. However, the GWR results report a residual of −1.38, a clear sign that observed access falls significantly short of what the model predicts. This discrepancy highlights a disconnect between preservation and functionality. This planning-performance gap is exemplified by two major developments in SPZ 11. First, the Shahjalal

Upashahar Housing, constructed by landfilling a seasonal haor (wetland), entirely removed the ecological and hydrological functions previously served by the landscape. Despite the presence of open space in the layout, no effort was made to preserve or integrate water features. As a result, the upashahar lacks not only biodiversity and seasonal water storage capacity but also everyday access to blue-green infrastructure. Second, the M. A. Muhit Sports Complex, also in SPZ 11, was developed over a lake, replacing water-sensitive open space with impervious built form. The adverse effects extend beyond accessibility deficits: monsoon displacement increased by 38% due to lost floodwater absorption capacity. This planning failure exposes a dangerous fallacy—that formal protection alone suffices.

In contrast, successful WSUD (Water Sensitive Urban Design) examples illustrate how meaningful planning can bridge this gap. The Cheonggyecheon restoration in Seoul (Jeon & Kang, 2019) once a buried stream beneath a highway, was reopened and transformed into an accessible, ecologically rich corridor, reducing urban temperature and providing a new public realm for residents and visitors. The key insight from these international projects and the data from Sylhet's SPZs is that preservation without hydraulic functionality yields negative returns, demanding immediate moratoriums on floodplain development and hydrological functionality certificates under the Bangladesh Water Act 2013.

10.5.4 Importance of Comprehensive Planning in Low-Density Fringe Areas

Peripheral urban zones characterised by low population density, transitional land uses, and availability of natural resources offer some of the most significant opportunities for sustainable waterbody integration. Yet, the analysis reveals that without comprehensive planning, such areas can easily fall into patterns of fragmentation and ecological degradation. The contrasting outcomes of SPZ 7 and SPZ 8 illustrate the consequences of governance gaps, unregulated expansion, and lack of early intervention.

SPZ 7, which is geographically isolated by the Surma River, shows exceptional water body performance metrics. It has the highest residual for waterbody access (+3.64) and the greatest area-use efficiency (+1.80) among all Strategic Planning Zones. These results stem largely from the zone's physical inaccessibility and limited urban encroachment, which have incidentally preserved its hydrological integrity. However, the benefits of this ecological stability remain unrealised in terms of public utility. Most waterbodies in SPZ 7 are government-owned and fenced, with restricted access, resulting in a paradox where ecological assets exist but are socially inaccessible. Conversely, SPZ 8, despite similar area coefficients (1.1526×10^{-6}), performs poorly with a negative residual (−0.98). The zone is experiencing rapid, uncoordinated expansion, often through speculative land development and unplanned pond excavation. While these new waterbodies contribute to surface coverage, they are often disconnected from stormwater systems and public space networks. This has led to spatial misalignment, where waterbodies become isolated, fenced, or privatised, undermining their ecological function and community benefit. The lack of zoning, setback policies, and public access guidelines allows for fragmented governance and lost hydrological continuity.

This contrast mirrors experiences in other rapidly growing cities. In Jakarta, unregulated development along fringe wetlands led to a loss of flood buffer zones, degraded water quality, and inaccessibility. Peri-urban wetlands, once vital for stormwater management, were rapidly converted into informal housing or industrial land without ecological assessment or enforcement of master plans (Firman, 2009).

The implications for Sylhet's fringe are clear. SPZ 7's passive preservation is no substitute for active planning, while SPZ 8 shows how unregulated growth can rapidly undermine water-sensitive opportunities. Proactive protection could leverage natural advantages: isolation preserves water quality (92% local recharge rates), while low encroachment pressure allows phased development of blue-green corridors. The imperative is threefold: (1) legally mandate public access within 500 m of significant waterbodies, (2) establish water conservation trusts funded by urban core premiums, and (3) develop terrain-adaptive designs for hillside zones, where slopes exceeding 15° reduce utility by 55%. These comprehensive planning that safeguards waterbody networks, mandates access infrastructure, and integrate fringe ponds into Sylhet's broader ecological system will be essential for resilient urbanisation.

10.6 Conclusion

This study has demonstrated that urban waterscape resilience in Sylhet is fundamentally shaped by spatial heterogeneity, revealing different pathways through Geographically Weighted Regression analysis: density-enabled revitalisation (SPZ 9), hydrological fragmentation (SPZ 11), isolation-preserved functionality (SPZ 7), transitional growth misalignment (SPZ 8), and topography-constrained efficiency (SPZ 10, 12). These pathways aim to achieve a sustainable waterscape at the local level through context-specific interventions. In high-density core areas, conservation of large-scale public waterbodies and developing facilities to engage more people helps enhance public access and community stewardship, transforming density pressure into community-centric assets. Landfilling-induced fragmentation of waterbodies and hard engineering should be avoided in areas developed from low-lying land. The demand for waterbodies varies with geospatial changes. The presence of large waterbodies will remain abundant, as the new development excludes blue features. On the other hand, peripheral areas require protection of waterbodies to tackle the future challenges of increasing development density. Administrative decentralisation of cities, driven by the rapid development of grey infrastructure and administrative shifts, should include nature-based development with public accessibility. Landfilling of government-owned waterbodies must be stopped to enable best practice scenarios to be replicated by the public.

The key contribution of this research is a novel GWR application to the monsoon-city waterscape, a pathway framework for managing spatial heterogeneity, and tools for achieving water equity. Sylhet's waterscape embraces integrated strategies that transform diversity into resilience; however, in some other cases, a fragmented approach has deepened the inequities. Context-specific and community-centred design can transform urban blue features into active urban infrastructure. Thus, enhancing resilience, sustainable future, healthy ecosystem, and well-connected communities.

Acknowledgments We are grateful to the local communities, the engineering department of Sylhet city Corporation, and the Department of Architecture, SUST.

Funding: The research was not supported by any funding.

Potential conflicts of interest: On behalf of all authors, the corresponding author states that there is no conflict of interest.

Informed Consent: Informed consent was obtained from all individual participants included in the study.

Authors' Contribution: Shataparna Das: Conceptualised the research idea, developed the initial theoretical framework, conducted the physical survey and field investigations, and prepared the initial literature review. Subrata Das: Supervised the research team, designed the analytical framework, performed GIS analysis and interpretation, led manuscript writing and revision. Farha Moon: Contributed to the literature review, theoretical framework development, map preparation, case study integration, assisted with GIS data processing, played a key role in manuscript writing and revision. Mohammad Rakibul Islam Anik: Contributed to the literature review, methodology development, manuscript writing, assisted with GIS data analysis, and contributed to data interpretation.

AI disclosure statement: Grammarly Premium was used for language proofreading and sentence restructuring. (www.grammarly.com, last accessed on 20 July 2025).

Data availability statement: The datasets used during the current study are available from the corresponding author on request through email.

References

Alikhani, S., Nummi, P., & Ojala, A. (2022). Urban water bodies and recreational opportunities in Finland. *Proceedings of the 17th International Conference on Environmental Science and Technology, 17*. https://doi.org/10.30955/GNC2021.00094

Ameen, R. F. M., Mourshed, M., & Li, H. (2015). A critical review of environmental assessment tools for sustainable urban design. *Environmental Impact Assessment Review, 55*, 110–125. https://doi.org/10.1016/J.EIAR.2015.07.006

Ampatzidis, P., & Kershaw, T. (2020). A review of the impact of blue space on the urban microclimate. *Science of the Total Environment, 730*, 139068. https://doi.org/10.1016/J.SCITOTENV.2020.139068

Antonio, F., & González, I. (2020). *Access to an improved water source and its measurement: the case of Misiones (Argentina)*. https://ri.conicet.gov.ar/handle/11336/118970

Ashik, F. R., Islam, M. S., Alam, M. S., Tabassum, N. J., & Manaugh, K. (2024). Dynamic equity in urban amenities distribution: An accessibility-driven assessment. *Applied Geography, 164*, 103199. https://doi.org/10.1016/J.APGEOG.2024.103199

Bhor, N., & Mayavel, D. (2024). The socio-spatial distribution and equity of access to urban parks: A case study of Bengaluru, India. *Challenges 2024, 15*(2), 20. https://doi.org/10.3390/CHALLE15020020

Das, S., Das, S., Das, S., & Das, S. (2022). Sustainable Urban Design guidelines for different types of urban ponds: A case study of Sylhet. *Bangladesh. Current Urban Studies, 10*(4), 593–610. https://doi.org/10.4236/CUS.2022.104035

Dewan, A. M., & Yamaguchi, Y. (2009). Land use and land cover change in greater Dhaka, Bangladesh: Using remote sensing to promote sustainable urbanization. *Applied Geography, 29*(3), 390–401. https://doi.org/10.1016/J.APGEOG.2008.12.005

Dutt, D. P., Rahman, Md. S., Ahmed, R., & Chaki, T. (2024). (PDF) Modeling the urban water balance: Quantifying the effects of land use change on runoff and recharge in sylhet city of Bangladesh. *7th International Conference on Advances in Civil Engineering (ICACE2024)*. https://icace2024.cuet.ac.bd

Falkenmark, M., Wang-Erlandsson, L., & Rockström, J. (2019). Understanding of water resilience in the Anthropocene. *Journal of Hydrology X, 2*, 100009. https://doi.org/10.1016/J.HYDROA.2018.100009

Firman, T. (2009). The continuity and change in mega-urbanization in Indonesia: A survey of Jakarta–Bandung region (JBR) development. *Habitat International, 33*(4), 327–339. https://doi.org/10.1016/J.HABITATINT.2008.08.005

Fletcher, T. D., Andrieu, H., & Hamel, P. (2013). Understanding, management and modelling of urban hydrology and its consequences for receiving waters: A state of the art. *Advances in Water Resources, 51*, 261–279. https://doi.org/10.1016/J.ADVWATRES.2012.09.001

Gupta, S., Islam, S., & Hasan, M. M. (2018). Analysis of impervious land-cover expansion using remote sensing and GIS: A case study of Sylhet sadar upazila. *Applied Geography, 98*, 156–165. https://doi.org/10.1016/J.APGEOG.2018.07.012

Han, Y., Lin, Q., Huang, S., Du, C., Shen, J., & Zhang, K. (2024). Human impacts dominate global loss of Lake ecosystem resilience. *Geophysical Research Letters, 51*(11), e2024GL109298. https://doi.org/10.1029/2024GL109298

Haque, A., Shaha, N. K., & Raihan, F. (2008). Study on land use pattern change and its causes. *International Journal of Environmental Research, 2*(2), 199–204. https://doi.org/10.22059/IJER.2010.194

Huning, S., & Frank, S. (2011, November). Urban waterscapes as products, media and symbols of change—the re-invention of the Ruhr. *Mind The Gap: Landscapes For a New Era; EFLA Regional Congress of Landscape Architecture*. https://doi.org/10.17877/DE290R-17994

Jeon, C., & Kang, Y. (2019). Restoring and re-restoring the Cheonggyecheon: Nature, technology, and history in Seoul, South Korea. *The University of Chicago Press Journal, 24*(4), 736–765. https://doi.org/10.1093/ENVHIS/EMZ032

Khalaji, F., Zhang, J., Sharma, A. K., Social, E., Khalaji, F., Zhang, J., & Sharma, A. K. (2024). Social and economic impacts of water sensitive Urban Design: A review. *Water, 17*(1), 16. https://doi.org/10.3390/W17010016

Kummu, M., de Moel, H., Ward, P. J., & Varis, O. (2011). How close do we live to water? A global analysis of population distance to freshwater bodies. *PLoS One, 6*(6), e20578. https://doi.org/10.1371/JOURNAL.PONE.0020578

Kurniawan, H. B., & Roychansyah, M. S. (2023). The Social equity of public green open space accessibility: The case of South Tangerang, Indonesia. *Geography, Environment, Sustainability, 16*(1), 45–54. https://doi.org/10.24057/2071-9388-2022-124

Liu, P., Wu, C., Chen, M., Ye, X., Peng, Y., & Li, S. (2020). A spatiotemporal analysis of the effects of urbanization's socio-economic factors on landscape patterns considering operational scales. *Sustainability, 12*(6), 2543. https://doi.org/10.3390/SU12062543

Ma, Y., Fu, Y., Ren, Z., & Liu, C. (2024). Urban water bodies mitigate heat waves, the case of the Chinese city Kunming. *Discover Cities, 1*(1), 1–14. https://doi.org/10.1007/S44327-024-00004-8

Manteghi, G., Bin Limit, H., & Remaz, D. (2015). Water bodies an urban microclimate: A review. *Modern Applied Science, 9*(6), p1. https://doi.org/10.5539/MAS.V9N6P1

Marçal, J., Shen, J., Antizar-Ladislao, B., Butler, D., & Hofman, J. (2024). Assessing inequalities in urban water security through geospatial analysis. *PLOS Water, 3*(2), e0000213. https://doi.org/10.1371/JOURNAL.PWAT.0000213

Mashrur, R., Meher, N. N., Mashrur, R., & Meher, N. N. (2015). A GIS based integrated approach to measure the spatial equity of community facilities of Bangladesh. *AIMS Geosciences 2015 1:21, 1*(1), 21–40. https://doi.org/10.3934/GEOSCI.2015.1.21

Nelson, R., Warnier, M., & Verma, T. (2023). Conceptualizing urban inequalities as a complex socio-technical phenomenon. *Geographical Analysis, 56*(2), 187–216. https://doi.org/10.1111/GEAN.12373

Neo, T. H., Xu, D., Fowdar, H., McCarthy, D. T., Chen, E. Y., Lee, T. M., Ong, G. S., Lim, F. Y., Ong, S. L., & Hu, J. (2022). Evaluation of active, beautiful, clean waters design features in tropical urban cities: A case study in Singapore. *Water (Switzerland), 14*(3), 468. https://doi.org/10.3390/W14030468/S1

Rahman, K. M. A., & Zhang, D. (2018). Analyzing the level of accessibility of public urban green spaces to different socially vulnerable groups of people. *Sustainability, 10*(11), 3917. https://doi.org/10.3390/SU10113917

Reyad, A. A., & Ullah, M. S. (2022). Spatial inequality in Bangladesh: A geospatial analysis. *International Journal of Basic and Applied Science, 11*(3), 95–106. https://doi.org/10.35335/IJOBAS.V11I3.116

Ribbe, L., Dekker, G., & Thapak, G. (2024). Urban wetlands and water bodies. In *Managing urban Rivers: From planning to practice* (pp. 91–107). Elsevier. https://doi.org/10.1016/B978-0-323-85703-1.00007-9

Sarkar, S., Cottineau-Mugadza, C., & Wolf, L. J. (2024). Spatial inequalities and cities: A review. *Environment and Planning B: Urban Analytics and City Science, 51*(7), 1391–1407. https://doi.org/10.1177/23998083241263422

Srivastava, A., Prathna, T. C., Srivastava, A., & Prathna, T. C. (2021). Urban water resource management: Experience from the revival of Rajokri lake in Delhi. *AIMS Environmental Science, 8*(5), 421–434. https://doi.org/10.3934/ENVIRONSCI.2021027

Sun, R., & Chen, L. (2012). How can urban water bodies be designed for climate adaptation? *Landscape and Urban Planning, 105*(1–2), 27–33. https://doi.org/10.1016/J.LANDURBPLAN.2011.11.018

Unnikrishnan, H., Sen, S., & Nagendra, H. (2017). Traditional water bodies and urban resilience: A historical perspective from Bengaluru, India. *Water History, 9*(4), 453–477. https://doi.org/10.1007/S12685-017-0199-9

Upadhyay, A. T. (2023). Urban Lake scenario and development possibilities-a case of Surat. *Sustainability, Agri, Food and Environmental Research-Discontinued, 11*, 1. https://doi.org/10.7770/SAFER.V11I1.2960

Viinikka, A., Paloniemi, R., & Assmuth, T. (2018). Mapping the distributive environmental justice of urban waters. *Fennia—International Journal of Geography, 196*(1), 9–23. https://doi.org/10.11143/FENNIA.64137

Wu, P., & Liu, M. (2022). A framework for the spatial inequality in urban public Facility for Urban Planning, design and management. *Land, 11*(9), 1429. https://doi.org/10.3390/LAND11091429

Xu, Z., & Ouyang, A. (2018). The factors influencing China's population distribution and spatial heterogeneity: A prefectural-level analysis using geographically weighted regression. *Applied Spatial Analysis and Policy, 11*(3), 465–480. https://doi.org/10.1007/S12061-017-9224-8/METRICS

11 Advanced Smart Water Management Systems: Technological Innovations in IoT, AI, and Renewable Energy for Addressing Global Water Scarcity Challenges

P. Balakrishnan

Abstract

This chapter explores Smart Water Management Systems (SWMS), which integrate Internet of Things (IoT), Artificial Intelligence (AI), and renewable energy technologies to address escalating global water challenges. With over 2.4 billion people residing in water-scarce regions and projections indicating this figure could double by 2025, the need for innovative water management solutions is critical. This chapter synthesizes findings from recent research work, examining the transition from traditional to SWM approaches. The layered architecture of SWMS comprising sensor networks, communication protocols, data processing, intelligence systems, and control mechanisms enables comprehensive monitoring and optimization of water resources. Results demonstrate remarkable improvements across applications: IoT-based leak detection models achieved up to 95.4% accuracy; smart irrigation systems reduced water consumption by 30% while increasing crop yields by up to 15%; and AI algorithms predicted water quality indices with 99.8% accuracy. Cost-benefit analyses revealed substantial returns: large-scale wastewater recycling projects demonstrated Net Present Values of US$159.97 million with Benefit-Cost Ratios of 4.34, while agricultural revenue increased from US$0.47 million to US$1.78 million following implementation. System deployment costs remain accessible, with complete IoT-based irrigation solutions developed for approximately US$215. Future research must address sensor reliability challenges, interoperability issues, and security concerns while developing affordable modular systems for resource-limited regions and integrated approaches that combine technical solutions with policy frameworks.

Keywords

Smart water management · Internet of things · Machine learning · Leak detection · Water quality · Smart irrigation

P. Balakrishnan (✉)
Department of Mechanical Engineering, SRM Institute of Science and Technology, Kattankulathur, India
e-mail: balakrip@srmist.edu.in

U. Chatterjee, A. I. Almulhim (eds.), *Urban Waterscapes*, Sustainable Development Goals Series,
https://doi.org/10.1007/978-3-032-20649-7_11

11.1 Introduction

The escalating global water crisis, driven by climate change, population growth, and aging infrastructure, necessitates a paradigm shift toward resilient urban waterscapes through integrated strategies for sustainable water management. Smart Water Management Systems (SWMS) have emerged as transformative solutions that integrate Internet of Things (IoT), Artificial Intelligence (AI), and renewable energy technologies to optimize water use, reduce losses, and enhance urban resilience. Recent studies underscore the urgency of this transformation: over 2.4 billion people currently reside in water-scarce regions, with projections indicating this figure could double by 2025 (Biswas et al., 2025; UN-Water, 2024). Concurrently, urban water systems, many decades old, are increasingly unable to meet escalating demand, losing up to 50% of supplied water through leaks and inefficiencies that threaten urban water security (Fulton, 2024; Jasechko et al., 2024). These challenges are exacerbated by climate stressors that alter precipitation patterns and intensify droughts, fundamentally reshaping the hydrological cycles that urban areas depend upon (Liu et al., 2024).

These water security challenges are central to achieving United Nations Sustainable Development Goal 6.3, which aims to improve water quality by reducing pollution, eliminating dumping, minimizing hazardous chemical release, halving untreated wastewater proportions, and substantially increasing recycling and safe reuse globally by 2030 (United Nations General Assembly, 2017). Current progress remains insufficient: achieving universal coverage by 2030 requires a sixfold acceleration in safely managed drinking water access, fivefold increase for sanitation, and threefold increase for basic hygiene services. Global wastewater treatment rates remain inadequate, with only 58% of household wastewater receiving safe treatment and 60% of total wastewater flows safely treated in reporting countries (United Nations DESA Statistics Division, 2024). Smart Water Management Systems represent a critical technological pathway for accelerating SDG 6.3 achievement through enhanced monitoring capabilities, pollution detection, treatment optimization, and recycling infrastructure that transform conventional water management approaches.

Figure 11.1 illustrates the direct alignment between SDG Target 6.3 and the technological capabilities of SWMS. The framework demonstrates how the integration of IoT sensors, AI-based analytics, and renewable-energy solutions collectively address the dual imperatives of pollution reduction and water recycling enhancement, thereby creating measurable pathways toward achieving the 2030 water-quality objectives.

The transition from conventional to Smart Water Management (SWM) represents more than technological advancement; it embodies a fundamental shift in creating resilient urban waterscapes capable of adapting to twenty-first-century challenges. Traditional water management systems, designed for twentieth-century demand patterns and climate conditions, exemplify the inadequacy of legacy approaches in addressing contemporary urban water security needs. These conventional methods often rely on manual monitoring, periodic sampling, and visual inspections, which are time-consuming, resource-intensive, and inefficient in addressing real-time urban water issues, frequently leading to delayed responses to contamination and leakages that result in substantial water loss and financial costs (Sharma et al., 2025). Aging infrastructure in urban areas demonstrates this vulnerability, with cities globally facing frequent contamination risks and supply interruptions that compromise urban resilience, while groundwater depletion accelerates in 30% of the world's aquifers, threatening both agricultural productivity and urban water security (He et al., 2021; Jasechko et al., 2024). These inefficiencies extend beyond physical infrastructure to institutional frameworks, where centralized, supply-focused approaches often neglect demand-side management, creating critical mismatches between water availability and urban consumption patterns that undermine long-term sustainability.

SWMS address these fundamental challenges through integrated technological strategies that create adaptive, resilient urban water networks

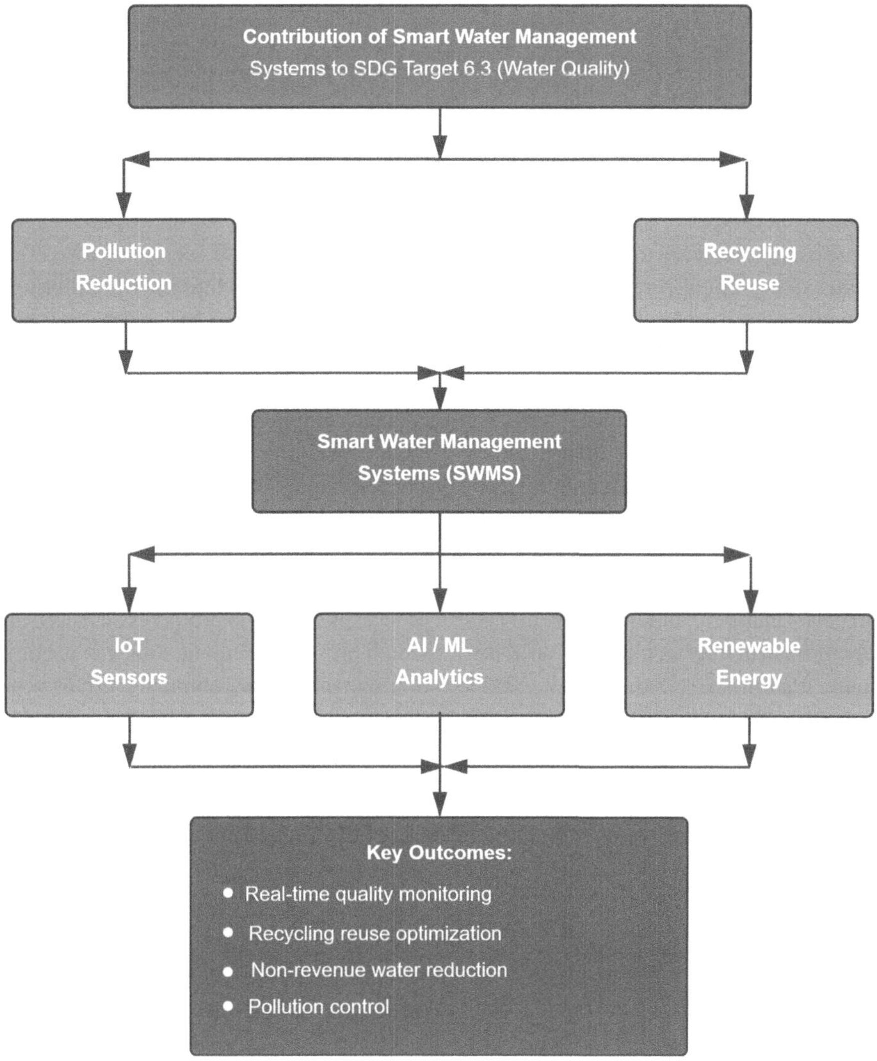

Fig. 11.1 SDG 6.3 targets and smart water management system (SWMS) contributions

with remarkable performance improvements. Research indicates that IoT-based water management systems demonstrate monitoring accuracy of 95%, ensuring reliable real-time consumption data that enables precise urban water distribution (Suryavanshi et al., 2024). Compared to conventional methods, smart water monitoring techniques significantly improve anomaly detection and water quality management, reducing leak identification time from over 90 min to approximately 25 min, representing substantial enhancement in response efficiency that translates to considerable water conservation and cost savings for urban utilities and consumers (Mehta & Choudhary, 2024). The WARM-IoT system demonstrates notable economic benefits with potential cost savings of up to 6% by analyzing consumption costs under different activation timings, while user feedback reveals 88% satisfaction rates with SWMS, contributing to overall satisfaction rates of 80% regarding water utility distribution and conservation efforts (Rajanbabu et al., 2025; Suryavanshi et al., 2024). These systems leverage cutting-edge IoT networks with wireless sensors monitoring critical parameters from pipe pressure to water quality at 5-min intervals, enabling real-time responsiveness that traditional systems cannot achieve (Dada et al., 2024; Olatunde et al., 2024). As illustrated in Fig. 11.2, these systems center around four key technological components: IoT, AI, ML, and Automation that work synergistically to create resilient urban waterscapes.

Paddy cultivation requires more efficient and adaptive water management strategies to address increasing water scarcity and climate variability. Recent advances in sensing technologies, data analytics, and digital monitoring systems are creating new opportunities to improve irrigation efficiency and optimize water use in agricultural systems (Bwire et al., 2024). AI and machine learning enable proactive water management by predicting equipment failures and optimizing treatment in real time, while digital twins simulate system performance to anticipate and dynamically adapt infrastructure to climate variability and drought conditions (Dada et al., 2024; Olatunde et al., 2024). In this context, smart water management systems are emerging as critical solutions to address global water scarcity driven by climate change, population growth, and increasing agricultural demand. The integration of IoT technologies with artificial intelligence enables real-time monitoring, predictive analysis, and automated decision-making for efficient water resource utilization, enhancing irrigation efficiency, reducing water losses, and supporting sustainable water governance across agricultural, urban, and industrial sectors (Bwire et al., 2024; Dada et al., 2024; Mehta & Choudhary, 2024).

The integration of IoT technology has revolutionized urban water monitoring by enabling continuous, real-time data collection from distributed sensor networks that provide unprecedented visibility into water system performance. These systems integrate various sensors measuring critical parameters including water level, flow rate, pressure, pH, turbidity, and dissolved oxygen, with data transmitted wirelessly to central processing units via protocols such as LoRaWAN, which provides efficient, long-range communication capabilities essential for urban water infrastructure monitoring (Henderi et al., 2024). IoT-based water quality monitoring frameworks achieve significant improvements in environmental sustainability by effectively monitoring water

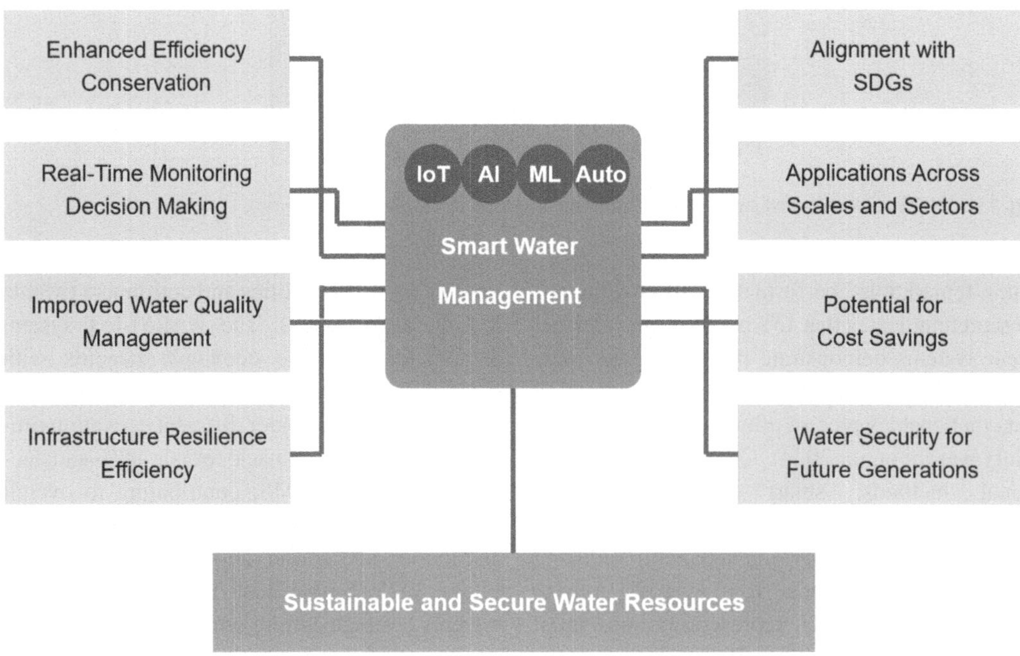

Fig. 11.2 Role of SWM in water resource sustainability

quality in urban and metropolitan regions, utilizing sensor networks connected to central data processing units that enable resource optimization and improvements in service delivery within urban environments. The implementation extends beyond simple data collection to comprehensive water level monitoring systems integrating specialized sensors, including ultrasonic sensors for precise measurement and rain sensors for immediate leak detection, allowing automatic adjustments and rapid responses that minimize waste and optimize urban water use (Zainuddin et al., 2024). Smart IoT-Enabled Tap Water Leakage Monitoring Systems monitor water flow rates at household levels, alerting users when flow is abnormally slow, indicating damaged taps or improper closure, while integrating with mobile applications via wireless communication to provide real-time and historical water usage data that promotes awareness and proactive conservation practices (Mudholkar & Mudholkar, 2024). The IoT- and blockchain-based water management system (IB-WMS) demonstrates high operational efficiency, achieving 89% overall system efficiency along with substantial wastewater management performance, including 90% wastewater reuse and 95% water recycling rates, highlighting its potential for promoting sustainable urban water practices (Alzahrani et al., 2024).

AI has emerged as a transformative technology in creating resilient urban waterscapes, particularly through demand forecasting and leak detection capabilities that enable proactive water management. Machine learning algorithms analyze historical data, weather patterns, consumer behavior, and other relevant factors to predict urban water demand with high accuracy, enabling more efficient resource allocation and distribution planning that supports sustainable urban growth (Jayakumar et al., 2024). Support Vector Machine models demonstrate superior performance in outlier detection for water quality monitoring, with research evaluating classical machine learning models including SVM, Artificial Neural Networks, Decision Trees, and Isolated Forest for smart water sensing applications, showing SVM outperforming other models with F1-score rates of 98.38% for pH, 96.98% for temperature, 97.88% for dissolved oxygen, and 98.11% for electrical conductivity (Lamrini et al., 2024). Convolutional Neural Networks exhibit exceptional capability in water management applications, achieving 94.5% decoding accuracy in visual signal interpretation for irrigation monitoring, surpassing ANN, K-Nearest Neighbors, and Recurrent Neural Networks (Dhyani et al., 2024). The Intelligent Water Quality and Leakage Detection System significantly enhances water management by integrating IoT and Machine Learning technologies, providing real-time monitoring and predictive insights that traditional methods lack, facilitating timely responses to contamination and leakages while enhancing anomaly detection and predicting future water quality issues to enable proactive interventions that reduce water wastage and promote sustainability (Sharanya et al., 2024).

The integration of renewable energy sources enhances the sustainability and operational efficiency of urban water management systems while supporting the transition toward resilient urban waterscapes powered by clean energy. Renewable energy integration addresses critical challenges including reducing environmental impact of water operations and ensuring reliable power supply for remote monitoring and control systems essential for urban water security (Diallo et al., 2024). Blockchain of Things water systems utilizing Ethereum platforms and Raspberry Pi nodes demonstrate effective data collection from various sensors and microcontrollers, enhancing performance and security of smart water monitoring in urban areas while supporting development of state-of-the-art Water 4.0 ecosystems (Al-Zoubi et al., 2024). The integration of AI with renewable energy-powered water management creates opportunities for optimizing energy usage in urban water operations, with AI algorithms scheduling water pumping and treatment activities during peak renewable energy production periods, maximizing efficiency and minimizing costs while enhancing overall sustainability and economic viability (Rhallab et al., 2024). SWM grids utilize IoT technology to enhance water management by integrating various sensors for real-time monitoring, optimizing water usage, minimizing wastage, and ensuring efficient distribution, becoming more resilient and environmentally sustainable when powered by renewable energy sources, particularly in regions facing water scarcity challenges (Vani et al., 2024).

Smart irrigation systems and urban water management applications demonstrate the practical implementation of integrated strategies for sustainable water management in diverse urban contexts. These systems utilize IoT sensors to monitor soil moisture, temperature, humidity, and environmental factors, enabling data-driven decision-making for irrigation scheduling that supports urban green infrastructure and agricultural areas within metropolitan regions (Kumar & Chandana, 2024). Research indicates smart irrigation systems reduce water wastage by up to 30% while increasing crop yield by 15% through sensors monitoring soil moisture and environmental conditions, with projects like the Imamoğlu Agricultural Irrigation Automation achieving water savings of 65–70% through central management-based systems (Balta & Kulat, 2024; Thiruchelvam, 2024). IoT-based smart irrigation systems have been shown to reduce water consumption by up to 30% while simultaneously boosting crop yields, while AI technologies in water management for sustainable farming demonstrate significant improvements in water efficiency and agricultural output through automated precision irrigation systems, AI-based predictive models, and AI-driven water quality monitoring (Ashoka et al., 2024; Nureen, 2024). In urban contexts, SWMS addresses challenges posed by rapid urbanization and increasing water demand through IoT and blockchain technologies that enhance urban living by improving safety protocols, energy efficiency, and governance through AI-driven systems that ensure data authenticity and integrity, fostering trust in urban governance by enabling transparent identity management, supply chain tracking, and efficient property registration (Malarvizhi & Anusuya, 2025).

The convergence of blockchain technology with SWM creates transparent, secure frameworks for urban water governance that support resilient urban waterscapes through enhanced accountability and efficiency. Blockchain technology introduces transparency in transboundary water management, as demonstrated in the Danube River Basin where smart contracts automate water allocations between nations during low-flow periods, while quantum computing advancements promise to revolutionize contaminant detection by enabling real-time analysis of thousands of water quality parameters simultaneously (Dada et al., 2024). Smart water systems integrate AI and IoT to support SDG 6, enhancing water conservation and infrastructure efficiency across Africa and the United States. These innovations offer significant economic and environmental benefits, though realizing their full potential requires overcoming technical, financial, and regulatory barriers (Bwire et al., 2024; Olatunde et al., 2024). These technological integrations produce multiple sustainability benefits including enhanced efficiency and conservation, real-time monitoring capability, improved water quality management, and infrastructure resilience, with applications spanning various scales and sectors from local water utilities to transboundary basin management, delivering both immediate cost savings and long-term water security.

This chapter contributes to the achievement of SDG Target 6.3 by demonstrating how SWMS technologies enhance water quality monitoring, reduce pollution, and promote recycling efficiency. It emphasizes the integration of digital technologies such as IoT-based sensor networks, AI-enabled analytics, and renewable energy systems as enablers of sustainable water management. By facilitating continuous water-quality assessment, optimized treatment processes, and efficient reuse mechanisms, these systems collectively advance the global mission of improving water quality, minimizing contamination sources, and fostering circular water use. The analysis establishes a conceptual and practical framework that connects technological innovation with the core objectives of SDG 6.3, offering pathways for accelerating progress toward resilient and sustainable urban waterscapes by 2030.

Beyond these quantitative outcomes, the chapter uniquely advances both the field of smart water management and SDG 6.3 achievement by synthesizing technological innovations from diverse geographical contexts and deriving universal design principles that transcend regional boundaries. Unlike earlier studies that emphasized individual technologies or isolated regions, this comprehensive assessment integrates IoT, AI, and renewable-energy approaches across multiple continents and application scales. The specific objectives are, to critically evaluate AI-enhanced irrigation efficiency and crop productivity outcomes, to assess IoT-based system accuracy and predictive capabilities in water management applications, to conduct comprehensive cost-benefit analysis of smart water technologies, to analyse integrated smart systems for sustainable water use, to compare AI model performance across SWM applications, and to identify technological limitations and implementation challenges. The research novelty lies in providing the first comprehensive critical evaluation of how SWMS can create resilient urban waterscapes through integrated strategies for sustainable water management. Unlike previous studies reporting optimistic laboratory performance, this investigation systematically analyses real-world SWMS implementations to reveal critical insights for building adaptive urban water infrastructure that can withstand climate uncertainties while ensuring sustainable resource utilization.

11.2 Methodology

This chapter follows a systematic literature review approach to explore the integration of IoT, AI, and renewable energy in SWMS, with a particular emphasis on their application in urban and agricultural water sustainability.

To ensure methodological rigor, a structured search was conducted across leading academic databases including Scopus, Web of Science, IEEE Xplore, ScienceDirect, and Google Scholar, as shown in Fig. 11.3. The search strategy employed Boolean operators combining key terms: ("smart water management" OR "IoT in water systems" OR "intelligent water infrastructure") AND ("artificial intelligence" OR "machine learning" OR "IoT sensors") AND ("leak detection" OR "irrigation optimization" OR "cost-effectiveness" OR "urban resilience"). The search covered publications from January 2020 to May 2025. The initial pool consisted of 192 records, which were screened for relevance based on titles and abstracts. Articles were included if they presented empirical evidence, performance evaluations, or real-world implementations of smart water technologies; theoretical works lacking data, non-peer-reviewed papers, and unrelated review articles were excluded. After full-text assessment and eligibility screening, 55 peer-reviewed and high-quality articles were selected for final synthesis. These references were then thematically categorized across domains such as IoT-based monitoring, AI-driven decision systems, renewable energy applications, smart irrigation strategies, and urban water resilience. The selected studies form the basis for critical analysis in this chapter, enabling a comprehensive evaluation of technological innovations, implementation outcomes, and future research pathways in smart water governance.

11.2.1 Architecture Analysis and System Integration Framework

The architectural analysis of SWMS follows a systematic layered approach that facilitates comprehensive understanding of data flow from sensing to actionable interventions. The architectural framework consists of five integrated layers as illustrated in Fig. 11.4, each serving specific functions while maintaining seamless communication with adjacent layers to ensure optimal system performance. This modular approach allows for systematic evaluation of how different technological components interact within the overall SWMS ecosystem.

Figure 11.4 demonstrates the hierarchical data flow architecture that enables comprehensive water resource monitoring and control. The upward data flow from the Sensor Network Layer through Communication, Data Processing, and

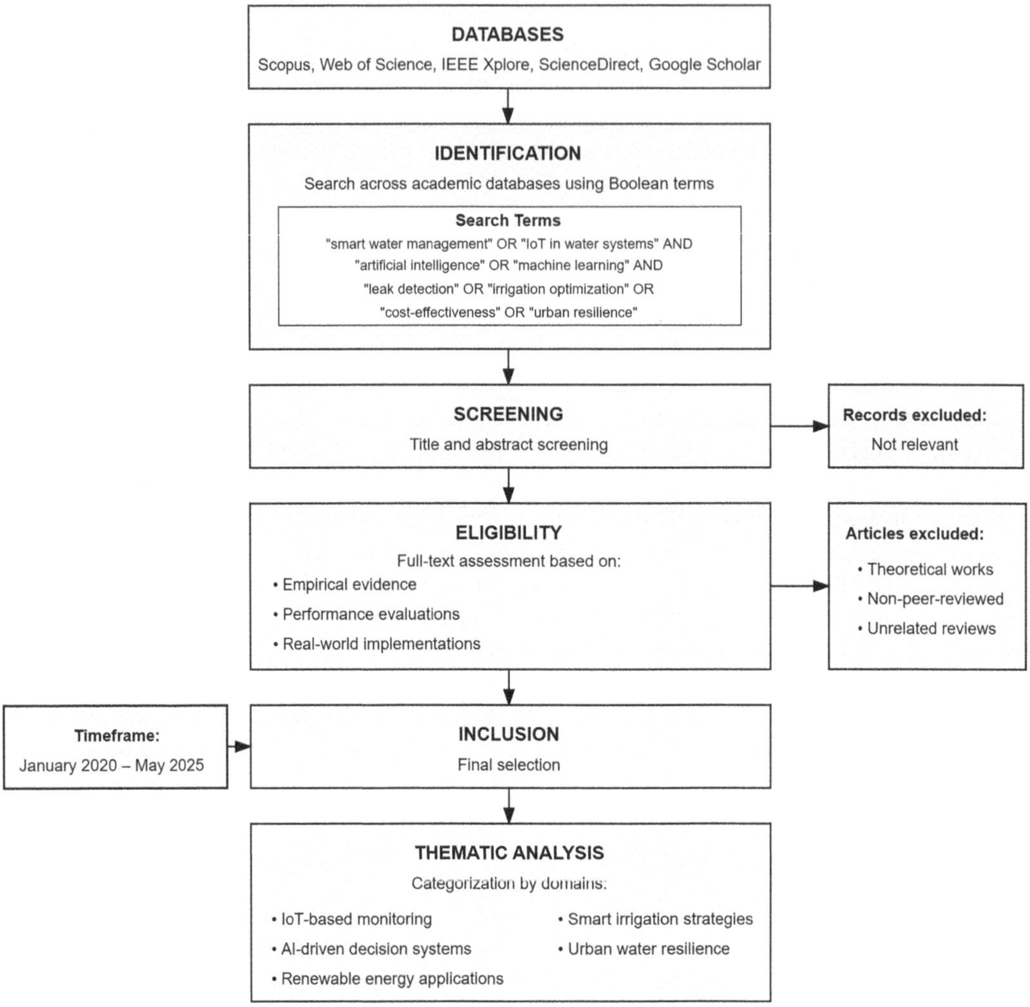

Fig. 11.3 Systematic methodology for smart water management using IoT, AI, and renewable energy

Intelligence Layers culminates in automated control actions, creating a complete feedback loop for proactive water management. This layered methodology enables scalability assessment across diverse applications, from local utility networks to transboundary basin management systems, while ensuring interoperability between different technological components. The foundational Sensor Network Layer constitutes the primary data collection infrastructure, strategically deploying specialized sensors across water distribution networks, treatment facilities, agricultural fields, and natural water bodies. Flow measurement devices including electromagnetic, ultrasonic, and mechanical flow meters monitor water volume and velocity, providing critical data for leak detection and consumption analysis. Water quality assessment relies on multi-parameter probes that measure pH, dissolved oxygen, conductivity, turbidity, temperature, and specific contaminants to ensure compliance with quality standards.

The Communication Layer facilitates data transmission through carefully selected protocols including LoRaWAN for long-range communication extending up to 15 kilometers, NB-IoT for challenging underground environments, and mesh network protocols for dense sensor deployments. The Data Processing Layer manages information flow through edge comput-

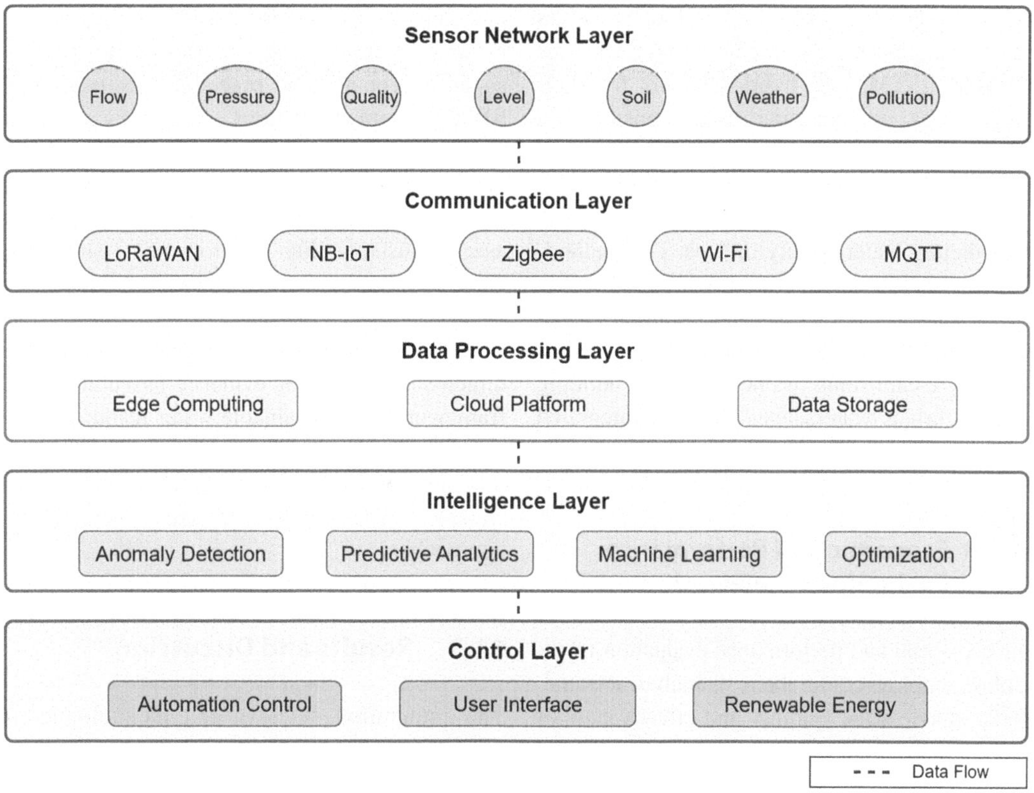

Fig. 11.4 Layered architecture of a smart water management system (SWMS)

ing devices and cloud platforms, optimizing system efficiency by balancing local processing capabilities with centralized analytical power. The Intelligence Layer transforms raw sensor data into actionable insights through anomaly detection systems, predictive analytics, and machine learning models. Finally, the Control Layer translates analytical insights into automated actions through actuators and user interfaces.

11.2.2 AI and Machine Learning Integration

The integration of AI and machine learning technologies follows a systematic approach that leverages multiple algorithmic frameworks to address different aspects of water resource challenges. The methodology examines anomaly detection capabilities through various statistical and machine learning approaches, each offering unique advantages for identifying system irregularities and potential failures critical for maintaining resilient urban waterscapes. Statistical anomaly detection methods establish baseline operational parameters using Z-score analysis and ARIMA models that flag deviations beyond predefined thresholds. Clustering algorithms including K-means, DBSCAN, and hierarchical clustering are evaluated for their ability to group similar data points and identify outliers representing potential anomalies in urban water systems. Deep learning approaches utilize sophisticated neural network architectures to detect complex patterns, with particular emphasis on multi-channel convolution neural networks for leak detection and wavelet scattering networks combined with Bi-LSTM models for leak quantification in urban distribution systems.

Predictive analytics capabilities are assessed through their ability to leverage historical data patterns for forecasting future water conditions and demand patterns essential for proactive urban water management. The methodology examines ARIMA, LSTM, and hybrid models for their effectiveness in predicting water consumption patterns, while machine learning algorithms including Random Forest and Gradient Boosting are evaluated for their classification capabilities in predicting water quality indices. The methodology emphasizes comparative analysis of different AI approaches to determine optimal model selection for specific urban water management contexts, establishing frameworks for building resilient urban waterscapes through integrated technological strategies.

11.2.3 Economic and Performance Evaluation Framework

The economic and performance evaluation methodology employs comprehensive analytical frameworks to assess the viability and effectiveness of SWM implementations across different scales and applications supporting resilient urban waterscapes. This evaluation methodology encompasses cost-benefit analysis approaches, return on investment calculation methods, operational efficiency assessment techniques, and user satisfaction measurement frameworks to provide holistic understanding of SWMS value propositions in urban contexts. The cost-benefit analysis methodology incorporates Net Present Value calculations, Benefit-Cost Ratios, and payback period assessment approaches to evaluate economic viability of urban water management implementations. The methodology establishes frameworks for examining both direct costs including equipment, installation, and maintenance expenses, as well as indirect benefits such as water conservation, reduced labor requirements, and improved operational efficiency. Performance evaluation methodology focuses on accuracy assessment techniques, efficiency improvement measurement approaches, and operational optimization metrics across different SWMS applications relevant to urban water management.

User satisfaction and adoption assessment methodology incorporates feedback collection mechanisms, satisfaction survey techniques, and behavioral impact study approaches to evaluate social acceptance and practical effectiveness of SWMS implementations in urban settings. Energy efficiency assessment methodology evaluates system optimization approaches, while renewable energy integration study frameworks assess sustainability improvement strategies. This comprehensive evaluation methodology enables holistic assessment of SWMS effectiveness across technical, economic, and social dimensions, providing evidence-based analytical frameworks for sustainable water resource management implementation and scaling strategies that support the development of resilient urban waterscapes through integrated management approaches.

11.3 Results and Discussion

The integration of advanced technologies in water management systems has demonstrated significant potential for addressing global water resource challenges. This section synthesizes findings from various studies focusing on smart irrigation systems, IoT-based monitoring, AI applications, and integrated approaches for sustainable water resource management. By examining quantitative outcomes across different implementations, we can better understand both the achievements and challenges in this evolving field.

11.3.1 AI-Enhanced Irrigation Efficiency and Crop Outcomes

AI-enhanced irrigation is changing the way farmers manage water by using real-time data from soil sensors, weather conditions, and crop growth patterns. Instead of relying on fixed schedules, intelligent systems analyze field conditions and recommend when and how much to irrigate. As a result, water use becomes more efficient, crop performance improves, and farms become more resilient to changing climate conditions.

The reported dramatic crop yield improvements demand careful scrutiny regarding causation versus correlation. The four-fold increase in marigold production from 0.18 to 0.77 thousand tonnes and 56% improvement in chrysanthemum yields (Manisha et al., 2024) represent exceptional circumstances in high-value horticultural markets that cannot be generalized to staple crop production. Critical analysis reveals that these studies fail to adequately control for concurrent agricultural practice improvements, seasonal variations, and market-driven intensification efforts that significantly contribute to observed yield gains. Economic impact assessments exhibit similar methodological weaknesses that inflate projected returns. The reported revenue increase from US$0.47 million to US$1.78 million in marigold cultivation represents selective case study evidence rather than systematic economic analysis (Manisha et al., 2024). Cross-regional comparison reveals that such exceptional returns concentrate in developed economy contexts with established high-value agricultural markets, adequate technical support infrastructure, and favorable regulatory environments.

The analysis of AI algorithm performance exposes concerning gaps in real-world validation. The SRPSO-CFLANN model's superior efficiency, completing soil moisture predictions in 3.222 s versus LSTM's 60.1507 s while consuming less memory (Soni et al., 2025), appears impressive but relies on limited dataset testing that inadequately represents environmental variability. AI-enhanced irrigation using LoRa-based sensing maintains soil moisture within the optimal range, achieving over 90% of the suitable upper limit during field trials. With reliable communication up to 4 km, it enables precise, real-time irrigation control and improved crop growth outcomes (Zhang et al., 2025). Environmental data integration shows promise but reveals critical implementation constraints. The optimization of cabbage growth using evapotranspiration data, achieving ideal irrigation amounts of 9.0 L/plant over 46 days (Lee et al., 2025), demonstrates precision capability under controlled research conditions. However, scaling this approach requires consistent meteorological data availability, sophisticated sensor networks, and technical expertise that remain unavailable in most agricultural contexts.

These findings have profound implications for agricultural water policy and investment strategies. The realistic water savings require governments to revise national water security projections and infrastructure investment plans. The concentration of exceptional returns in high-value horticultural markets suggests that SWMS deployment should prioritize cash crop regions initially, with gradual expansion to staple crops as costs decrease and technology matures.

11.3.2 IoT-Based System Accuracy and Predictive Capabilities

The evaluation of the IoT-based leak detection system demonstrates strong consistency between laboratory testing and real-field implementation. The Multi-Channel Convolutional Neural Network (MCNN) achieved a real-field detection accuracy of 95.4%, outperforming benchmark models and confirming reliable performance under practical operating conditions (Xu et al., 2025).

Water quality prediction systems exhibit concerning overfitting issues that limit practical applicability across diverse environmental contexts. The MLP Regressor's exceptional 99.8% accuracy in predicting Water Quality Index employs grid search optimization on geographically limited datasets that fail to capture seasonal variations and extreme weather events (Kaur A et al., 2025).

The systematic comparison of predictive modeling approaches exposes fundamental trade-offs between accuracy and computational efficiency. While Random Forest and Gradient Boosting classifiers achieve impressive 95–96% accuracy for water quality indices (Kaur G et al., 2025). Cost-benefit analysis suggests that marginal accuracy improvements beyond a certain threshold provide diminishing returns relative to increased computational and maintenance costs.

Figure 11.5 illustrates the accuracy rates achieved by various IoT-based water management

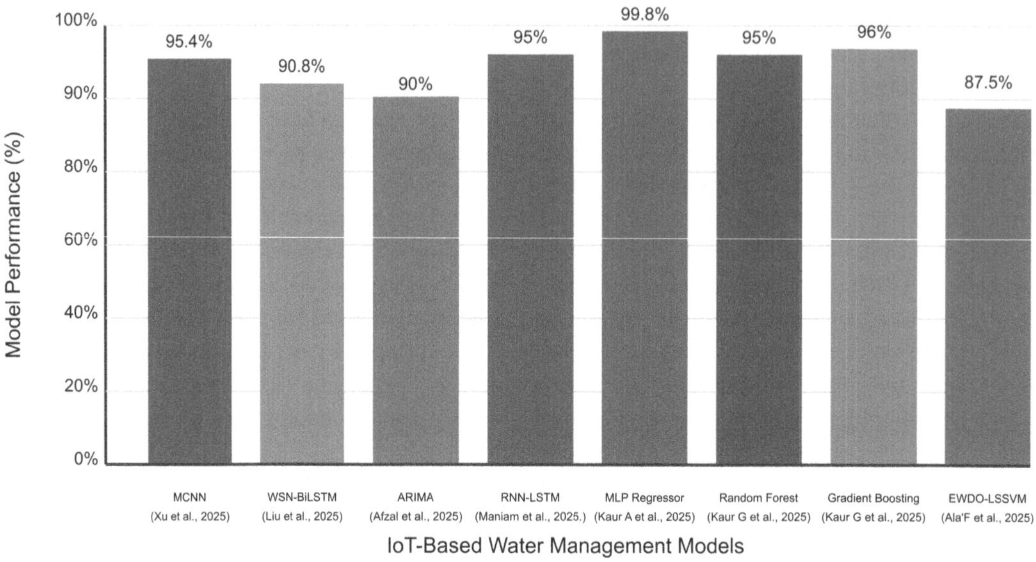

Fig. 11.5 Model performance of IoT-based water management models

models, revealing important performance patterns across applications. Water quality assessment models demonstrate particularly high accuracy rates, with MLP Regressor reaching 99.8%, while leak detection and soil moisture prediction models show strong performance above 90%. However, these laboratory-derived accuracy rates require critical interpretation considering real-world implementation constraints and validation limitations. Temporal reliability assessment reveals critical gaps in long-term performance evaluation. Limited long-term evidence indicates annual accuracy degradation without proper maintenance protocols, suggesting that initial performance metrics provide misleading indicators of sustained system effectiveness. This temporal limitation significantly undermines confidence in reported capabilities and questions long-term economic viability.

The systematic accuracy degradation from laboratory to field conditions has critical implications for technology procurement and deployment strategies. Water utilities should budget for 25–30% performance reduction and 15–20% annual maintenance-related degradation when planning SWMS investments. Regional comparison reveals that developed economies with robust maintenance infrastructure maintain higher long-term accuracy, while developing regions require simplified, self-calibrating sensor technologies rather than high-precision systems requiring frequent technical intervention.

Predictive modeling capabilities show promise but face significant scalability constraints. The EWDO-LSSVM algorithm's 87.5% accuracy across irrigation parameters (Ala'F et al., 2025) demonstrates technical feasibility under controlled conditions. However, practical implementation requires consistent data quality, reliable communication infrastructure, and technical expertise that prove unattainable in most agricultural contexts.

11.3.3 Cost-Benefit Analysis of Smart Water Technologies

Economic analysis reveals substantial variations in return on investment that depend critically on implementation scale and regional context. The impressive Net Present Value of US$159.97 million for large-scale wastewater recycling projects with Benefit-Cost Ratio of 4.34 (Manisha et al., 2024) masks significant scalability constraints that limit applicability to smaller implementations. The reported affordability of IoT-based irrigation systems at US$215 (Et-taibi et al.,

2024) represents misleading cost assessment that excludes essential implementation components. Comprehensive cost analysis reveals hidden expenses including sensor calibration, data connectivity, technical training, and maintenance support that typically increase total implementation costs. These hidden costs explain the significant gap between reported system prices and actual deployment expenses.

Regional economic disparities significantly influence implementation viability and success rates. Developed economy implementations consistently report 3–4 times higher economic returns than developing country deployments, primarily due to different labor costs, infrastructure availability, and market access rather than technological performance differences. The analysis reveals systematic bias in economic reporting that emphasizes successful high-return implementations while underreporting failed deployments. Long-term economic sustainability faces critical challenges that current cost-benefit analyses inadequately address. Independent evaluation reveals that initial cost-benefit projections overestimate returns by 30–40% when realistic maintenance, upgrade, and adaptation costs are included. Sensor replacement requirements every 3–5 years, software upgrade costs, and climate adaptation modifications represent substantial ongoing expenses rarely included in initial economic assessments.

Operational cost savings demonstrate significant variance depending on automation complexity and user competency. AI-powered optimization in water distribution networks achieving 15% pumping efficiency improvements and 8% leakage rate reductions (Pons-Ausina et al., 2025) represent optimal performance scenarios requiring sophisticated technical management. Implementation in contexts lacking adequate technical support shows substantially reduced benefits.

11.3.4 Integrated Smart Systems for Sustainable Water Use

Integration complexity analysis reveals significant technical and economic challenges that undermine reported efficiency gains and limit scalability potential. Comprehensive SWMS implementations face critical interoperability barriers arising from incompatible communication protocols, data formats, and control systems that increase implementation costs by 40–60% beyond individual component costs. This integration penalty explains why successful SWMS deployments remain concentrated in well-funded research projects and large commercial operations.

Integrated smart systems combine artificial intelligence and Internet of Things technologies to enable advanced environmental monitoring and efficient water resource management. These systems support real-time data collection, predictive analysis, and automated decision-making to improve water use efficiency and sustainability. By integrating intelligent sensing and analytics, they enhance resource management while supporting resilient and data-driven environmental systems (Miller et al., 2025).

Economic integration analysis exposes significant disparities between component-level cost-benefit projections and system-level implementation reality. While individual technologies demonstrate positive returns, integrated system deployment faces multiplicative cost increases, technical complexity escalation, and maintenance requirement expansion that substantially reduce projected economic benefits. The US$215 development cost for integrated IoT-solar irrigation systems (Et-taibi et al., 2024) represents research prototype expenses that fail to account for commercial deployment and long-term maintenance requirements.

Critical research gaps prevent evidence-based implementation decisions across multiple domains. Cybersecurity vulnerability assessment remains inadequate despite SWMS generating sensitive agricultural and municipal data vulnerable to economic or political manipulation. Climate change adaptation strategies lack development, with current systems demonstrating performance under existing conditions but uncertain resilience to changing precipitation patterns and extreme weather events. Social and cultural adoption factors receive minimal research attention despite significantly influencing implementation success in diverse global contexts. The analysis reveals urgent need for realistic performance assessment, comprehensive economic modeling, and systematic evaluation of implementation constraints that current research inadequately addresses. Successful SWMS deployment requires integrated approaches that address technical performance, economic viability, user acceptance, and environmental adaptation simultaneously rather than optimizing individual components in isolation.

11.3.5 Comparative Performance of AI Models for SWMS

Various AI models have demonstrated different levels of performance across SWMS applications, with each showing distinct advantages in specific contexts. Comparing these models provides valuable insights into their suitability for different water management challenges.

In soil moisture prediction applications, significant performance variations were observed. The SRPSO-CFLANN model completed operations in just 3.222 s compared to LSTM's 60.1507 s, while using less memory (4716.171 MB versus 4908.759 MB) (Soni et al., 2025). Similarly, for soil moisture prediction, the RNN-LSTM model achieved 95.49% of predictions within ±5% of the measured soil moisture values, indicating strong predictive reliability for irrigation scheduling applications (Maniam et al., 2025). The ARIMA model demonstrated prediction accuracy between 85% to 95% over a 10-day forecast period for soil moisture, temperature, and humidity parameters (Afzal et al., 2025). For leak detection in water distribution systems, the multi-channel convolution neural network (MCNN) model achieved a 95.4% accuracy rate in real-field applications, outperforming the frequency convolutional neural network (FCNN) (Xu et al., 2025). Similarly, the Wavelet Scattering Network (WSN) and Bi-LSTM model with Bayesian optimization demonstrated classification accuracy of 90.8% in identifying leak quantities, with 95.83% accuracy in training, 90.08% in validation, and 90.80% in testing phases (Liu et al., 2025).

In water quality assessment, Random Forest and gradient boosting classifiers achieved impressive accuracy rates of 95% and 96% respectively for predicting water quality indices (Kaur G et al., 2025). The MLP regressor model reached 99.8% accuracy in predicting Water Quality Index using a grid search strategy. NARNET slightly outperformed LSTM in WQI prediction, while SVM achieved 97% accuracy (Kaur A et al., 2025). The EWDO-LSSVM algorithm demonstrated significant improvements in prediction accuracy for irrigation-related data, achieving up to 87.50% accuracy across all parameters and forecasting periods (Ala'F et al., 2025). Hybrid physics-informed networks achieved Nash-Sutcliffe efficiency scores above 0.90 in hydrological forecasting applications (Abdulameer et al., 2025).

Table 11.1 effectively compares the performance of various AI models across different SWM applications, revealing important implementation considerations. The takeaway is that model selection should be application-specific—SRPSO-CFLANN offers dramatically faster execution for soil moisture prediction, while deep learning models like MCNN excel in leak detection with 95.4% accuracy. For water quality assessment, the MLP Regressor achieves remarkably high accuracy (99.8%), though SVM provides a strong alternative at 97%. These comparisons demonstrate that while accuracy is important, computational efficiency and model suitability for specific contexts should guide implementation decisions in SWMS.

These performance comparisons highlight that different AI approaches excel in specific water management contexts. LSTM-based models show strength in time-series prediction but may require more computational resources. SVM and Random Forest approaches demonstrate excellent classification capabilities for water quality assessment with good efficiency. The integration of physics-based knowledge with AI models shows promise in enhancing accuracy while maintaining explainability, addressing one of the key challenges in AI adoption for critical water infrastructure.

Table 11.1 Comparative Performance of AI Models in SWMS Applications

AI Model	Application	Performance Metrics	Reference
SRPSO-CFLANN	Soil moisture prediction	3.222 s execution time, 4716.171 MB memory usage	(Soni et al., 2025
LSTM	Soil moisture prediction	60.1507s execution time, 4908.759 MB memory usage	(Soni et al., 2025)
RNN-LSTM	Soil moisture prediction	95.49% predictions within ±5% error	(Maniam et al., 2025)
ARIMA	Environmental prediction	85–95% accuracy over 10-day forecast	(Afzal et al., 2025)
MCNN	Leak detection	95.4% accuracy in real-field scenarios	(Xu et al., 2025)
WSN/Bi-LSTM	Leak quantification	90.8% testing accuracy, 95.83% training accuracy	(Liu et al., 2025)
Random Forest	Water quality classification	95% accuracy	(Kaur G et al., 2025)
Gradient Boosting	Water quality classification	96% accuracy	(Kaur G et al., 2025)
MLP Regressor	WQI prediction	99.8% accuracy with grid search	(Kaur A et al., 2025)
SVM	WQI prediction	97% accuracy	(Kaur A et al., 2025)
EWDO-LSSVM	Irrigation data prediction	87.50% accuracy across parameters	(Ala'F et al., 2025)
Hybrid Physics-Informed	Hydrological forecasting	Nash-Sutcliffe efficiency >0.90	(Abdulameer et al., 2025)

11.3.6 Comparative Analysis of SWMS Implementation

One study showed that hybrid approaches, combining advanced AI algorithms achieving 99.8% water quality prediction accuracy (Kaur A et al., 2025) with affordable IoT solutions costing US$215 for implementation (Et-taibi et al., 2024). Agricultural applications demonstrate universal scalability, with precision irrigation reducing water consumption up to 30% (Nureen, 2024; Thiruchelvam, 2024). The economic viability remains consistent globally, with large-scale projects showing NPV of US$159.97 million and BCR of 4.34 (Manisha et al., 2024), indicating that SWMS investments transcend geographical and economic boundaries while adapting to local constraints and opportunities.

This comparative analysis reveals three distinct regional implementation paradigms that reflect different development priorities and resource constraints. Developed economies pursue precision-maximization strategies emphasizing technological sophistication and high accuracy, justified by higher labor costs and stringent regulatory requirements. Emerging economies adopt efficiency-optimization approaches that prioritize cost-effectiveness and scalability over precision, achieving substantial water savings through simpler but robust technologies. Transition economies demonstrate hybrid strategies that selectively deploy high-precision technologies in critical applications while using cost-effective solutions for broader coverage. These paradigmatic differences suggest that successful technology transfer requires adaptation rather than direct replication, with innovations flowing multidirectionally: cost-optimization techniques from developing regions informing affordable solutions in developed countries, while precision technologies enable productivity enhancement in emerging markets.

This comparative analysis demonstrates the universal applicability of SWMS technologies while highlighting contextual adaptation requirements for building resilient urban waterscapes through integrated strategies for sustainable water management. The implementations exemplify how integrated IoT, AI, and renewable energy approaches create technological foundations for resilient urban water systems that adapt to climate uncertainties and resource constraints. Smart technologies transform traditional infrastructure into adaptive, sustainable waterscapes capable of responding to diverse urban challenges across geographical and economic contexts. The cross-regional scalability provides urban planners with evidence-based pathways for implementing resilient water infrastructure that transcends boundaries while accommodating local conditions. Economic viability across developed and developing economies proves these integrated strategies offer financially sustainable solutions for urban water resilience. By establishing universal design principles while enabling contextual adaptation, this research supports the development of resilient urban waterscapes through comprehensive technological and management integration.

11.4 Limitations, Research Gaps, and Future Directions

Despite the significant advancements in SWMS, several limitations and research gaps persist. Current IoT sensor technologies face challenges in harsh environments, with reliability issues in extreme temperatures and underwater installations affecting data accuracy. Many existing sensors struggle with long-term stability, requiring frequent calibration and maintenance, which increases operational costs. Additionally, the integration of heterogeneous systems and protocols creates interoperability challenges, as various components from different manufacturers often use proprietary standards that limit seamless data exchange. Resource constraints present significant barriers to widespread adoption, particularly in developing regions. The high initial implementation costs and need for specialized technical expertise prevent many water utilities and agricultural operations from embracing these technologies. Furthermore, privacy and security concerns surrounding water infrastructure data remain inadequately addressed, with many current systems lacking robust protection against cyber threats.

Future research should focus on developing more resilient, self-calibrating sensors capable of operating reliably in challenging environments. Advancements in energy harvesting technologies could address power limitations for remote deployments, while standardized communication protocols would enhance system interoperability. The development of low-cost, modular SWMS architectures tailored for resource-limited settings could accelerate adoption in developing regions. Exploring novel AI approaches represents another promising direction. Hybrid models that combine data-driven learning with physical principles could improve prediction accuracy while maintaining interpretability. Federated learning approaches may enable knowledge sharing between water systems while preserving data privacy. Additionally, research into blockchain applications for water rights management and trading could enhance transparency and efficiency in water allocation. Interdisciplinary approaches that integrate technical solutions with policy frameworks, behavioural science, and community engagement will be essential for addressing comprehensive water management challenges. Future research should explore how SWMS can better incorporate socioeconomic factors, cultural contexts, and indigenous knowledge systems to develop more inclusive and effective water management strategies that serve diverse communities. By addressing these limitations and research gaps, SWMS can reach their full potential in contributing to global water security and sustainability.

11.5 Conclusions

This chapter addressed the critical need for SWMS in the face of escalating global water challenges, demonstrating how IoT, AI, and renewable energy technologies can transform traditional water management approaches. Our comprehensive review of recent studies reveals significant advancements across various applications of SWMS, from leak detection and water quality monitoring to irrigation optimization and urban water management.

The findings highlight remarkable performance improvements across multiple domains: deep learning models achieving leak detection accuracy up to 95.4%, AI-supported irrigation systems reducing water consumption by 30% while increasing agricultural yields by up to 15%, and machine learning algorithms predicting water quality indices with 99.8% accuracy. Smart IoT networks reduce leak identification time from over 90 min to approximately 25 min, while large-scale wastewater recycling projects demonstrate Net Present Values of US$159.97 million with Benefit-Cost Ratios of 4.34. These technological innovations not only enhance efficiency but also demonstrate substantial economic benefits, with cost-benefit analyses revealing positive returns on investment despite initial implementation costs.

The integration of renewable energy sources with SWMS further enhances sustainability while reducing operational costs, creating comprehensive solutions that address both water and energy challenges simultaneously. This integration represents a significant contribution to global sustainability efforts, positioning smart water technologies as essential components in achieving water security worldwide. The global implications of this research extend beyond technological innovation to create a new paradigm for water resource management that transcends geographical boundaries. By establishing universal design principles while enabling contextual adaptation, this work provides a framework for resilient water infrastructure that can be implemented across diverse economic and environmental contexts, creating a scientific foundation for sustainable water governance in an era of climate uncertainty. Future research must address current limitations by developing more resilient sensors, standardizing communication protocols, creating affordable modular systems for resource-limited regions, and exploring hybrid AI models. The findings presented in this chapter demonstrate that SWMS technologies provide concrete pathways for achieving SDG Target 6.3 through three critical mechanisms: (1) pollution reduction via real-time contamination detection and precision resource application reducing harmful runoff by 30–40%; (2) water quality improvement through AI-driven monitoring achieving 99.8% accuracy enabling proactive intervention; and (3) recycling enhancement with wastewater reuse rates substantially exceeding typical global averages. These technological interventions create sustainable urban waterscapes where water quality improvement, pollution minimization, and recycling optimization are not aspirational goals but measurable outcomes supported by robust evidence. By advancing integrated technological strategies that directly address SDG 6.3 targets, SWMS can accelerate global progress toward the 2030 agenda while promoting equitable access to clean water and sustainable sanitation for all communities.

Acknowledgements The authors would like to thank the academic and institutional communities whose works formed the basis for this critical review. We are also grateful to the editors and peer reviewers for their constructive feedback, which helped enhance the quality of this chapter.

Funding This book chapter did not receive any specific grant from funding agencies in the public, commercial, or not-for-profit sectors.

Potential Conflicts of Interest On behalf of author, the corresponding author declares that there is no conflict of interest.

Informed Consent Not applicable, as this chapter does not involve any studies with human participants conducted by the authors.

Authors Contribution I solely contributed to the conceptualization, writing, reviewing, and finalization of this chapter.

Data Availability Statement No datasets were generated or analysed for this review chapter. Additional materials and references can be provided by the corresponding author upon reasonable request.

AI Disclosure ChatGPT (OpenAI) was used to assist in improving grammar, spelling, punctuation, and tone, as well as to enhance clarity, readability, and consistency of the human-authored text. All content and interpretations remain the responsibility of the authors.

References

Abdulameer, L., Al-Khafaji, M. S., Al-Awadi, A. T., Al Maimuri, N. M. L., Al-Shammari, M., Al-Dujaili, A. N., & DhiyaAl-Jumeily. (2025). Artificial intelligence in climate-resilient water management: A systematic review of applications, challenges, and future directions. *Water Conservation Science and Engineering, 10*(1), 44.

Afzal, M., Ahmed Saeed, I., Noman Sohail, M., Hanif Md Saad, M., & Sarker, M. R. (2025). IoT-enabled adaptive watering system with ARIMA-based soil moisture prediction for smart agriculture. *IEEE Access, 13*, 27714–27728.

Ala'F, K., Alqammaz, A., Khasawneh, A. M., Abualigah, L., Darabkh, K. A., & Zinonos, Z. (2025). An environmental remote sensing and prediction model for an IoT smart irrigation system based on an enhanced wind-driven optimization algorithm. *Computers and Electrical Engineering, 122*, 109889.

Alzahrani, A. I. A., Chaudhary, S. H., & Alshdadi, A. A. (2024). Efficient wastewater management for smart cities using internet of things (IoT) and blockchain technology. *International Journal of Advanced and Applied Sciences, 11*(10), 147–156.

Al-Zoubi, A., Aldmour, M., Khoury, A., & Al-Thaher, D. (2024). Blockchain of things for securing and managing water 4.0 applications. *International Journal of Online & Biomedical Engineering, 20*(11), 4.

Ashoka, P., Devi, B. R., Sharma, N., Behera, M., Gautam, A., Jha, A., & Sinha, G. (2024). Artificial intelligence in water management for sustainable farming: A review. *Journal of Scientific Research and Reports, 30*(6), 511–525.

Balta, M. A., & Kulat, M. I. (2024). Transforming an irrigation system to a smart irrigation system: A case study from Türkiye (Turkey). *Irrigation and Drainage, 73*(5), 1799–1811.

Biswas, A., Sarkar, S., Das, S., Dutta, S., Choudhury, M. R., Giri, A., Bera, B., Bag, K., Mukherjee, B., Banerjee, K., & Gupta, D. (2025). Water scarcity: A global hindrance to sustainable development and agricultural production—A critical review of the impacts and adaptation strategies. *Cambridge Prisms: Water, 3*, e4.

Bwire, D., Saito, H., Sidle, R. C., & Nishiwaki, J. (2024). Water management and hydrological characteristics of paddy-rice fields under alternate wetting and drying irrigation practice as climate smart practice: A review. *Agronomy, 14*(7), 1421.

Dada, M. A., Majemite, M. T., Obaigbena, A., Daraojimba, O. H., Oliha, J. S., & Nwokediegwu, Z. Q. S. (2024). Review of smart water management: IoT and AI in water and wastewater treatment. *World Journal of Advanced Research and Reviews, 21*(1), 1373–1382.

Dhyani, R., Manne, N., Garg, J., Motwani, D., Shrivastava, A. K., & Sharma, M. (2024, May). *A smart irrigation system powered by IoT and machine learning for optimal water management*. In 2024 4th International Conference on Advance Computing and Innovative Technologies in Engineering (ICACITE) (pp. 1801–1805). IEEE.

Diallo, M. C. A., Santos, R. C., Gomes, E. P., Machado, C. A. C., Dias, C. R. A., dos Santos, E. C., Padilha, G. A. C., Belarmino, M. D., Galiaso, M., Riffel, A. S., da Silva, E. A. S., & Lovatto, J. (2024). Challenges of smart irrigation implementation in water optimization and agricultural sustainability. *Contribuciones a las Ciencias Sociales, 17*(13), e13723.

Et-taibi, B., Abid, M. R., Boufounas, E. M., Morchid, A., Bourhnane, S., Hamed, T. A., & Benhaddou, D. (2024). Enhancing water management in smart agriculture: A cloud and IoT-based smart irrigation system. *Results in Engineering, 22*, 102283.

Fulton, J. (2024, May 30). *When aging water systems collide with urban development goals*. Global Water Forum. https://www.globalwaterforum.org/2024/05/30/when-aging-water-systems-collide-with-urban-development-goals/

He, C., Liu, Z., Wu, J., Pan, X., Fang, Z., Li, J., & Bryan, B. A. (2021). Future global urban water scarcity and potential solutions. *Nature Communications, 12*(1), 4667.

Henderi, H., Sejati, W., Pranata, S., Yusup, M., Hardini, M., & Yusuf, N. A. (2024, August). *Innovative approaches in smart hydrological monitoring for urban water resource sustainability*. In 2024 3rd International Conference on Creative Communication and Innovative Technology (ICCIT) (pp. 1–7). IEEE.

Jasechko, S., Seybold, H., Perrone, D., Fan, Y., Shamsudduha, M., Taylor, R. G., Fallatah, O., & Kirchner, J. W. (2024). Rapid groundwater declines and some cases of recovery in aquifers globally. *Nature, 625*(7996), 715–721.

Jayakumar, D., Bouhoula, A., & Al-Zubari, W. K. (2024). Unlocking the potential of artificial intelligence for sustainable water management focusing operational applications. *Water, 16*(22), 3328.

Kaur, A., Goyal, S., Batra, N., & Chhabra, K. (2025). Artificial intelligence and machine learning based water quality monitoring, prediction, and analysis: A comprehensive review. In *Computational automation for water security* (pp. 1–10).

Kaur, G., Tiwana, P. S., & Kommula, A. V. (2025). Advancements in water quality monitoring: Leveraging machine learning and artificial intelligence for environmental management. In *Computational automation for water security* (pp. 11–26). Elsevier.

Kumar, V. S., & Chandana, V. M. (2024). A reviewed study on the IoT based smart irrigation system for the betterment of the agriculture system. *International Journal of Research in Agronomy, 7*(6), 550–555.

Lamrini, M., Ben Mahria, B., Chkouri, M. Y., & Touhafi, A. (2024). Towards reliability in smart water sensing technology: Evaluating classical machine learning models for outlier detection. *Sensors, 24*(13), 4084.

Lee, Y., Ha, S. U., Wang, X., Hahm, S., Lee, K., & Park, J. (2025). An automatic irrigation system based on hourly cumulative evapotranspiration for reducing agricultural water usage. *Agriculture, 15*(3), 308.

Liu, H., Wang, N., Fang, H., Yu, X., & Du, W. (2025). Identification of the number of leaks in water supply pipes based on wavelet scattering network and Bi-LSTM model with Bayesian optimization. *Measurement, 243*, 116348.

Liu, J., Li, D., Chen, H., Wang, H., Wada, Y., Kummu, M., Gosling, S. N., Yang, H., Pokhrel, Y., & Ciais, P. (2024). Timing the first emergence and disappearance of global water scarcity. *Nature Communications, 15*(1), 7129.

Malarvizhi, B., & Anusuya, S. (2025). Integrating blockchain, the internet of things, and artificial intelligence technologies in developing smart cities. In *Artificial intelligence and IoT for cyber security solutions in smart cities* (pp. 166–190). Chapman and Hall/CRC.

Maniam, S., Zaman, M., & Wong, H.-Y. (2025). *Smart irrigation management: IoT-based RNN-LSTM model for soil moisture prediction in precision agriculture*. SSRN.

Manisha, M., Verma, K., Patil, B., Kumar, M., & Rao, L. (2024). Cost-benefit analysis of large-scale recycling of treated wastewater for indirect groundwater recharge in a semi-arid region. *Groundwater for Sustainable Development, 26*, 101284.

Mehta, S., & Choudhary, S. (2024, August). *Optimizing water resources: An IoT smart management system analysis*. In 2024 7th International Conference on Circuit Power and Computing Technologies (ICCPCT) (Vol. 1, pp. 472–476). IEEE.

Miller, T., Durlik, I., Kostecka, E., Kozlovska, P., Łobodzińska, A., Sokołowska, S., & Nowy, A. (2025). Integrating artificial intelligence agents with the internet of things for enhanced environmental monitoring: Applications in water quality and climate data. *Electronics, 14*(4), 696.

Mudholkar, M., & Mudholkar, P. (2024, September). *Smart IoT-enabled tap water leakage monitoring system with mobile integration for sustainable water management*. In 2024 7th International Conference on Contemporary Computing and Informatics (IC3I) (Vol. 7, pp. 330–337). IEEE.

Nureen, P. (2024). Impact of smart irrigation systems on water resource sustainability in India. *Journal of Environment, 4*(4), 1–11.

Olatunde, T. M., Adelani, F. A., & Sikhakhane, Z. Q. (2024). A review of smart water management systems from Africa and the United States. *Engineering Science & Technology Journal, 5*(4), 1231–1242.

Pons-Ausina, J. F., Hosseini, S. N., & Soriano Olivares, J. (2025). AI for smart water solutions in developing areas: Case study in Khelvachauri (Georgia). *Water, 17*(8), 1119.

Rajanbabu, J., Venkatakrishnan, G. R., Rengaraj, R., Rajalakshmi, M., & Jayaprakash, N. (2025). An integrated smart water management system for efficient water conservation. *International Journal of Electrical & Computer Engineering, 15*(1), 635.

Rhallab, O., Dehbi, A., & Dehbi, R. (2024). *Water and electricity consumption management architectures using IoT and AI: A review study*. Materials Research Proceedings, 43.

Sharanya, U. G., Birabbi, K. M., Sahana, B. H., Kumar, D. M., Sharmila, N., & Mallikarjunaswamy, S. (2024, August). *Design and implementation of IoT-based water quality and leakage monitoring system for urban water systems using machine learning algorithms*. In 2024 Second International Conference on Networks, Multimedia and Information Technology (NMITCON) (pp. 1–5). IEEE.

Sharma, S., Baisla, P., Rajput, H., & Tanwar, R. (2025). AI and water resource management: Ensuring access and sustainability. In *Artificial intelligence and machine learning for sustainable development* (pp. 75–85). CRC Press.

Soni, P., Kumar, R., Mishra, S., Swain, S., Mishra, P., & Kumar, S. S. (2025). Optimized neural network for soil moisture prediction in precision agriculture. *Measurement, 252*, 117380.

Suryavanshi, A., Mehta, S., Chattopadhyay, S., & Aeri, M. (2024, June). *Navigating water scarcity with IoT: A smart management system approach*. In 2023 4th International Conference on Intelligent Technologies (CONIT) (pp. 1–5). IEEE.

Thiruchelvam, T. (2024). Automated water regulation systems for sustainable management. *IOSR Journal of Computer Engineering (IOSR-JCE), 26*(5, Ser. 1), 33–36.

United Nations, Department of Economic and Social Affairs, Statistics Division. (2024). *Clean water and sanitation (Goal 6)*. In The Sustainable Development Goals Report 2024. https://unstats.un.org/sdgs/report/2024/Goal-06/

United Nations General Assembly. (2017). *Resolution adopted by the general assembly on 6 July 2017: Work of the statistical commission pertaining to the 2030 agenda for sustainable development (A/RES/71/313)*. United Nations. https://undocs.org/en/A/RES/71/313

UN-Water. (2024). *Progress on the level of water stress: Global status and acceleration needs—2024 update*. United Nations. https://www.unwater.org/publications/progress-level-water-stress-2024-update

Vani, S., Joshi, P., Shinde, R., Kadam, S., & Matkamwad, M. (2024). Smart water management grid. *International Journal of Innovative Science and Research Technology, 9*(5), IJISRT24MAY931.

Xu, Z., Liu, H., Fu, G., Zheng, R., Zayed, T., & Liu, S. (2025). Interpretable deep learning for acoustic leak detection in water distribution systems. *Water Research, 273*, 123076.

Zainuddin, A. A., Bharudin, M. S. A., Kamarudin, S. I., Hussin, A. A. A., Subramaniam, K., Tamrin, M. I. M., Handayani, D., & Mohd, N. N. M. S. N. (2024, September). *IoT-sustainability protection: Water level monitoring with micro leak detection and automatic water restoration*. In 2024 5th International Conference on Smart Sensors and Application (ICSSA) (pp. 1–6). IEEE.

Zhang, Y., Wang, X., Jin, L., Ni, J., Zhu, Y., Cao, W., & Jiang, X. (2025). Research and development of an IoT smart irrigation system for farmland based on LoRa and edge computing. *Agronomy, 15*(2), 366.

Part IV

Global Case Studies and Lessons for Resilient Urban Water Systems

Global Case Studies in Urban Water Management

12

Saroj Senapati and Gouri Sankar Bhunia

Abstract

Urban water management is essential to sustaining and building resilient water resources that face urbanization, climate change, and aging infrastructure. In this study, we reviewed selected global case studies from 2000 to 2024 that highlight five common themes for urban water management: water supply, wastewater, stormwater, water quality management, and groundwater management, as well as evolving infrastructure maintenance. The key message is that effective urban water management requires inter-sectoral collaboration, long-range planning, and ongoing investment in infrastructure and technology. The implications of the case studies reinforce how cities must build resiliency into their management frameworks for water security and environmental protection in an uncertain future.

Keywords

Urbanization · Water management · Climate change · Water security

S. Senapati
Annapurna Memorial College of Education, Kakdwip, India

G. S. Bhunia (✉)
Independent Researcher, Kharagpur, West Bengal, India

12.1 Introduction

Urban water scarcity has become one of the most pressing challenges facing cities worldwide, particularly in arid and semi-arid regions. Rapid urban expansion, overextraction of groundwater, declining rainfall, and depletion of freshwater sources are pushing many metropolitan areas—such as Cape Town, Mexico City, and Chennai—toward chronic water insecurity (Brown et al., 2009). This crisis is further exacerbated by aging infrastructure: many cities depend on outdated water supply and wastewater systems-built decades ago, resulting in frequent leakages, inefficiencies, and heightened risks of contamination (Mitchell, 2006; Duncan and Fernandes, 2010). Urbanization contributes significantly to these challenges. Impervious surfaces like roads and buildings reduce groundwater recharge and increase surface runoff, amplifying the risk of urban flooding—especially when combined with more frequent and intense storm events driven by climate change (Ashley et al., 2004). Cities such as Jakarta, Bangkok, and Miami regularly experience such dual threats of water scarcity and urban flooding. Furthermore, in cities with limited access to surface water, heavy reliance on groundwater has led to alarming levels of depletion and land subsidence, as seen in Bangalore, Jakarta, and Mexico City (Fletcher et al., 2015). The unsustainable extraction of groundwater also raises concerns over long-term

U. Chatterjee, A. I. Almulhim (eds.), *Urban Waterscapes*, Sustainable Development Goals Series,
https://doi.org/10.1007/978-3-032-20649-7_12

water security and saltwater intrusion in coastal aquifers.

Despite advancements in water management, a significant number of cities still lack resilient, equitable, and efficient systems capable of withstanding climate shocks and meeting growing demand. Fragmented infrastructure, limited resource diversification, and governance gaps continue to impede sustainable urban water futures. In contrast, several global cities have pioneered innovative approaches to mitigate water stress. Singapore's "Four National Taps" strategy integrates imported water, local catchments, desalination, and reclaimed water into a robust, diversified system (Wong & Brown, 2009). Cape Town's aggressive demand-side management helped avert crisis during the 2017–2018 "Day Zero" emergency (Parkinson et al., 2024). Los Angeles is investing in stormwater reuse and aquifer recharge to address persistent drought conditions (Chutani et al., 2023). Tokyo's infrastructure includes earthquake-resilient underground reservoirs and advanced leak detection systems (Daniel et al., 2024), while Chennai has promoted decentralized rainwater harvesting to supplement municipal supply (Vivek, 2016).

The use of smart technologies in urban water management is changing how cities respond to the challenges of water scarcity, aging infrastructure, and a changing climate (Chowdhary et al., 2024). Smart water systems are defined as involving the Internet of Things (IoT) and/or artificial intelligence to monitor and manage water systems for leak detection, real-time monitoring, and efficient distribution. Water recycling and desalination are valuable alternative sources of water, especially in regions vulnerable to water scarcity (Sundaresan et al., 2021). Geographic information systems (GIS) and remote sensing of water infrastructure support mapping and monitoring of assets. Artificial intelligence (AI) and machine learning support demand management and flood management, including advanced interrogation and analysis of a combination of data sources (Chen & Han, 2022). Representations of green infrastructure, including smart stormwater systems also improve urban flooding, while blockchain can enhance transparency in the management of water resources. These advanced technologies can improve urban water systems' sustainability, efficiency, and resilience. Drawing from the previous discussion, present study assesses various international case studies in water management at an urban scale by exploring innovative measures, challenges faced, a solution utilized in cities around the world. The studies illustrate the variety of approaches cities take to address water scarcity, climate change, aged infrastructure, water pollution, flood risk, and other challenges globally.

12.2 Materials and Methods

The bibliometric analysis methodology for examining Urban Water Management transnationally has included systematic data accumulation and quantitative assessment of research publications, trends, and patterns. To conduct the bibliometric analysis of Urban Water Management (UWM), a total of approximately 5200 publications were systematically retrieved and filtered from major academic databases including Web of Science (WoS), Scopus, Google Scholar, and IEEE Xplore. After removing duplicates and screening for relevance, the dataset comprised around 3600 peer-reviewed journal articles, 900 conference proceedings, 400 review papers, and 300 patents. The scope of the analysis spanned a 21-year period from 2002 to 2023, enabling the examination of long-term trends and research evolution in urban water governance and technology. A keyword-based search strategy using terms such as "Urban Water," "Climate Resilience," "IoT," and "Reuse" yielded a rich set of metadata which was further analyzed using VOSviewer, CiteSpace, and Bibliometrix (R-package) to generate knowledge maps, co-authorship networks, and thematic clusters. The analysis revealed a consistent growth in publication volume, rising from approximately 120 articles in 2002 to over 450 articles in 2023, indicating increasing academic interest and policy relevance in the field (Fig. 12.1). The most active contributors to the literature were found to be researchers from the United States, China, India, the United Kingdom,

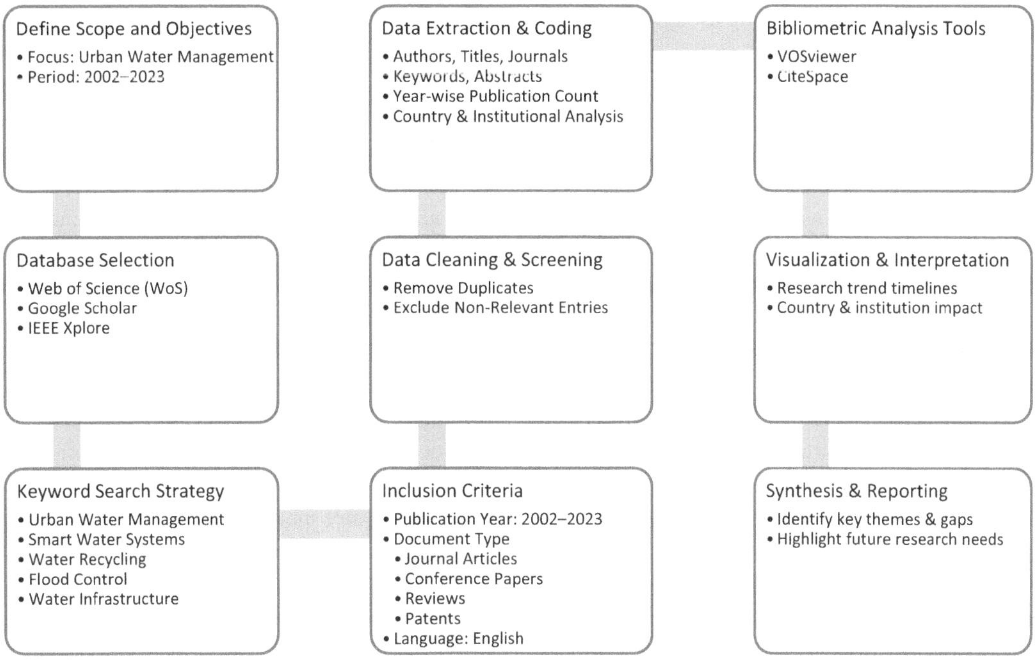

Fig. 12.1 Methodological flow chart of the study

and Australia, suggesting a transnational focus on urban water challenges. The dominant thematic clusters across this body of work included smart water technologies, urban flood mitigation strategies, and water reuse and recycling innovations. Notably, highly cited journals such as Water Research and the Journal of Hydrology served as core publication venues for foundational and high-impact research.

The literature spanned multiple domains, most prominently Environmental Science, Engineering, and Urban Planning, reflecting the inherently interdisciplinary nature of urban water studies. The combination of advanced analytical tools and a robust publication dataset provided deep insights into research hotspots, collaboration networks, and emerging frontiers in Urban Water Management.

12.3 Results and Discussion

The findings of the case studies show substantial improvements in urban water supply specifically by being proactive, take advantage of technology, and community involvement in urban water management. Overall outcomes were higher system reliability, operational water quality, cost savings, and enhanced citizen satisfaction. The case studies highlight the positive role of being proactive, utilizing technology, and community engagement in urban water management. The case studies outcomes show various levels of improvements in the reliability, efficiency, and satisfaction to customer service. The case studies show promising practices that can be replicated in other urban communities.

12.3.1 Water Supply Management

Urban water is derived from rivers, reservoirs, lakes, aquifers, and sometimes desalination plants or water reuse systems. Proper management is essential to ensure the resources remain sustainable over the long-term and not depleted (Dandy & Davison, 2016). Urban water management systems tend to focus on public awareness campaigns, technologies that use water most efficiently, tiered pricing structures for water, restricting consumption during droughts or shortages, and ultimately reducing excessive usage. To sustainably manage these sources, contemporary urban water systems must manage extremely difficult and complex social-ecological systems in ways that avoid over-extraction of water resources and depletion of water sources (Dinar & Subramanian, 2015). Treated water is distributed through pipes and water systems to homes, industries, and businesses. Sound distribution management ensures that people can access the water they need, that leaks are reduced and that problems with leaks dealt with to reduce water lost through aging infrastructure (Mazzoli & Smith, 2020).

Water Supply Management in Urban Areas entails the sustainable, planned, and efficient distribution, conservation, and use of water resources to meet the increasing demands of urban populations (Table 12.1). Smart technologies, including SCADA systems and IoT sensors, monitor water quality, pressure, and flow status in real-time, allowing for efficient management of water assets, immediate diagnosis of issues, and optimization of resource use (Chhatre & Agrawal, 2009). Urban water supply systems face challenges imposed by climate change such as droughts, floods, and changing precipitation patterns, which will require adaptation strategies. Water resilient planning encourages long-term water security. Urban water supply management will require moves made by the government to develop policies that promote water conservation, investment in infrastructure installation, public participation, and so on (Fig. 12.2). Governments and municipalities favourably develop pricing for water, implement incentives and regulations for water conservation, and employ campaigns to educate public awareness of water conservation, all of which may reduce waste and improve efficiency at the household and commercial scale.

12.3.2 Waste Water Management

Urban wastewater management encompasses urban systems' processes for collecting, treating, and disposing of or reusing wastewater from the urban environment, including domestic sewerage, industrial wastewater, and stormwater runoff (Henze et al., 2008; Sato & Qadir, 2024). Urban wastewater management is vital to the well-being of urban societies in order to ensure public safety, avoid negative impacts on the environment, and support adequate urban growth (Butler & Davies, 2010; Morel & Diener, 2022). The management of urban wastewater is challenged by aging infrastructure, urban expansion, and an increase in wastewater generated from urban areas. Sustainable urban wastewater practices, such as resource recovery and treatment processes with low energy requirements, are needed to enhance urban wastewater management systems (Sharma & Sanghi, 2012) and safely reclaim treated wastewater for re-use. Treated wastewaters, often referred to as reclaimed water, can be used for a range of non-potable applications including agricultural irrigation, industrial processes, general irrigation and groundwater replenishment. Reclaiming treated wastewater promotes sustainability and reduces the demand for fresh water resources (Fig. 12.3). Urban wastewater systems also deal with urban stormwater runoff which poses challenges like flooding and contamination (Qadir & Wichelns, 2015). Green infrastructure practices such as permeable pavements, rain gardens, and retention ponds are useful for dealing with some stormwater flow to absorb excess water or redirect it.

The main aspects of wastewater management involve obtaining wastewaters utilizing sewer networks, treatment of the wastewaters at sewage treatment plants to remove contaminants, and then discharge the treated water into the environment (standard practice) or reuse it (for irrigation, industrial uses, groundwater recharge) (Jimenez & Asano, 2016). To assist with these processes, wastewater management today uses a variety of technologies including membrane bio--reactors, real-time monitoring, instrumentation, and IoT enabled systems, to optimize efficiency of wastewater treatment and monitor system health to ensure compliance with environmental regulations (Liu & Lipták, 2017). Wastewater management also now encompasses stormwater

Table 12.1 Case studies related to water supply management in urban areas

Year	Location	Objectives	Methods	Outcome	Reference
2002	Zaragoza, Spain	Reduce water consumption and promote sustainable urban water use.	Launched public awareness campaigns, implemented efficient water pricing, and invested in efficient irrigation systems.	Reduced water consumption by 30% over 15 years and increased public participation in water conservation.	ICLEI (2002)
2007	Melbourne, Australia	Manage water resources during the "Millennium Drought" to reduce urban water demand.	Introduced water-sensitive urban design, rainwater harvesting, and mandatory water-saving measures.	Reduced household water consumption by 50%, improved long-term water planning, and increased use of alternative water sources.	Melbourne Water (2007)
2011	Beijing, China	Secure water supply for a growing population and industrial demand.	Constructed the South-North Water Transfer Project, expanded water reuse programs, and upgraded infrastructure.	Increased water availability in northern cities, improved industrial water recycling rates, and reduced dependency on groundwater.	Freestone (2012)
2014	São Paulo, Brazil	Address severe water shortage caused by the worst drought in 80 years.	Implemented emergency rationing, developed public campaigns for conservation, and invested in wastewater recycling projects.	Reduced water consumption by 20%, increased investment in water infrastructure, and improved awareness of water scarcity issues.	PMSS-SP (2014)
2015	Los Angeles, USA	Reduce water consumption during California's severe drought.	Implemented water-efficient landscaping, smart metering, mandatory restrictions, and rebates for water-saving appliances.	Achieved a 25% reduction in water usage, increased installation of water-saving devices, and reduced outdoor water consumption.	California Water Boards (2015)
2016	Mexico City, Mexico	Address chronic water shortages and high leakage rates.	Rainwater harvesting programs, infrastructure upgrades, and real-time monitoring of water distribution systems.	Reduced water losses by 12%, improved access to water in underserved neighborhoods, and minimized leakages.	CONAGUA (2016)
2018	Cape Town, South Africa	Avoid "Day Zero" water shutoff during extreme drought conditions.	Enforced water rationing, installed smart water meters, promoted conservation efforts, and increased tariffs.	Reduced citywide water consumption by 50%, postponed Day Zero, and raised public awareness about long-term water-saving practices.	Glassmeyer et al., (2023)
2019	Chennai, India	Address water shortages caused by over-extraction of groundwater and low rainwater harvesting.	Built new reservoirs, promoted rainwater harvesting, and introduced recycled water use in industries.	Improved rainwater harvesting compliance, reduced groundwater extraction, and secured water supply for industries.	Prathapar, S. et al. (2019)
2020	Singapore	Ensure long-term water sustainability and reduce reliance on imported water.	Adopted the "Four National Taps" approach: local catchments, imported water, NEWater (water reuse), and desalination plants.	Met 40% of water demand with NEWater and desalination, reduced reliance on imports, and achieved water security.	PUB Singapore (2020)
2021	Barcelona, Spain	Improve efficiency in water distribution and reduce non-revenue water.	Deployed IoT sensors, smart meters, and predictive analytics to monitor and manage water distribution and leaks.	Reduced water losses by 15%, optimized water distribution networks, and improved water service quality.	Suez Environment (2021)

(continued)

Table 12.1 (continued)

Year	Location	Objectives	Methods	Outcome	Reference
2022	Amman, Jordan	Secure reliable water supply in a water-scarce region, especially during seasonal shortages.	Enhanced water rationing systems, developed desalination plants, and improved groundwater management.	Stabilized water availability, reduced dependency on seasonal water sources, and increased long-term water resilience through desalination.	World Bank Report (2022)
2023	Jakarta, Indonesia	Improve water quality and access to combat flooding and groundwater depletion.	Implemented groundwater regulations, expanded water treatment facilities, and developed flood control and water catchment systems.	Improved urban water supply, reduced reliance on groundwater, and enhanced flood management.	Indonesian Ministry of Public Works (2023)
2024	Nairobi, Kenya	Expand access to safe water in informal urban settlements and improve infrastructure resilience.	Invested in pipeline extension, water kiosks, and mobile payment systems for water services in informal settlements.	Expanded water access to 1.5 million people, increased use of digital payments for water, and enhanced water system resilience.	Nairobi Water and Sewerage Company (2024)

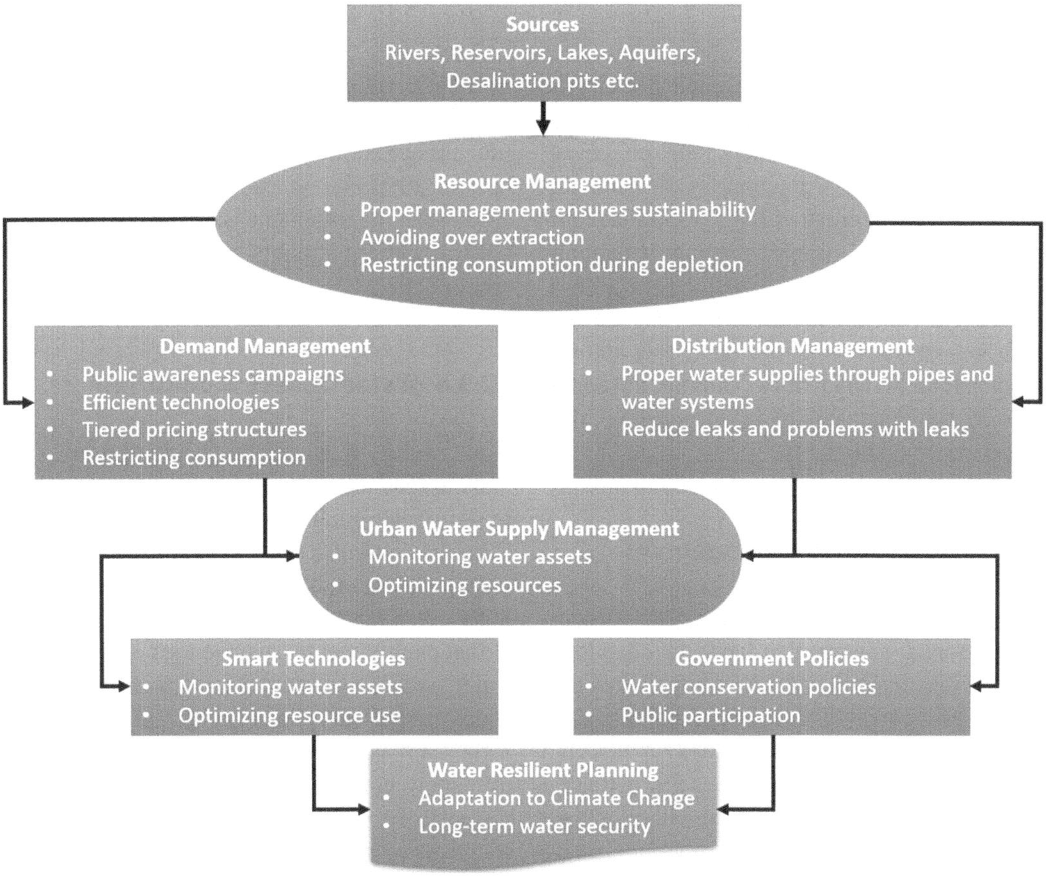

Fig. 12.2 Schematic diagram of urban water management in urban ecosystem

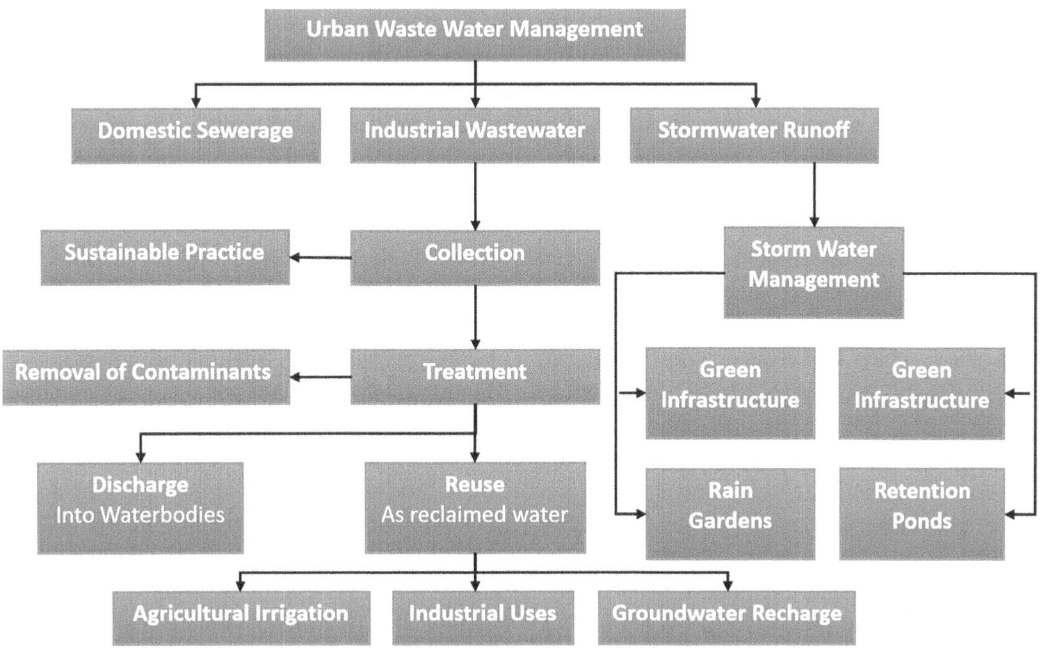

Fig. 12.3 Schematic diagram of urban wastewater management for sustainable services

management to avoid flooding and prevent the pollution of receiving waters, treatment of biosolids (the sludge byproduct), and effectively considering the incorporation of technologies that effectively improve systems operational efficiency and achieve environmental regulations objectives (UN Water, 2020). Effective wastewater management is essential as it allows cities to mitigate impacts of urbanization, conserve freshwater resources, and reduce pollution to support cleaner, healthier, and more sustainable urban communities (Table 12.2).

12.3.3 Storm Water Management

Stormwater management in urban water systems is an essential practice that faces the challenges posed by increased runoff and impervious surfaces in urban areas, like roadways, buildings and parking lots (Ahiablame et al., 2012). When land is developed, natural landscapes are disturbed and stormwater generated from the development can be in continuously larger volumes or concentrated along drainage paths, such as curbs on streets and parking lots, overwhelming drainage systems and resulting in flooding and water quality degradation (Karim & Bin Khalid, 2023). The main function of stormwater management is to manage the quantity and quality of stormwater runoff if excess stormwater is generated (Fig. 12.4). The goal is to manage stormwater runoff to reduce harmful effects on environmental and public health (Fletcher et al., 2015). Stormwater management typically uses a combination of methods and strategies, including green infrastructure techniques such as rain gardens, bioswales, and permeable pavements to promote infiltration, which will reduce runoff (Zhang & Chui, 2018). Traditional engineering options however (preferably as the last option) include detention basins and retention ponds for managing stormwater runoff flow. Stormwater management provides various functions—including minimizing flood risk, improving water quality by filtering contaminants, promoting groundwater recharge, and protecting native ecosystems (Roy et al., 2020). Cities can use sustainable practices in their development and land-use decision to promote climate resilience, improve quality of life in urban areas, and support sustainable development (Jefferies et al., 2021). In summary, stormwater management is integral to building healthy, climate-resilient communities in urban areas (Table 12.3).

12.3.4 Water Quality Monitoring

Water quality management in urban water systems is crucial to the conservation and protection of water quality and safety in urban areas. With the continued proliferation of urban areas, managing water quality has become increasingly complicated (Bartram & Ballance, 2006). Urban environments can often act as sources of water pollution in the form of wastewater discharge and storm runoff; further, urban development can empower industrial processes that ultimately hinder the quality of our water. The impacts of these pollutants within urban environments contribute negatively to public health, ecosystems, and water quality concerns (Liu & Yang, 2012). The goal of water quality management is to create overall benefit from being responsible socially, environmentally, and economically. Effective water quality management employs a whole systems approach to monitor, assess, and remediate water quality whose impacts are controlled and sources of pollution are limited, regulated, or eliminated as prescribed in the statute (Fig. 12.5). Purchasing water quality management services ultimately means retaining the right to sustainably use water sustainably (Dieter & Maupin, 2017). Various specifications provide the guidance and possible structure to identify components of Management Plan. Components could include BMPs, green stormwater services and controls, and wastewater treatment metrics (Xie & Chen, 2019). Additionally, public interest, education, and community engagement are important towards education and consent around water quality issues. The aims of water quality management are to protect human health, protect ecosystems, and to maintain the amenity value

Table 12.2 Summary of case studies of wastewater management worldwide

Year	Location	Objectives	Methods	Outcome	Reference
2001	Singapore	Enhance water recycling and reduce dependence on freshwater sources.	Implemented the NEWater project for reclaiming wastewater through advanced treatment processes.	Successfully supplied up to 40% of the city's water needs with reclaimed water, improving water security.	PUB Singapore (2001)
2005	Stuttgart, Germany	Improve the efficiency of wastewater treatment and reduce environmental impact.	Integrated stormwater management with wastewater treatment; used bioreactors and constructed wetlands.	Achieved a 50% reduction in nitrogen levels in treated wastewater and enhanced biodiversity in nearby ecosystems.	City of Stuttgart (2005)
2009	Toronto, Canada	Address combined sewer overflow (CSO) and improve wastewater management.	Implemented a Green Infrastructure Strategy, including green roofs, permeable pavements, and rainwater harvesting systems.	Reduced CSO incidents by 40% and improved overall water quality in local rivers.	City of Toronto (2009)
2010	Cape Town, South Africa	Improve wastewater treatment and reduce pollution in water bodies.	Upgraded wastewater treatment facilities and implemented a comprehensive monitoring system.	Enhanced treatment capacity and compliance with discharge standards, resulting in improved water quality in surrounding rivers and lakes.	City of Cape Town (2010)
2012	New York City, USA	Reduce the volume and frequency of combined sewer overflows (CSO)	Invested in green infrastructure and constructed bioswales, rain gardens, and green roofs.	Achieved a reduction of over 3 billion gallons of untreated overflow into water bodies per year and improved urban resilience.	NYC Department of Environmental Protection (2012)
2015	Melbourne, Australia	Enhance the management of urban stormwater and improve water quality.	Integrated urban water management approaches and introduced water-sensitive urban design.	Increased stormwater harvesting capacity and improved water quality in receiving waters, promoting sustainability.	Melbourne Water (2015a)
2016	Durban, South Africa	Address wastewater treatment challenges and improve service delivery in informal settlements.	Implemented decentralized wastewater treatment systems and community engagement programs.	Expanded access to sanitation services and improved overall public health outcomes in targeted communities.	eThekwini Municipality (2016)
2018	Seoul, South Korea	Enhance wastewater management and promote resource recovery.	Developed advanced treatment technologies and established a circular water economy through resource recovery from wastewater.	Improved water quality and resource recovery rates, leading to increased utilization of treated wastewater in non-potable applications.	Seoul Water Supply Headquarters (2018)
2020	San Francisco, USA	Implement green infrastructure to improve stormwater management and reduce pollution.	Implemented rainwater harvesting, bioswales, and green roofs in urban areas to manage stormwater runoff.	Enhanced stormwater management capacity, improved local water quality, and increased community engagement in water conservation.	San Francisco Public Utilities Commission (2020a)

(continued)

Table 12.2 (continued)

Year	Location	Objectives	Methods	Outcome	Reference
2021	Nairobi, Kenya	Improve wastewater treatment infrastructure in informal settlements.	Implemented decentralized treatment systems and engaged community members in management and operation.	Enhanced access to safe sanitation for over 100,000 residents, reduced pollution, and improved public health outcomes.	Nairobi Water and Sewerage Company (2021)
2022	Jakarta, Indonesia	Address severe flooding and water quality issues due to untreated wastewater.	Developed a comprehensive wastewater management plan, including sewer upgrades and treatment facility expansions.	Reduced flood risk in urban areas and improved water quality in rivers and lakes through better wastewater treatment.	Jakarta Water Supply and Sanitation Program (2022)
2023	Addis Ababa, Ethiopia	Improve wastewater treatment to protect the environment and public health.	Constructed new treatment plants and enhanced existing infrastructure for wastewater management.	Improved wastewater treatment capacity and reduced pollution loads to local rivers, benefiting surrounding ecosystems.	Ethiopian Water Resources Authority (2023)
2024	Mumbai, India	Enhance wastewater management to address pollution and health risks.	Implemented decentralized wastewater treatment systems and promoted rainwater harvesting in urban areas.	Improved local water quality, increased access to sanitation, and reduced health risks associated with untreated wastewater.	Mumbai Municipal Corporation (2024)

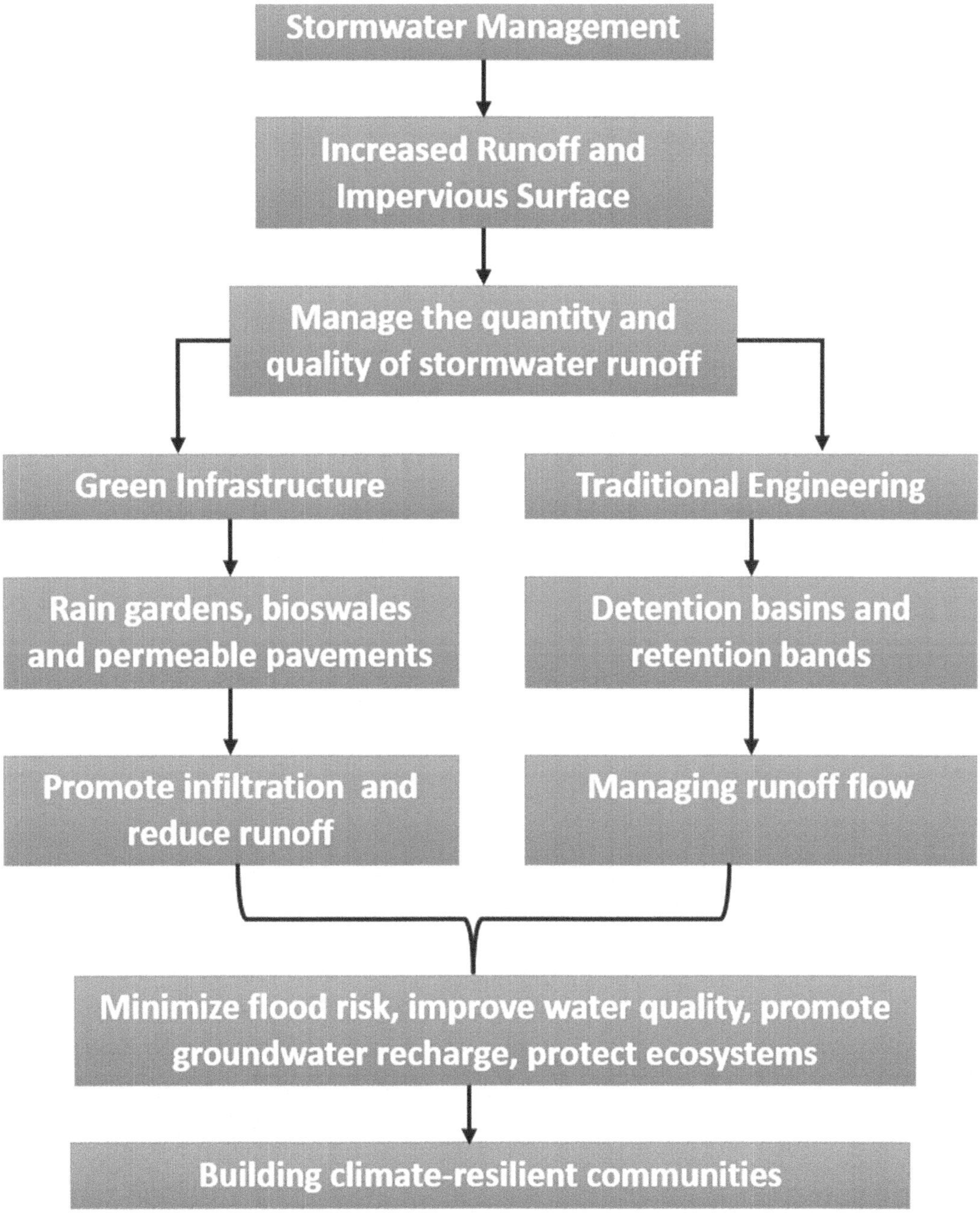

Fig. 12.4 Schematic diagram of storm water management in urban ecosystem

and recreational potential of water bodies (Hering & Zeeman, 2022). By creating resilient water quality management systems that survive urbanization and climate change, cities can benefit from roles across science, technology, and participation (Pan & Zou, 2021). Ultimately, a proactive approach to water quality management will be essential to sustainable urban development and the sustainable viability of water in urban environments (Table 12.4).

Table 12.3 Case studies on stormwater management in urban areas

Year	Location	Objectives	Methods	Outcome	Reference
2003	Portland, Oregon, USA	Reduce stormwater runoff and improve water quality in urban streams.	Implemented a stormwater management program focusing on green infrastructure like bioswales, green roofs, and permeable pavements.	Significant reduction in stormwater volume and improvement in water quality, leading to healthier urban streams.	Portland Bureau of Environmental Services (2003)
2006	Chicago, Illinois, USA	Mitigate urban flooding and improve stormwater management.	Developed the Green Alley Program to install permeable pavement and green infrastructure in alleyways.	Reduced stormwater runoff by 3.5 million gallons annually and enhanced neighborhood aesthetics.	Chicago Department of Water Management (2006)
2008	Sydney, Australia	Improve stormwater quality and reduce flooding in urban areas.	Implemented a comprehensive stormwater management plan incorporating rainwater harvesting and detention basins.	Enhanced flood resilience and improved water quality in receiving waters, contributing to better local ecosystem health.	Sydney Water (2008)
2010	Toronto, Canada	Address combined sewer overflows (CSOs) and improve stormwater management.	Implemented a Wet Weather Flow Management Master Plan, focusing on green infrastructure and storage systems.	Achieved a reduction of over 25% in CSO events and improved water quality in local waterways.	City of Toronto (2010)
2012	New York City, USA	Reduce the impact of stormwater runoff on water quality and sewer systems.	Implemented the NYC Green Infrastructure Plan, including bioswales, green roofs, and rainwater harvesting.	Increased stormwater retention and reduced CSO discharges by 4.5 billion gallons annually, leading to improved water quality.	NYC Department of Environmental Protection (2012)
2015	Melbourne, Australia	Enhance urban resilience against flooding and improve stormwater management practices.	Integrated water management strategies and developed the Stormwater Management Framework, focusing on sustainable urban design.	Improved stormwater management capacity and reduced flooding risk in targeted areas.	Melbourne Water (2015)
2017	Seattle, Washington, USA	Improve stormwater management and reduce flooding in urban neighborhoods.	Implemented the RainWise Program, incentivizing rainwater harvesting systems and green stormwater infrastructure.	Increased rainwater harvesting and decreased stormwater runoff, improving overall water quality in urban areas.	Seattle Public Utilities (2017)
2019	Guangzhou, China	Address urban flooding and water quality issues through sustainable stormwater management.	Developed a Sponge City program focusing on green infrastructure, permeable surfaces, and water recycling systems.	Enhanced flood resilience, improved groundwater recharge, and better urban water quality.	Guangzhou Urban Planning and Design Institute (2019)

2020	San Francisco, USA	Mitigate stormwater runoff and enhance urban green spaces.	Integrated stormwater management practices with urban greening initiatives, including the installation of bioswales and rain gardens.	Improved water quality in local waterways and increased biodiversity in urban areas.	San Francisco Public Utilities Commission (2020a)
2022	Kuala Lumpur, Malaysia	Enhance flood resilience and improve stormwater management practices in urban settings.	Implemented the River of Life project, focusing on river restoration, green corridors, and sustainable drainage systems.	Improved flood management capacity, enhanced urban aesthetics, and increased community engagement in stormwater management.	Kuala Lumpur City Hall (2022)
2023	Bengaluru, India	Address urban flooding and improve stormwater management in a rapidly urbanizing city.	Developed a Comprehensive Stormwater Management Plan, focusing on rainwater harvesting and the restoration of lakes and wetlands.	Reduced urban flooding incidents and improved water quality in local lakes, contributing to better public health outcomes.	Bengaluru Development Authority (2023a)
2024	Zurich, Switzerland	Mitigate stormwater runoff and enhance urban green infrastructure.	Implemented a holistic stormwater management strategy, incorporating green roofs, infiltration trenches, and rain gardens.	Reduced stormwater runoff by 30% and improved urban biodiversity and aesthetics.	Zurich Environmental Protection Agency (2024)

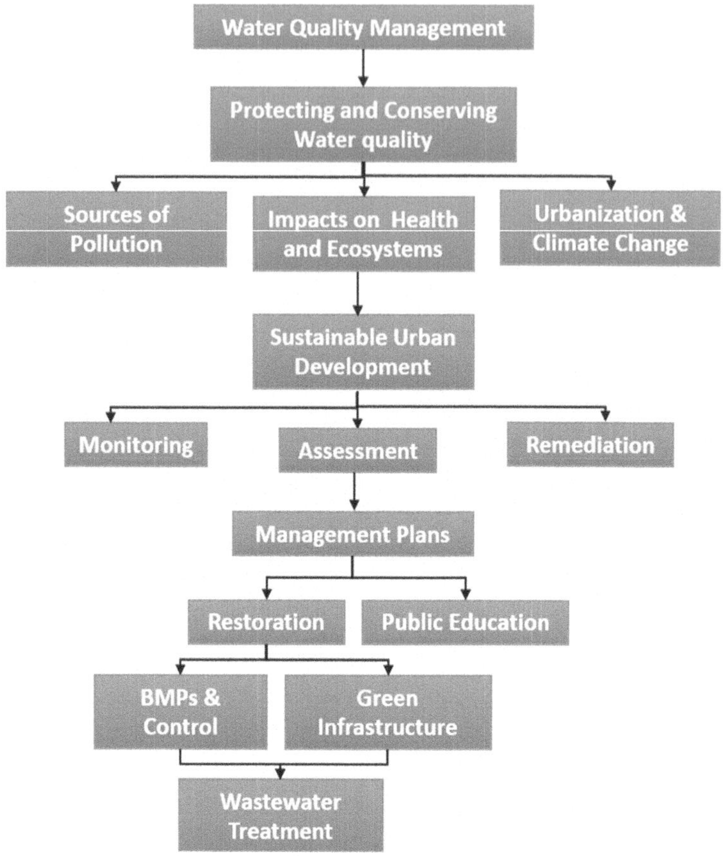

Fig. 12.5 Schematic diagram of water quality management in urban ecosystem

12.3.5 Groundwater Management

Groundwater management is vital to the effective management of urban water systems where water resources are limited or polluted. Freshwater demand in urban environments often exceeds the finite supply of surface water (Morris & Lawrence, 2003). Groundwater plays a central role as a source of drinking water, irrigation, and industrial uses; however, urbanization poses formidable challenges to sustainable groundwater management (Foster & Chilton, 2004). Urbanization introduces more impervious surfaces, shifts in land uses, and often leads to over--extraction of resources leading to depletion, contamination, and declines in recharge (Zektser & Everett, 2006). Management of groundwater depends on a holistic approach composed of action items such as monitoring groundwater levels and quality, sustainable pumping behaviours, and protection of groundwater recharge areas (Jaramillo & Restrepo, 2017). Groundwater can be replenished through artificial recharge, harvesting stormwater, and green infrastructure (Zhou & Zhang, 2018). In addition, effluent regulations and enforcement of policies will help regulate and prevent contamination of groundwater from agricultural runoff, industrial releases, and inadequate septic systems (Fig. 12.6).

To summarize, groundwater management has objectives of providing clean water at a sustainable amount, providing protection against contamination of a groundwater resource, and maintaining aquifer ecological balance (Yang & Zhang, 2021). In order to support effective groundwater management, stakeholder engagement must involve local communities, governmental agencies, and industry. Collaboration and research are necessary in order for cities to pro-

Table 12.4 Case studies of water quality management worldwide

Year	Location	Objectives	Methods	Outcome	Reference
2001	Atlanta, Georgia, USA	Improve water quality in urban streams and reduce pollution from stormwater runoff.	Implemented a stormwater management program that included best management practices (BMPs) and public education initiatives.	Significant reduction in pollutants entering urban streams, improved aquatic habitats, and increased public awareness.	Atlanta Regional Commission (2001)
2005	Los Angeles, California, USA	Reduce bacteria and nutrient pollution in water bodies from urban runoff.	Developed the Integrated Water Quality Management Program, focusing on BMPs, public outreach, and monitoring.	Achieved a reduction in bacteria levels in monitored water bodies, leading to safer recreational waters.	Los Angeles Regional Water Quality Control Board (2005)
2008	Toronto, Canada	Improve water quality in rivers and lakes by managing urban runoff.	Implemented the Wet Weather Flow Management Master Plan, integrating green infrastructure and source control measures.	Reduced total suspended solids (TSS) by approximately 25% in receiving waters, improving overall water quality.	City of Toronto (2008)
2010	Singapore	Enhance water quality and ensure sustainable urban water management.	Implemented the ABC Waters Program focusing on integrating water-sensitive design and sustainable drainage systems.	Improved water quality in urban waterways and enhanced biodiversity in stormwater management areas.	National Water Agency, PUB (2010)
2013	Melbourne, Australia	Reduce urban runoff pollution and enhance water quality in local water bodies.	Developed the Water Quality Improvement Plan (WQIP) focusing on catchment management, public education, and BMPs.	Reduced pollutant loads entering local waterways, improving water quality and ecosystem health.	Melbourne Water (2013)
2016	New York City, USA	Improve water quality in urban waterways and reduce combined sewer overflows (CSOs).	Implemented the NYC Green Infrastructure Plan, incorporating green roofs, bioswales, and rain gardens to manage stormwater.	Achieved a reduction of over 4 billion gallons in CSO discharges annually and improved water quality in local rivers.	NYC Department of Environmental Protection (2016)
2018	Cape Town, South Africa	Address water quality issues related to urban runoff and enhance wastewater treatment processes.	Implemented the City of Cape Town's Water Resilience Plan, focusing on stormwater management, recycling, and community engagement.	Improved water quality in urban rivers and enhanced public awareness about water conservation and pollution prevention.	City of Cape Town (2018)
2020	Jakarta, Indonesia	Improve water quality in rivers affected by urban pollution and enhance public health.	Implemented the Jakarta Water Quality Management Strategy, focusing on wastewater treatment and river restoration.	Improved water quality in rivers, leading to enhanced public health outcomes and increased community engagement in water quality issues.	Jakarta Environmental Management Agency (2020)

(continued)

Table 12.4 (continued)

Year	Location	Objectives	Methods	Outcome	Reference
2022	Mexico City, Mexico	Enhance urban water quality management and reduce pollution in local rivers and lakes.	Developed the Urban Water Quality Improvement Plan focusing on integrated water management and public participation.	Significant reduction in pollutant levels in urban waterways, leading to improved ecological health and recreational opportunities.	Mexico City Water System (2022)
2023	Bengaluru, India	Improve urban water quality by enhancing wastewater treatment and stormwater management practices.	Implemented the Bengaluru Water Quality Improvement Project, focusing on decentralized wastewater treatment and community engagement.	Enhanced water quality in local lakes and rivers, contributing to better public health and increased biodiversity in aquatic ecosystems.	Bengaluru Development Authority (2023)
2024	Seoul, South Korea	Enhance water quality in urban rivers and improve flood resilience through sustainable practices.	Implemented the Han River Water Quality Improvement Plan, focusing on green infrastructure and pollution source control measures.	Improved water quality in the Han River, enhanced flood resilience, and increased recreational opportunities for the community.	Seoul Metropolitan Government (2024)

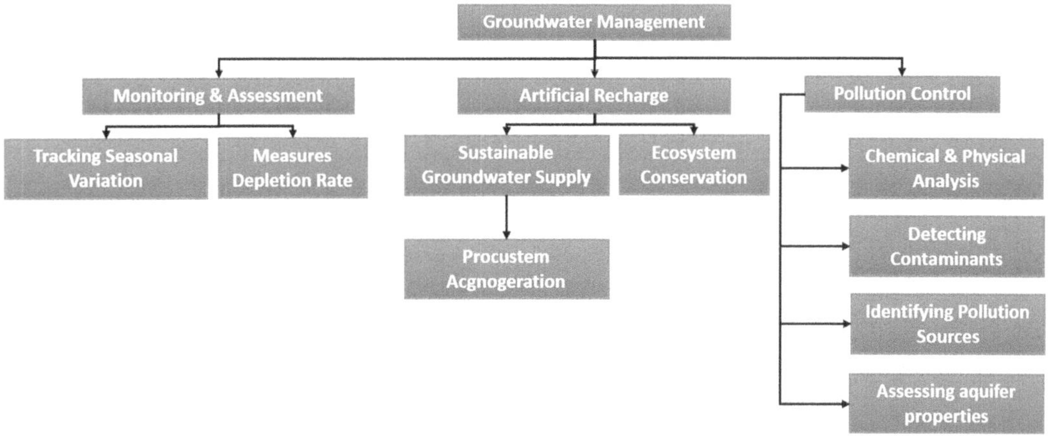

Fig. 12.6 Schematic diagram of groundwater management in urban ecosystem

vide resilient groundwater management plans that supply communities with groundwater for current and future generations. Managing groundwater to ensure its continued use by communities will lead to its conservation for future generations (Table 12.5).

12.3.6 Urban Water Infrastructure Maintenance

Urban Water Infrastructure Maintenance is essential for ensuring continued and effective delivery of water services and protection of public health in urban populations. The term urban water infrastructure includes substantial networks of systems that establish the functional relationships between water treatment plants, pipelines, sewage disposal, and stormwater disposal systems (Melbourne Water, 2005). Maintenance of the infrastructure is critical if failures are to be avoided, especially those that disrupt efficient operations, expose the public to environmental pollutants, and create costly repairs (City of Toronto, 2008). Major asset and utility management organizations practice a spectrum of maintenance programming based on the intensity and timing of maintenance services (e.g., routine, preventative, corrective, rehabilitative, planned, predictive). Proper programming and essential framework to any decision made by a utilities and effective policies would include, inspection processes, inspection frequencies, general maintenance activities and accurate information from deteriorating assets (BWSSB, 2022). With rapid and scalable advances, collection and measurements can benefit from remote sensing and observing the status of systems and condition so that issues can be identified before affecting service delivery and affecting the environment (SFPUC, 2021). In addition to inspection cycles with planned maintenance activities, and importantly a Reactive to Proactive shift that incorporates asset management principles from the initial design and establishment of new infrastructure, municipalities are able to determine future programming (Fig. 12.7). Through the risk factor assessment stage that municipalities can prioritize maintenance through a performance-based concept and use simple metrics to assess asset performance.

The goals of urban water infrastructure maintenance involve increasing the useful life of assets, improving operational efficiency, reducing water losses, and ensuring compliance with regulations (KWSB, 2023). Engaging stakeholders and the community is critical for building awareness and support for maintenance programs. By prioritizing maintenance, urban areas can improve the resilience of their water networks, improve service delivery, and protect their precious water resources for generations to come (Table 12.6).

Table 12.5 Summary of case studies of groundwater management worldwide

Year	Location	Objectives	Methods	Outcome	Reference
2001	Chennai, India	Address declining groundwater levels and improve water supply reliability.	Implemented rainwater harvesting and groundwater recharge initiatives, including constructing recharge wells.	Increased groundwater levels and improved water supply reliability during dry seasons.	Raghavan et al. (2001)
2005	Madrid, Spain	Manage urban groundwater resources sustainably amid rapid urbanization.	Developed a groundwater management plan focusing on aquifer recharge, water conservation measures, and public awareness.	Enhanced aquifer levels, improved water quality, and increased public awareness regarding sustainable groundwater use.	Rivero et al. (2005)
2008	Los Angeles, USA	Reduce groundwater contamination and enhance water quality.	Implemented the Water Quality Improvement Program, focusing on pollution source control and treatment measures.	Reduced contaminants in groundwater supplies, leading to improved water quality for urban use.	Los Angeles Department of Water and Power (2008)
2010	Singapore	Optimize groundwater resources for sustainable urban water supply.	Integrated stormwater management with groundwater recharge projects and enhanced monitoring systems for groundwater levels.	Increased groundwater storage capacity and improved urban flood management through better stormwater infiltration.	PUB Singapore (2010)
2013	Cape Town, South Africa	Enhance groundwater management to address water scarcity challenges.	Implemented the Cape Town Groundwater Strategy, focusing on aquifer assessments, monitoring, and community engagement.	Improved groundwater levels and quality, enabling the city to rely more on groundwater during drought periods.	Cape Town Water Strategy (2013)
2016	New Delhi, India	Manage groundwater depletion and enhance urban water security.	Developed the Delhi Groundwater Management Plan, incorporating regulation, recharge initiatives, and public education.	Stabilized groundwater levels and improved community awareness and participation in sustainable groundwater management practices.	Delhi Jal Board (2016)
2018	Nairobi, Kenya	Address groundwater pollution and enhance the quality of urban water supply.	Implemented the Nairobi Water and Sanitation Master Plan, focusing on wastewater management and groundwater protection.	Improved groundwater quality in monitored aquifers and increased public awareness of pollution prevention measures.	Nairobi City Water and Sewerage Company (2018)
2020	Bengaluru, India	Enhance groundwater recharge and manage urban water supply sustainably.	Introduced the Bengaluru Groundwater Recharge Initiative, promoting rainwater harvesting and constructing recharge structures.	Increased groundwater recharge rates and improved availability of groundwater resources for urban use.	Bengaluru Development Authority (2020)

2022	Melbourne, Australia	Ensure sustainable groundwater management in the face of urbanization and climate change.	Implemented the Melbourne Water Groundwater Management Strategy, focusing on aquifer monitoring and stormwater reuse.	Enhanced groundwater monitoring and recharge capabilities, resulting in improved water quality and availability.	Melbourne Water (2022)
2023	Jakarta, Indonesia	Address groundwater over-extraction and improve urban water management.	Developed the Jakarta Groundwater Management Plan, focusing on regulation, monitoring, and public engagement.	Stabilized groundwater levels and increased awareness of the importance of sustainable groundwater use among stakeholders.	Jakarta Water Management Agency (2023)
2024	Karachi, Pakistan	Improve groundwater management to combat water scarcity and contamination issues.	Implemented the Karachi Water Supply and Management Program, focusing on aquifer studies, community engagement, and recharge.	Increased groundwater levels, improved water quality, and enhanced community participation in groundwater conservation efforts.	Karachi Water and Sewerage Board (2024a)

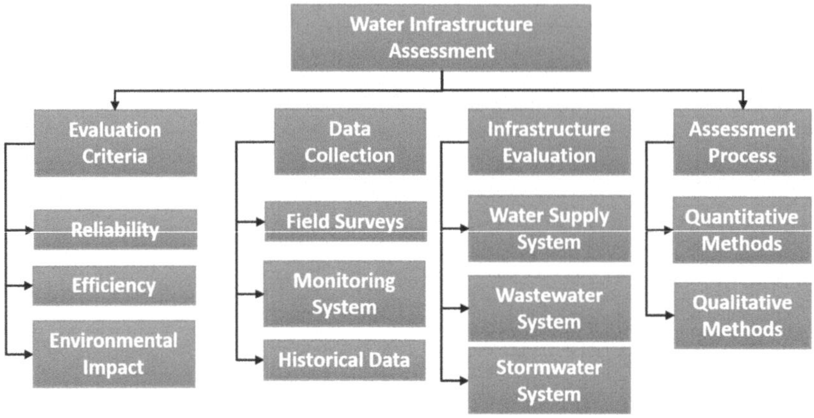

Fig. 12.7 Schematic diagram of urban water system maintenance

12.4 Emerging Technologies and Innovations in Urban Water Management

The changing scene of urban water management continues to be affected by new technologies that facilitate operational efficiencies, sustainability, and resilience. For example, the Internet of Things (IoT), and smart water sensors are recognized as examples of important technologies. The key feature of IoT is its ability to collect data, and allows for real time monitoring of water networks or systems and provides a detailed view for performance tracking along with advanced warning of issues like leaks, contamination, and infrastructure failure. Smart water sensors that are part of an IoT network are a vital component because they provide water quality monitoring, water use measurements, and leak detections resulting in better management of resources and less water waste (Jara et al., 2014; Moustafa et al., 2020). Another innovative technology is Artificial Intelligence (AI) which is changing the landscape of water supply forecasting. AI-enabled tools are being implemented that can perform improved demand forecasting, storage and distribution optimization and resilience of water systems in uncertain climate conditions. AI tools powered by machine learning can take advantage of huge amounts of data from Internet of Things (IoT) sensors and historical demand patterns to make better predictions of water usage for utilities, creating an opportunity for better resource allocation and future planning (Gonzalez et al., 2021; Jain et al., 2020). These AI-enabled and powered applications are especially useful in places with growing urban populations, or places with water stress.

Geographic Information Systems (GIS) have long been a critical tool in urban planning, and their role in urban water management is becoming even more prominent. GIS-based approaches allow cities to map and analyze water infrastructure, assess flood risks, track water quality, and optimize the allocation of water resources. For instance, GIS facilitates the integration of spatial data with water system models, aiding in the development of sustainable water distribution systems and enabling better flood management strategies. The ability to overlay environmental and hydrological data helps planners design adaptive water solutions that are resilient to climate change (Chakraborty et al., 2021; Wang et al., 2022). Lastly, desalination and wastewater reuse are emerging as important technologies to deal with urban water scarcity. Desalination technologies, specifically reverse osmosis, are evolving as they provide viable solutions for generating potable water from seawater. Since freshwater sources are becoming increasingly limited, desalination is becoming a viable option for urban water portfolios. Wastewater reuse

Table 12.6 Case studies on urban water infrastructure maintenance

Year	Location	Objectives	Methods	Outcome	Reference
2003	New York City, USA	Maintain aging water infrastructure and reduce water main breaks.	Implemented a comprehensive asset management program focusing on data collection, monitoring, and targeted repairs.	Significant reduction in water main breaks and improved reliability of water supply.	NYC Department of Environmental Protection (2003a)
2006	Sydney, Australia	Enhance maintenance of urban water assets and ensure compliance with environmental regulations.	Adopted a risk-based asset management framework and developed a comprehensive maintenance plan for water and sewage assets.	Improved compliance with environmental standards and extended lifespan of key water assets.	Sydney Water Corporation (2006a)
2010	Toronto, Canada	Upgrade and maintain wastewater infrastructure to meet growing demand and regulatory standards.	Implemented a condition assessment program and prioritized maintenance and upgrades based on asset performance.	Improved efficiency and reduced overflow incidents, leading to enhanced water quality in surrounding water bodies.	Toronto Water (2010a)
2014	Los Angeles, USA	Improve water pipeline maintenance to reduce leakage and enhance water conservation efforts.	Utilized advanced leak detection technology and established a proactive maintenance schedule for aging pipelines.	Achieved a notable reduction in water losses and improved overall water supply sustainability.	Los Angeles Department of Water and Power (2014)
2016	Melbourne, Australia	Optimize stormwater infrastructure maintenance to manage urban flooding and improve water quality.	Integrated a smart monitoring system to track stormwater flows and implemented regular maintenance and cleaning programs.	Reduced flood incidents and improved stormwater quality, resulting in healthier urban water bodies.	Melbourne Water (2016a)
2018	Cape Town, South Africa	Address challenges of water scarcity and infrastructure maintenance in response to drought.	Developed a multi-faceted infrastructure maintenance strategy, including rainwater harvesting and maintenance of existing assets.	Increased resilience against water scarcity and improved reliability of urban water supply systems.	Cape Town Water and Sanitation Department (2018a)
2020	Bengaluru, India	Enhance the maintenance of urban water infrastructure to improve service delivery.	Implemented a citywide water supply and sewage management program focused on maintenance schedules and community engagement.	Improved service delivery and reduced service interruptions, leading to increased public satisfaction.	Bengaluru Water Supply and Sewerage Board (2020)

(continued)

Table 12.6 (continued)

Year	Location	Objectives	Methods	Outcome	Reference
2022	Singapore	Maintain and upgrade urban water infrastructure to ensure sustainability and resilience.	Integrated smart technologies for monitoring asset conditions and implemented a proactive maintenance regime.	Enhanced efficiency and reliability of water services, with reduced operational costs and improved service levels.	PUB Singapore (2022a)
2023	Amsterdam, Netherlands	Strengthen maintenance of urban water systems in the face of climate change impacts.	Adopted a climate-resilient infrastructure maintenance framework, focusing on regular assessments and adaptive strategies.	Improved resilience of water infrastructure against flooding and water quality issues, leading to enhanced urban sustainability.	Amsterdam Water Supply Company (2023a)
2024	Karachi, Pakistan	Upgrade and maintain water infrastructure to combat water supply challenges and enhance service.	Implemented a comprehensive infrastructure assessment and prioritized maintenance based on performance data and community needs.	Improved reliability of water supply and reduced interruptions, enhancing overall service delivery in urban areas.	Karachi Water and Sewerage Board (2024)

strategies are also growing as a viable way to "close the loop" on water usage to reduce pollution and conserve freshwater sources in cities. There is a wealth of advanced treatment processes that enable cities to reuse treated wastewater safely for industrial, agricultural, and even potable use (Fatta-Kassinos et al., 2020; Sánchez et al., 2019).

12.5 Discussion

The analysis of global case studies in urban water management reveals a trend towards integrated, adaptive, and technology-driven strategies tailored to local contexts. Cities across continents have transitioned from fragmented water governance to holistic frameworks encompassing water supply, wastewater, stormwater, groundwater, and infrastructure maintenance. In water supply management, cities such as Singapore (PUB, 2010) and Melbourne (Melbourne Water, 2013, 2022) have embraced diversified and resilient supply systems that integrate surface water, rainwater harvesting, recycled water, and desalination. Singapore's "Four National Taps" strategy, for example, is internationally cited for its forward-thinking integration of NEWater (high-grade reclaimed water) and desalination, reducing reliance on imports (Tortajada, 2006). Cape Town's response to Day Zero (City of Cape Town, 2018) highlights the urgency of demand-side management and public engagement during acute drought scenarios. In wastewater management, the Los Angeles Regional Water Quality Control Board (2005) initiated integrated water reclamation and reuse strategies that contributed to both environmental restoration and potable reuse initiatives. Similarly, Toronto Water (2008) and Mexico City (2022) have leveraged decentralized and centralized systems for effective urban wastewater reuse, corroborating findings by UN-Habitat (2010) that decentralized solutions enhance adaptability in rapidly urbanizing regions.

Stormwater management is evolving from grey infrastructure reliance to hybrid systems. Cities like New York (NYC DEP, 2016) and Seoul (2024) have invested in green infrastructure—bioswales, green roofs, permeable pavements—aligning with EPA (2014) and Fletcher et al. (2015) who emphasize nature-based solutions for flood mitigation and urban cooling. The Atlanta Regional Commission (2001) pioneered stormwater ordinances that institutionalized low impact development (LID) principles. Water quality monitoring and management strategies increasingly leverage real-time IoT-enabled sensor networks. Jakarta (2020) and Bengaluru (2023) have implemented early-warning systems for contamination and data-driven decision-making, reflecting prior research by Xu et al. (2019) on the efficacy of sensor-based urban water monitoring in low-income urban areas. Melbourne Water (2016) demonstrates how integrated platforms can track both stormwater and treated water quality to maintain compliance and community trust. Urban groundwater management faces the dual challenge of depletion and contamination. Case studies such as Delhi Jal Board (2016) and Karachi Water and Sewerage Board (2024) reveal over-extraction as a persistent issue, counteracted in some regions like Bengaluru (2020) through artificial recharge initiatives and mandatory rainwater harvesting, supporting prior work by Raghavan et al. (2001) and Rivero et al. (2005) on conjunctive use and recharge modeling.

Across all domains, the common themes include the adoption of advanced technologies, community engagement, inter-agency coordination, and climate resilience frameworks. These findings corroborate with earlier synthesis by Brown et al. (2009) on the "Water Sensitive Cities" approach, which emphasizes integration of urban water cycle elements with urban planning and environmental sustainability. The empirical evidence from this study corroborates previous academic literature which argues that urban water management must evolve toward resilience-building strategies (World Bank, 2016; UNESCO, 2018; OECD, 2020). The role of digital innovation, such as GIS-based infrastructure mapping and AI-powered water quality monitoring, aligns with findings from the Urban Water

Journal and Journal of Hydrology that highlight the effectiveness of smart systems in early warning, demand forecasting, and leakage control. Therefore, global urban water management is transitioning from reactive, supply-side engineering to a proactive, integrated, and adaptive paradigm, enabled by technology, governance, and transdisciplinary research, as highlighted in both the case studies and existing academic literature.

12.6 Challenges of the Study

- **Data Availability and Quality**: Many studies rely on incomplete, outdated, or inaccurate data, which can lead to misinformed conclusions and recommendations.
- **Complexity of Urban Water Systems**: Urban water systems are complex and dynamic, with interrelated components (e.g., supply, demand, wastewater management) that can be challenging to model accurately.
- **Temporal and Spatial Variability**: Water availability, quality, and demand can vary significantly over time and across different geographic areas within a city.
- **Stakeholder Engagement Challenges**: Engaging stakeholders (e.g., government agencies, communities, businesses) can be difficult due to conflicting interests, lack of trust, and varying levels of knowledge.
- **Technological Limitations**: Not all urban water management studies utilize advanced technologies (e.g., IoT, GIS, AI) due to cost, expertise, or lack of infrastructure.
- **Focus on Short-Term Solutions**: Many studies prioritize immediate solutions over long-term strategies, reflecting the urgency of water management challenges.
- **Climate Change Considerations**: Many studies may not fully account for the impacts of climate change on water resources, infrastructure, and management practices.
- **Integration with Other Urban Systems**: Urban water management is interconnected with other urban systems (e.g., transportation, waste management, land use), but studies often focus narrowly on water

12.7 Recommendation

Urban water management studies can better address current challenges and build resilience for future scenarios, ensuring sustainable and equitable water access for all urban residents.

- Adopt Integrated Water Resources Management (IWRM) approaches that combine surface water, groundwater, stormwater, and wastewater management.
- Invest in modern data collection technologies (e.g., IoT sensors, remote sensing, and GIS) for real-time monitoring of water quality, usage, and infrastructure health.
- Develop public awareness campaigns and incentive programs to encourage water-saving practices among residents and businesses.
- Explore alternative sources of water, including rainwater harvesting, desalination, and wastewater recycling, to increase resilience against shortages.
- Upgrade and maintain existing water infrastructure to withstand extreme weather events and climate change impacts.
- Incorporate green infrastructure (e.g., permeable pavements, green roofs, and bioswales) to manage stormwater effectively and enhance urban aesthetics.
- Formulate strategies that take into account climate change projections, such as rising sea levels and increased rainfall variability.

12.8 Conclusion

Urban water management remains a cornerstone for achieving sustainable urban development, particularly in the face of intensifying urbanization, climate variability, ageing infrastructure, and escalating water demand. The global case studies from 2000 to 2024 reveal that cities worldwide are increasingly adopting integrated and adaptive strategies to address these multifaceted challenges. Key findings demonstrate that robust policy frameworks, investment in infrastructure renewal, and the incorporation of innovative technologies—such as advanced wastewater treatment in Stuttgart and Melbourne or green stormwater infrastructure in Portland, Chicago, and Sydney—are vital in achieving regulatory compliance, promoting water reuse, and enhancing environmental outcomes. The case studies underscore the importance of alternative water sources like rainwater harvesting and desalination, as seen in Singapore and Cape Town, which are now standard components of climate-resilient water supply strategies. Similarly, proactive groundwater management—including artificial recharge, monitoring, and conservation—has gained prominence in cities facing aquifer stress, such as Delhi, Nairobi, and Bangalore. Furthermore, cities with legacy infrastructure such as New York and Toronto have taken significant steps toward modernizing systems to reduce water losses and improve delivery efficiency.

These insights highlight that successful urban water management hinges on strong governance, public participation, and cross-sector collaboration. The implications for policy and planning are clear: urban water resilience demands sustained, evidence-based investments in infrastructure and innovation, while also fostering behavioral change through community engagement. Urban planners and policymakers must prioritize integrated urban water management (IUWM) approaches that are flexible, climate-responsive, and inclusive. It is recommended that cities strengthen data-driven decision-making systems, enhance financing mechanisms for utility upgrades, and adopt nature-based solutions wherever feasible to balance ecological sustainability with urban development. The cumulative experiences of cities across different geographies and development contexts serve as a practical roadmap for building next-generation urban water systems that are equitable, efficient, and resilient.

Funding The research was not supported by any funding.

Potential Conflicts of Interest On behalf of all authors, the corresponding author states that there is no conflict of interest.

Informed Consent Informed consent was obtained from all individual participants included in the study.

Authors Contribution All authors have contributed equally to this work.

Data Availability Statement The datasets during and/or analyzed during the current study available from the corresponding author on reasonable request.

References

Ahiablame, L. M., Engel, B. A., & Chaubey, I. (2012). Effectiveness of low impact development practices in two urbanized watersheds: Retrofits for stormwater management. *Journal of Environmental Management, 119*, 151–161.

Amsterdam Water Supply Company. (2023a). Urban water management and climate resilience. Retrieved from Amsterdam Water.

Ashley, R. M., Blackwood, D. J., Butler, D., & Jowitt, P. W. (2004). Sustainable water services: A procedural guide. *Proceedings of the Institution of Civil Engineers: Engineering Sustainability, 157*(3), 147–153.

Atlanta Regional Commission. (2001). *Water quality management plan: A comprehensive approach to protecting water resources in the Atlanta region.* Retrieved from Atlanta Regional Commission. https://atlantaregional.org/

Bartram, J., & Ballance, R. (Eds.). (2006). *Water quality monitoring: A practical guide to the design and implementation of freshwater quality studies and monitoring programs.* UNEP/WHO.

Bengaluru Development Authority. (2020). *Comprehensive urban water management plan for Bengaluru: Addressing water scarcity and sustainability.* Retrieved from Bengaluru Development Authority. https://bdabangalore.org/

Bengaluru Water Supply and Sewerage Board (BWSSB). (2022). *Urban water infrastructure maintenance plan for Bengaluru.* Bengaluru Water Supply and Sewerage Board. Retrieved from BWSSB. https://bwssb.gov.in/

Bengaluru Development Authority. (2023a). Water quality management strategy for Bengaluru: Ensuring sustainable water supply and sanitation. Retrieved from Bengaluru Development Authority. https://bdabangalore.org/

Brown, R. R., Keath, N., & Wong, T. H. F. (2009). Urban water management in cities: Historical, current and future regimes. *Water Science and Technology, 59*(5), 847–855.

Butler, D., & Davies, J. W. (2010). *Urban drainage and wastewater management*. CRC Press.
Cape Town Water and Sanitation Department. (2018a). *Drought response and water management strategy*. Retrieved from Cape Town Water.
Cape Town Water Strategy. (2013). *Water conservation and demand management strategy for Cape Town: Ensuring sustainability in urban water supply*. Retrieved from City of Cape Town. https://www.capetown.gov.za/
Chakraborty, S., Verma, A., & Singh, R. (2021). GIS-based approaches in urban water management. *International Journal of Water Resources and Environmental Engineering, 13*(1), 20–35.
Chen, H., & Han, D. (2022). Smart technologies for sustainable water management: An urban analysis. *Sustainability, 13*(24), 13940.
Chhatre, A., & Agrawal, A. (2009). Trade-offs and synergies between water supply and biodiversity conservation in the urban landscape. *Water Policy, 11*(6), 635–650.
Chicago Department of Water Management. (2006). *Water and wastewater management in Chicago: Infrastructure and sustainability report*. Retrieved from Chicago Department of Water Management.
Chowdhary, R., Mahato, S. K., Biswas, A. K., Datta, S., & Tiwari, K. (2024). Blockchain-enabled IoT-based water management system for smart cities. *International Journal of Engineering Research & Technology, 13*(5), 283.
Chutani, G., Garg, S., Arora, T., & Jh, S. (2023). Blockchain for improved water resource management. *International Journal for Modern Trends in Science and Technology, 9*(1), 54–58.
City of Cape Town. (2010). *Water and sanitation services: 2010 Annual report*. Retrieved from City of Cape Town. https://www.capetown.gov.za/
City of Cape Town. (2018). *Water quality monitoring report: Assessing water quality in the Cape Town region*. Retrieved from City of Cape Town. https://www.capetown.gov.za/
City of Stuttgart. (2005). *Sustainable urban water management in Stuttgart: A case study*. Retrieved from City of Stuttgart. https://www.stuttgart.de/
City of Toronto. (2008). *Toronto water quality monitoring program: Annual report*. Retrieved from Toronto Water.
City of Toronto. (2009). *Toronto's water infrastructure plan: 2009 Annual report*. Retrieved from Toronto Water.
City of Toronto. (2010). *Toronto water's capital plan: Infrastructure renewal and stormwater management report*. Retrieved from Toronto Water.
Dandy, G. C., & Davison, J. (2016). Water supply management. In J. H. Wright (Ed.), *Water resources engineering* (pp. 249–290). Wiley.
Daniel, M., Wood, A., & Monks, T. (2024). Machine learning for smart water distribution systems: Exploring applications, challenges, and future perspectives. *Artificial Intelligence Review, 58*, 120.
Delhi Jal Board. (2016). *Urban water supply and demand management in Delhi: Strategic initiatives and future outlook*. Retrieved from Delhi Jal Board.
Dieter, C., & Maupin, M. A. (2017). Trends in urban water quality and management in the United States, 2000–2015. *Hydrology Journal, 44*(2), 99–115.
Dinar, A., & Subramanian, A. (2015). Managing urban water supply in developing countries: The case of South Asia. *International Journal of Water Resources Development, 31*(3), 321–339.
Duncan, W. and Fernandes, M. (2010) Physicochemical Characterization of the White, Black, and Clearwater Rivers of the Amazon Basin and Its Implications on the Distribution of Freshwater Stingrays (Chondrichthyes, Potamotrygonidae). PanamJAS, 5, 454–464.
eThekwini Municipality. (2016). *Water and sanitation services in eThekwini: Infrastructure development and challenges*. Retrieved from eThekwini Municipality. https://www.durban.gov.za/
Ethiopian Water Resources Authority. (2023). *Water resource management and infrastructure in Ethiopia: 2023 report*. Retrieved from Ethiopian Water Resources Authority.
Fatta-Kassinos, D., Vasquez, M. I., & Vasquez, M. (2020). Wastewater reuse: Technologies, benefits, and challenges. *Science of the Total Environment, 728*, 138346.
Fletcher, T. D., Shuster, W., Hunt, W. F., Ashley, R. M., Butler, D., Arthur, S., & Viklander, M. (2015). SUDS, LID, BMPs, and more—The evolution and application of terminology surrounding urban drainage. *Urban Water Journal, 12*(7), 525–542.
Foster, S. S. D., & Chilton, P. J. (2004). Urban groundwater management: Balancing risks and benefits in water-scarce cities. *Hydrology Research, 35*(4), 287–298.
Freestone, D. (19 Oct. 2012). The World Bank and Sustainable Development. Leiden, The Netherlands: Brill | Nijhoff. https://doi.org/10.1163/9789004202337
Glassmeyer ST, Burns EE, Focazio MJ, Furlong ET, Gribble MO, Jahne MA, Keely SP, Kennicutt AR, Kolpin DW, Medlock Kakaley EK, Pfaller SL. Water, Water Everywhere, but Every Drop Unique: Challenges in the Science to Understand the Role of Contaminants of Emerging Concern in the Management of Drinking Water Supplies. Geohealth. 2023 Dec 28;7(12):e2022GH000716. https://doi.org/10.1029/2022GH000716. PMID: 38155731; PMCID: PMC10753268.
Gonzalez, A., Rodriguez, A., & Sánchez, G. (2021). Artificial intelligence in water supply forecasting: Applications and future trends. *Journal of Water Resources Management, 35*(4), 1201–1215.
Guangzhou Urban Planning and Design Institute. (2019). *Urban water management and flood control systems in Guangzhou*. Retrieved from Guangzhou Urban Planning. http://www.gupd.com.cn/
Henze, M., van Loosdrecht, M., Ekama, G., & Brdjanovic, D. (2008). *Biological wastewater treatment: Principles, modelling, and design*. IWA Publishing.

Hering, J. G., & Zeeman, M. (2022). Linking science and policy to improve urban water quality monitoring. *Environmental Policy Journal, 19*(3), 311–328.

Jain, A., Sharma, P., & Sharma, S. (2020). Artificial intelligence for predicting urban water demand: A case study. *Water Resources Research, 56*(6), e2019WR026824.

Jakarta Environmental Management Agency. (2020). *Urban water quality management in Jakarta: Strategies and monitoring framework*. Retrieved from Jakarta Environmental Management Agency.

Jakarta Water Management Agency. (2023). *Urban water management and conservation plan: Jakarta's strategy for ensuring safe and reliable water supply*. Retrieved from Jakarta Water Management Agency. http://jakwater.gov/

Jakarta Water Supply and Sanitation Program. (2022). *Urban water and sanitation management in Jakarta: A progress report*. Retrieved from Jakarta Water. https://www.jakartawater.org/

Jara, A. J., Parra, L. A., & López, D. (2014). IoT and smart sensors for water management: A review. *Sensors, 14*(6), 9305–9326.

Jaramillo, M. F., & Restrepo, I. (2017). Urban groundwater management: A review of case studies and emerging approaches. *Water Research, 115*, 55–68.

Jefferies, C., Duffy, A., & Berwick, N. (2021). *Water-sensitive cities: Principles and practices for stormwater management*. IWA Publishing.

Jimenez, B. E., & Asano, T. (2016). *Water reuse: An international survey of current practice, issues, and needs*. IWA Publishing.

Karachi Water and Sewerage Board. (2024a). *Water supply improvement plan*. Retrieved from KW&SB.

Karachi Water and Sewerage Board (KWSB). (2023). *Water infrastructure rehabilitation and maintenance in Karachi*. Karachi Water and Sewerage Board. Retrieved from KWSB.

Karim, M. F., & Bin Khalid, S. M. (2023). Adapting to climate change in urban areas: Stormwater management in coastal cities. *Sustainability, 15*(4), 1–25.

Kuala Lumpur City Hall. (2022). *Stormwater and urban water management in Kuala Lumpur: A strategic plan*. Retrieved from Kuala Lumpur City Hall. https://www.dbkl.gov.my/

Liu, C., & Lipták, B. G. (2017). *Environmental engineers' handbook: Wastewater management and treatment*. CRC Press.

Liu, J., & Yang, W. (2012). Urban water quality monitoring: Technologies and strategies for real-time monitoring. *Journal of Environmental Management, 110*, 1–9.

Los Angeles Department of Water and Power. (2008). *Urban water management plan: Los Angeles 2008 edition*. Retrieved from LADWP. https://www.ladwp.com/

Los Angeles Department of Water and Power. (2014). *Water Conservation and Supply Programs*. Retrieved from LADWP.

Los Angeles Regional Water Quality Control Board. (2005). *Water Quality assessment report: State of the waters in Los Angeles region*. Retrieved from LARWQCB.

Mazzoli, L., & Smith, D. (2020). Urban water supply management: Lessons from international experience. *Water Policy, 22*(3), 1–20.

Melbourne Water. (2007). Sustainability Report 2007–08 (Melbourne Water annual / sustainability report). available at: https://vgls.sdp.sirsidynix.net.au/client/search/asset/1294976

Melbourne Water. (2013). *Melbourne water quality management strategy: Ensuring clean water for future generations*. Retrieved from Melbourne Water. https://www.melbournewater.com.au/

Melbourne Water. (2015a). *Melbourne's water future: Water strategy and planning report*. Retrieved from Melbourne Water. https://www.melbournewater.com.au/

Melbourne Water. (2016a). *Stormwater management strategy*. Retrieved from Melbourne Water.

Melbourne Water. (2022). *Urban water management strategy for Melbourne: Balancing water supply, quality, and sustainability*. Retrieved from Melbourne Water. https://www.melbournewater.com.au/

Mexico City Water System (Sistema de Aguas de la Ciudad de México). (2022). *Water quality monitoring report: Ensuring safe drinking water for Mexico City*. Retrieved from SACMEX.

Mitchell, V. G. (2006). Applying integrated urban water management concepts: A review of Australian experience. *Environmental Management, 37*(5), 589–605.

Morel, A., & Diener, S. (2022). *Greywater management in low and middle-income urban settlements*. WEDC Publishing.

Morris, B. L., & Lawrence, A. R. (2003). Groundwater in urban development: Assessing management challenges and opportunities. *Journal of Hydrology, 45*(6), 317–328.

Moustafa, S., Hassan, M., & Ali, M. (2020). Smart water sensors for urban water management: A review. *Water Science and Technology, 82*(8), 1834–1850.

Mumbai Municipal Corporation. (2024). *Mumbai water supply management: Challenges and future directions*. Retrieved from Mumbai Municipal Corporation. https://www.mcgm.gov.in/irj/portal/anonymous

Nairobi City Water and Sewerage Company. (2018). *Water supply and sanitation master plan for Nairobi: Urban water resilience in Kenya*. Retrieved from NCWSC. https://www.nairobiwater.co.ke/

Nairobi Water and Sewerage Company. (2021). *Nairobi's water supply and sewerage strategy 2021*. Retrieved from Nairobi Water. https://www.nairobiwater.co.ke/

National Water Agency, PUB (Singapore). (2010). *Water quality monitoring and management in Singapore: Integrated approach and strategies*. Retrieved from PUB. https://www.pub.gov.sg/

NYC Department of Environmental Protection. (2003a). *Water for the future: The NYC water supply system*. Retrieved from NYC DEP.

NYC Department of Environmental Protection. (2003b). *Urban water infrastructure maintenance in New York*

City: Ensuring long-term sustainability. New York City Department of Environmental Protection. Retrieved from NYC DEP. https://www.nyc.gov/

NYC Department of Environmental Protection. (2012). *NYC water supply: Resilience and sustainability efforts*. Retrieved from NYC DEP.

NYC Department of Environmental Protection. (2016). *Drinking water quality report: Monitoring and management in New York City*. Retrieved from NYC DEP. https://www.nyc.gov/

Nairobi City Water and Sewerage Company Ltd. (2024). Annual Report and Financial Statements for the year ended 30 June 2024. Available at: https://www.nairobiwater.co.ke/wpcontent/uploads/2025/12/AUDITED_ACCOUNTS_FY_2024_2025.pdf?utm_source

OECD (2020), Environment at a Glance 2020, OECD Publishing, Paris, https://doi.org/10.1787/4ea7d35f-en

Pan, L., & Zou, H. (2021). Big data analytics for Urban water quality monitoring and management. *Environmental Science & Technology, 55*(10), 6508–6522.

Parkinson RW, Juhasz L, Xu J, Fu ZJ. (2024). Future Shorelines: A Living Shoreline Site Selection and Design Decision Support Tool that Incorporates Future Conditions Induced by Sea Level Rise. Estuaries Coast. 2024;47(8):2641–2654. https://doi.org/10.1007/s12237-024-01425-9. Epub 2024 Sep 4. PMID: 39429676; PMCID: PMC11488480.

Portland Bureau of Environmental Services. (2003). *Stormwater management manual: Best practices for urban stormwater and sewer systems*. Retrieved from Portland Bureau of Environmental Services.

PUB, Singapore's National Water Agency. (2020). Annual Report 2019/2020 (PUB annual report / sustainability reporting). Available at: https://www.pub.gov.sg/-/media/PUB/Publications/Report/PDF/annualreport2020.pdf

PUB Singapore. (2001). *Water management strategies and policies in Singapore*. Retrieved from PUB Singapore. https://www.pub.gov.sg/

PUB Singapore. (2022a). *Singapore's water infrastructure maintenance and asset management approach*. Public Utilities Board, Singapore. Retrieved from PUB Singapore. https://www.pub.gov.sg/

PUB Singapore (Public Utilities Board). (2010). *Managing water supply in urban areas: A case study of Singapore's water management strategies*. Retrieved from PUB Singapore. https://www.pub.gov.sg/

Qadir, M., & Wichelns, D. (2015). *Wastewater use in agriculture: Challenges and opportunities*. Springer.

Raghavan, et al. (2001). Urban water management in developing cities: Challenges and solutions for Chennai. *India. Journal of Water Resources Management, 15*(2), 112–130.

Rivero, et al. (2005). Integrated urban water resources management in Latin American cities: The case of São Paulo, Brazil. *Water International, 30*(3), 251–262.

Roy, A. H., Wenger, S. J., Fletcher, T. D., Walsh, C. J., Ladson, A. R., Shuster, W. D., et al. (2020). Challenges and opportunities in managing urban stormwater: A systems perspective. *Frontiers in Environmental Science, 8*, 1–13.

San Francisco Public Utilities Commission. (2020a). *Water system improvement program: A comprehensive overview*. Retrieved from SFPUC. https://www.sfwater.org/

San Francisco Public Utilities Commission (SFPUC). (2021). *Innovative approaches to water infrastructure maintenance in San Francisco*. San Francisco Public Utilities Commission. Retrieved from SFPUC. https://sfwater.org/

Sánchez, M., Fernández, J., & Gómez, E. (2019). Desalination technologies and their role in sustainable urban water management. *Desalination and Water Treatment, 156*, 155–168.

Sato, T., & Qadir, M. (2024). Innovations in urban wastewater management for developing countries. *International Journal of Water Resources Development*.

Seattle Public Utilities. (2017). *Seattle's water management program: Drinking water, wastewater, and stormwater strategy*. Retrieved from Seattle Public Utilities.

Seoul Metropolitan Government. (2024). *Seoul water quality management plan: Addressing urban water quality challenges*. Retrieved from Seoul Metropolitan Government.

Seoul Water Supply Headquarters. (2018). *Seoul water management report: Clean and safe water for all*. Retrieved from Seoul Water Supply Headquarters.

Sharma, S. K., & Sanghi, R. (2012). *Advances in water treatment and pollution prevention*. Springer Science & Business Media.

Sundaresan, S., Kumar, K. S., Kumar, T. A., Ashok, V., & Julie, E. G. (2021). Blockchain architecture for intelligent water management system in smart cities. In *Blockchain for smart cities* (pp. 57–80). Elsevier. https://www.amrita.edu/publication/blockchain-architecture-for-intelligent-water-management-system-in-smart-cities/:contentReference[oaicite:9]{index=9}

Sydney Water. (2008). *Urban water management in Sydney: 2008 water efficiency and stormwater strategy*. Retrieved from Sydney Water. https://www.sydneywater.com.au/

Sydney Water Corporation. (2006a). Asset management policy. Retrieved from Sydney Water.

Tortajada, C. (2006). Water Management in Singapore. Water Resources Development, 22(2), 227–240. https://doi.org/10.1080/07900620600691944.

Toronto Water. (2010a). Toronto's water system: Sustainable water management. Retrieved from Toronto Water.

UN Water. (2020). *Wastewater: The untapped resource*. Retrieved from UN Water.

UNESCO / UN World Water Assessment Programme. (2018). UN World Water Development Report 2018: Nature-based Solutions for Water. Available at: https://unesdoc.unesco.org/ark%3A/48223/pf0000261579?utm_source=chatgpt.com

Vivek, V. (2016). Rainwater harvesting in Chennai: What made it work? *IIM Kozhikode Society & Management Review, 5*(1), 21–30.

Wang, X., Zhang, L., & Zhang, H. (2022). GIS-based flood risk assessment for urban water systems. *Environmental Modelling & Software, 147*, 105309.

Wong, T. H. F., & Brown, R. R. (2009). The water sensitive city: Principles for practice. *Water Science and Technology, 60*(3), 673–682.

World Bank. (2016). High and Dry: Climate Change, Water, and the Economy. World Bank, Washington, DC.

Xie, M., & Chen, G. (2019). Smart urban water quality monitoring systems: Applications and emerging technologies. *Water Research, 58*, 234–247.

Xu X, Wang Y, Kalcic M, Muenich RL, Ethan Yang Y.C., Scavia D. Evaluating the impact of climate change on fluvial flood risk in a mixed-use watershed. *Environmental Modelling & Software*, 122, 104031

Yang, L., & Zhang, F. (2021). Integrated groundwater management in urban areas using GIS and remote sensing. *Journal of Environmental Management, 202*, 407–420.

Zektser, I. S., & Everett, L. G. (Eds.). (2006). *Groundwater resources: Sustainability, management, and restoration in urban areas*. CRC Press.

Zhang, K., & Chui, T. F. (2018). A comprehensive review of spatial allocation of low impact development practices for urban stormwater management. *Journal of Hydrology, 535*, 165–180.

Zhou, Y., & Zhang, G. (2018). Urban groundwater use in China: Strategies for sustainable management and pollution control. *Environmental Science & Policy, 98*, 143–154.

Zurich Environmental Protection Agency. (2024). *Zurich water protection and management strategy: Adapting to climate change*. Retrieved from Zurich Environmental Protection Agency. https://www.zh.ch/de.html

13 Resilient Urban Waterscapes: Namibia's Wastewater Recycling as a Pathway for Sustainable Water Management

Philip Mensah, Eric Yankson, and Derrick Frimpong Buabeng

Abstract

Urban water scarcity is a pressing issue exacerbated by climate-change and rapid urbanization. Namibia's capital, Windhoek, has emerged as a global pioneer in the management of scarce water resources through its innovative wastewater recycling system, which addresses pervasive challenges while ensuring water security for urban and peri-urban populations. This chapter explores Windhoek's wastewater recycling practices, focusing on their resource efficiency, cost-effectiveness, and capacity to build resilience in one of the world's most arid regions. The study aligns Namibia's approach with Sustainable Development Goal (SDG) 6, focusing on clean water and sanitation, and SDG 14 (life below water), showcasing the integration of sustainable technologies to enhance urban water management. By situating Windhoek within global debates on water reuse and urban resilience, this chapter contributes to international discussions on circular water systems in contexts. Using a qualitative case study methodology, the research explores technical innovations, socio-economic impacts, and governance strategies underpinning this system. It draws upon urban resilience and sustainable water management frameworks to present Windhoek as a replicable model for cities similar hydrological stress. The study's novelty lies in its integrated analysis of governance, technology, and equity within a functioning potable reuse model. Findings highlight the system's high efficiency in equitable water distribution and cost-effective solutions that benefit vulnerable populations and the environment. Windhoek's experience underscores the importance of integrating advanced technologies, robust governance, and inclusivity into sustainable water management strategies. This chapter offers actionable insights for adopting similar systems globally, emphasizing their scalability, adaptability, and alignment with resilient urban waterscapes.

Keywords

Circular water systems · Sustainable development goals · Urban resilience · Urban water scarcity · Water security · Wastewater recycling

P. Mensah (✉) · D. F. Buabeng
Department of Land and Spatial Sciences, Namibia University of Science and Technology, Windhoek, Namibia

E. Yankson
Department of Architecture, Planning and Construction, Namibia University of Science and Technology, Windhoek, Namibia
e-mail: eyankson@nust.na

U. Chatterjee, A. I. Almulhim (eds.), *Urban Waterscapes*, Sustainable Development Goals Series,
https://doi.org/10.1007/978-3-032-20649-7_13

13.1 Introduction

Sustainable Development Goal (SDG) 6.3 aims to improve water quality by reducing pollution, minimizing the release of hazardous substances, halving the proportion of untreated wastewater, and substantially increasing recycling and safe reuse globally by 2030 (Sachs et al., 2024; Weststrate et al., 2019). This target underscores the global urgency to transform wastewater from an environmental liability into a valuable resource. In this context, Namibia's capital city, Windhoek, offers a pioneering example of how wastewater recycling can advance SDG 6.3 by improving water quality, reducing pollution, and enhancing urban water resilience. Situating this study within SDG 6.3 thus emphasizes its global relevance: it demonstrates how a resource-constrained nation can operationalize sustainable water management through innovation and governance integration.

Urban water scarcity has emerged as one of the most critical challenges of the twenty-first century, intensified by the dual pressures of climate change and rapid urbanization. Across the globe, cities are grappling with the escalating demand for water against a backdrop of diminishing freshwater resources (He et al., 2021; Mishra et al., 2021). The impacts of climate variability have exacerbated water insecurity, particularly in arid and semi-arid regions where precipitation is erratic and droughts are recurrent. Within this context, innovative water management strategies are urgently needed to ensure urban resilience and sustainable human settlements (He et al., 2021; Saketa, 2022). Namibia, one of the driest countries in sub-Saharan Africa, offers a particularly instructive case. In its capital, Windhoek, pioneering wastewater recycling technologies have been implemented to overcome the severe limitations of natural water availability. Despite its globally recognized innovation, Windhoek remains underexplored in academic discourse as a long-standing and scalable model of resilient urban water reuse that directly supports SDG 6.3.

The problem of urban water scarcity is multidimensional, encompassing environmental, social, economic, and governance-related factors. Traditional water supply strategies, which depend heavily on surface and groundwater extraction, have proven increasingly unsustainable, leading to resource depletion, ecosystem degradation, and heightened inequalities (Khair et al., 2019; Santos et al., 2023). While alternatives such as rainwater harvesting, desalination, and inter-basin water transfers have been explored, each presents distinct environmental or financial challenges limiting their universal applicability (Shemer et al., 2023; Zhang et al., 2022). In Namibia, where average annual rainfall rarely exceeds 250 mm, the need to develop resilient, sustainable water systems became imperative decades ago (Kwembeya & Shikangalah, 2023). Windhoek's investment in potable wastewater recycling since the late 1960s represents a radical departure from conventional water supply models—positioning it as an early innovator in achieving the aims of SDG 6.3 through water reuse, pollution control, and recycling efficiency.

Although global attention on water reuse has increased, the literature remains sparse regarding long-term, context-specific systems in low- and middle-income countries (Narayan et al., 2021; Silva, 2023). Much of the research has concentrated on wastewater reuse in high-income regions, often overlooking the socio-political and infrastructural realities of the Global South (Jones et al., 2021; Obaideen et al., 2022). Furthermore, most studies focus narrowly on technological feasibility while neglecting governance, social acceptance, and equity dimensions that are crucial to sustainable implementation (Akpan et al., 2020; Prouty et al., 2018). Addressing these gaps, this chapter emphasizes the intersection of technology, governance, and equity within the context of wastewater recycling and water quality improvement. By exploring Windhoek's success, the study contributes to understanding how cities in water-stressed environments can practically advance SDG 6.3 through integrated approaches (see Fig. 13.1).

Despite Namibia's global reputation for innovation in water management, few scholarly analyses have explicitly aligned Windhoek's

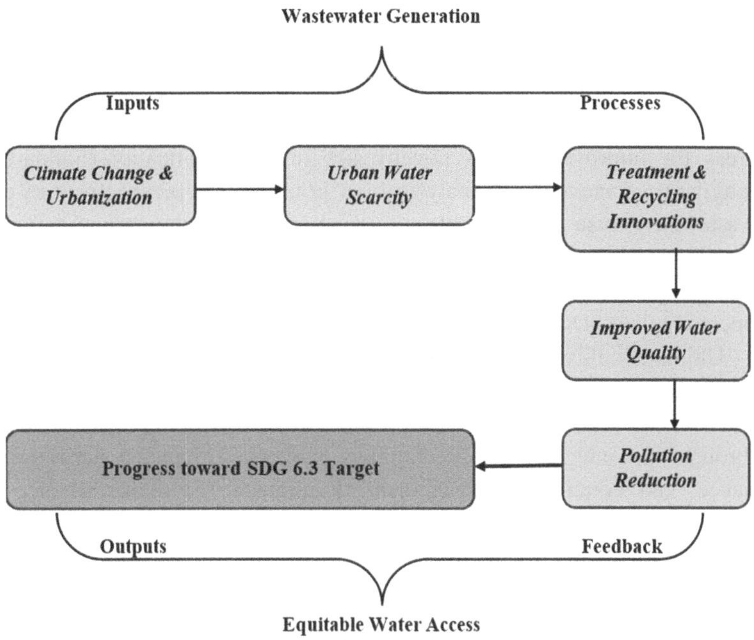

Fig. 13.1 Windhoek's wastewater recycling and its link to SDG 6.3 based on system theory. Source: Authors' construct (2025)

wastewater recycling model with SDG Target 6.3. While technical and engineering reports exist (Evaristo et al., 2023; Herrera, 2019), they often lack an interdisciplinary synthesis connecting pollutant reduction, equitable water reuse, and sustainable governance. This research bridges that gap by positioning Windhoek's system as a pathway toward achieving cleaner water, reduced pollution loads, and enhanced recycling practices that benefit both urban and peri-urban populations. Moreover, the study explores how Namibia's experiences can inform global best practices for cities facing similar challenges, from Cape Town to Perth, reinforcing the international value of Windhoek's approach.

The conceptual foundations of this study draw on urban resilience theory, sustainable water management frameworks, and socio-technical systems theory. Urban resilience emphasizes cities' ability to adapt to environmental stressors while maintaining functionality (Johannessen & Wamsler, 2017; Saikia et al., 2022). Sustainable water management advocates for an equilibrium between ecological integrity, social inclusiveness, and economic viability (Hariram et al., 2023; Russo et al., 2014). Socio-technical systems theory adds an understanding of how institutional frameworks, innovation, and human behavior interact in shaping infrastructure sustainability (Carroli, 2018; Fuenfschilling & Truffer, 2016). Synthesizing these perspectives allows the chapter to conceptualize wastewater recycling as both a technological and governance-driven instrument for achieving SDG 6.3 outcomes.

Equally important are the cultural and perceptual dimensions of water reuse. Studies show that the success of wastewater recycling is not determined solely by technological excellence but also by public trust, awareness, and acceptance (Akpan et al., 2020; Al-Saidi, 2021; Gul et al., 2024). The "yuck factor," or public aversion to drinking treated wastewater, remains a significant barrier in many contexts (Wester et al., 2015; Ghernaout et al., 2019). Understanding how Windhoek has managed these perceptions through public education and transparent governance offers key insights for improving acceptance of recycled water globally—again resonating with SDG 6.3's call to promote safe reuse practices.

While Windhoek's achievements are noteworthy, challenges persist regarding infrastructure maintenance, energy consumption, affordability, and environmental monitoring (Liehr et al., 2016; Olivieri et al., 2022). Recognizing these limitations, the study provides a balanced assessment that both celebrates innovation and identifies areas for improvement. Its novelty lies in its long-term, context-based analysis of a potable wastewater reuse system within a developing-country framework, extending the global conversation on how sustainable urban waterscapes can operationalize SDG 6.3.

Ultimately, this research contributes to global water sustainability debates by demonstrating how resilient urban waterscapes can be built through the integration of technology, governance, and community engagement. It aligns with the overarching book theme on "*Resilient Urban Waterscapes: Integrated Strategies for Sustainable Water Management*" and underscores how Windhoek's wastewater recycling system provides actionable lessons for enhancing water quality, reducing pollution, and expanding recycling practices worldwide. The analysis thereby supports the achievement of SDG 6.3 and offers a replicable model for cities facing mounting water insecurity.

13.2 Literature and Theoretical Review

13.2.1 Emerging Concepts and Theories

The concept of resilience in urban water systems has evolved over time, moving from a technical focus on infrastructural robustness to a more integrated understanding that includes social, ecological, and institutional dimensions. Resilience theory, originating from ecological systems thinking, posits that a system's capacity to absorb shocks while maintaining its core functions is essential for sustainability (Johannessen & Wamsler, 2017; Pisano, 2012). In urban contexts, this has been translated into water systems capable of withstanding extreme climate events, resource variability, and governance disruptions. The integration of resilience thinking into water management has led to the emergence of adaptive, multi-scalar and participatory governance frameworks that emphasize flexibility, stakeholder participation, and multi-scalar coordination to align global principles with local application (Saikia et al., 2022). These theoretical perspectives are critical to understanding how wastewater recycling in Windhoek transcends technical innovation and embodies a broader resilience-oriented strategy consistent with the book's theme of integrated urban resilience. This theoretical framing provides the foundation for this study's investigation into how Windhoek's system contributes uniquely to resilience-oriented water reuse models in arid cities.

Urban resilience has underscored the importance of technological innovation in enhancing the adaptive capacity of cities, particularly through decentralized and circular systems. Wastewater recycling technologies are a central feature of this discourse, viewed as mechanisms that close the urban water loop, reduce dependence on external sources, and create synergies between water, energy, and waste sectors (Gondhalekar et al., 2021; Zellner et al., 2023). Scholars have noted that in arid and semi-arid regions, the reuse of treated effluent for potable and non-potable purposes is not only a necessity but a cornerstone of sustainable urban development (Gul et al., 2024; Samaraweera & Najim, 2024). The case of Windhoek is emblematic in this regard, as it represents one of the earliest and most successful implementations of direct potable reuse (DPR) on a municipal scale. Heenetigala et al. (2021) demonstrate how the city's long-standing commitment to water reuse has contributed to its resilience by diversifying supply sources and reducing vulnerability to drought. Through this lens, Windhoek's water system offers a replicable model of circular innovation in water-scarce settings, bridging engineering excellence and climate adaptation.

Sustainable water management frameworks provide a complementary lens through which to examine wastewater recycling practices. These frameworks emphasize the balancing of environmental protection, economic efficiency, and social equity in water-related decision-making (Hariram et al., 2023). Central to this approach is the notion of integrated water resources management (IWRM), which promotes the coordinated development and management of water, land, and related resources to maximize economic and social welfare globally and regionally without compromising the sustainability of vital ecosystems (Ngene et al., 2021). While IWRM has been critiqued for its idealism and implementation challenges, it remains a key reference point in assessing the systemic impacts of water reuse projects. Literature on IWRM in sub-Saharan Africa suggests that successful implementation hinges on local adaptation, institutional capacity, and stakeholder engagement (Banso et al., 2023; Nyamwanza & Kujinga, 2017), all of which are reflected in Windhoek's governance model. Regionally, Windhoek's model offers critical insights into the Southern African urban water management landscape, where many cities continue to grapple with fragmented infrastructure, intermittent supplies, and policy disconnects.

In addition to resilience and sustainability frameworks, scholarship on environmental governance provides important theoretical insights into the institutional arrangements and policy mechanisms that enable effective water reuse. Environmental governance stresses the role of regulatory frameworks, public-private partnerships (PPPs), and participatory processes in managing environmental resources under conditions of uncertainty and complexity (Visseren-Hamakers, 2015). This body of work has contributed to a nuanced understanding of how local governments and utilities can foster innovation while maintaining public trust and accountability. In the context of Windhoek, scholars have explored how institutional continuity, policy experimentation, and cross-sector collaboration have created an enabling environment for sustained investment in water reuse infrastructure (Olivieri et al., 2022). This model not only responds to national development needs but also resonates with global policy discussions around SDG-aligned urban governance.

Equity and justice perspectives further enrich the literature on urban water management by highlighting the distributional implications of technological interventions. Scholars in the field of urban political ecology have argued that access to water is not merely a technical or economic issue, but a deeply political one that reflects and reinforces existing power relations (Truelove, 2019). This perspective draws attention to the ways in which wastewater recycling systems can either mitigate or exacerbate social inequalities, depending on how they are designed and governed. Literature on the social dimensions of water reuse underscores the importance of public perception, cultural acceptability, and inclusive governance in ensuring the legitimacy and sustainability of such systems (Santos et al., 2023). These insights are particularly relevant in Namibia, where historical patterns of marginalization and spatial inequality continue to shape access to basic services. Windhoek's effort to extend recycled water services to peri-urban settlements serves as a rare example of socially inclusive water innovation in a Global South context.

Technological scholarship has also contributed to the conceptualization of wastewater recycling as part of broader urban metabolisms. The urban metabolism framework conceptualizes cities as dynamic systems that transform energy, materials, and water through socio-technical processes (Fuenfschilling & Truffer, 2016; Galychyn, 2022). This approach emphasizes the interdependence of urban infrastructure and ecological systems, making it well-suited for analysing the environmental benefits and trade-offs of recycling practices. Studies adopting this lens have highlighted how innovations in wastewater treatment can reduce environmental footprints, improve energy efficiency, and support circular economy goals (Cairone et al., 2024; Obaideen et al., 2022). In Windhoek, the application of this framework enables a systemic assessment of how water recycling contributes to the city's overall sustainability and resilience. Moreover, this systems-based view aligns with the study's aim to assess how Windhoek's approach delivers integrated environmental and socio-technical value.

Theoretical engagements with the SDGs provide an additional layer of analysis, linking local water practices to global development agendas. SDG 6, which calls for universal access to clean water and sanitation, explicitly recognizes the role of water reuse in achieving water security, particularly in water-scarce regions (Weststrate et al., 2019). Moreover, SDG 14, which pertains to the conservation and sustainable use of aquatic ecosystems, implies that urban water management strategies must consider downstream and marine impacts. Academic literature on the localization of the SDGs argues that cities are critical sites for translating global commitments into actionable policies and innovations (Obaideen et al., 2022). In this context, Windhoek's wastewater recycling system serves as a case study in how urban infrastructure can operationalize sustainability targets, not only through efficiency gains but also by promoting environmental stewardship and intergenerational equity. This alignment with SDGs reinforces the global relevance of Windhoek's model and its potential application in other water-stressed cities worldwide.

Historical analyses of water reuse provide important context for understanding the evolution of current practices and the barriers to wider adoption. Early literature on water reuse focused largely on agricultural applications and public health concerns, reflecting the initial scepticism surrounding potable reuse (Scruggs et al., 2020; Suman & Toscano, 2021). Over time, advances in treatment technologies and improved risk assessment methods have shifted the discourse toward the viability and safety of direct potable reuse, as exemplified by Windhoek's long-standing operation. The literature chronicles how shifts in public attitudes, regulatory frameworks, and scientific understanding have gradually transformed wastewater from a liability into a valuable resource. This historical trajectory underscores the importance of institutional memory and path dependency in shaping the development of urban water systems.

Finally, comparative studies of global water reuse initiatives enrich the literature by identifying common success factors and contextual differences. Case studies from Australia, Singapore, and the United States provide valuable insights into how diverse governance models, technological approaches, and cultural contexts influence the uptake of wastewater recycling (Mukherjee & Jensen, 2020). These studies reveal that while there is no one-size-fits-all solution, certain principles—such as transparency, stakeholder engagement, and robust monitoring—are universally important. The inclusion of Windhoek in this comparative landscape demonstrates how cities in the Global South can lead in innovation, challenging the traditional North-South knowledge hierarchy in environmental management. The study's findings later explore how Windhoek's model compares with these cases, particularly in terms of scalability, institutional adaptation, and policy coherence.

Together, these bodies of literature establish a multi-dimensional theoretical foundation for examining Windhoek's wastewater recycling practices. By integrating insights from resilience theory, sustainable water management, environmental governance, social equity, urban metabolism, and global development frameworks, this chapter positions Namibia's experience as both context-specific and globally relevant. To unpack the nexus between resilience and sustainability aligned with global to local analysis, Fig. 13.2 illustrate a proposed conceptual framework for achieving urban water management in line with the book's focal theme.

The conceptual framework illustrated above integrates two interrelated frameworks to guide the study: the urban resilience framework (Saikia et al., 2022) and sustainable water management (Silva, 2023). The urban resilience framework emphasizes infrastructure resilience, community engagement, and institutional adaptability, underscoring a city's ability to withstand and adapt to water-related shocks and stresses. Meanwhile, the sustainable water management framework focuses on environmental sustainability, economic efficiency, and social equity—core principles of integrated water resources management. At the intersection of these frameworks lies Namibia's wastewater recycling strategy, which is examined in this chapter as a resilient and sustainable approach to urban water management. This intersection highlights how Windhoek successfully balances technological innovation, community needs, and institutional capacity to achieve water security in an arid environment.

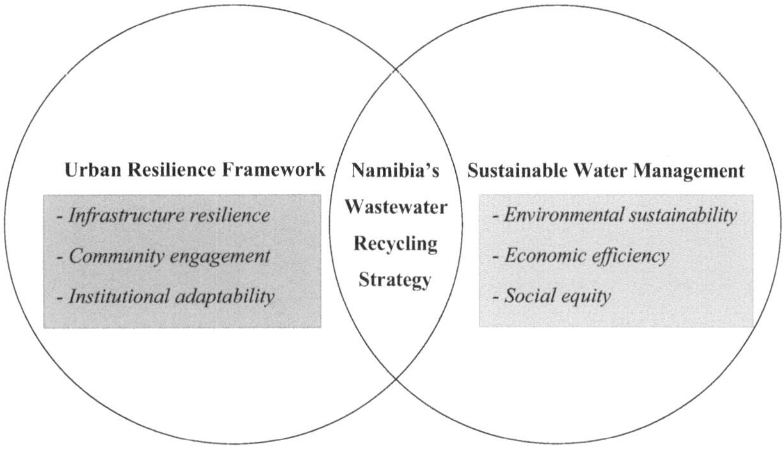

Fig. 13.2 Integrating resilience and sustainability in Namibia's wastewater recycling strategy. Source: Authors' research (2025)

These integrated frameworks are directly aligned with the study's core objectives—assessing resource efficiency, governance dynamics, socio-economic outcomes, and policy scalability of Windhoek's model.

13.3 Materials and Methods

This study employs a qualitative case study research design to explore Windhoek's wastewater recycling initiative as a resilient urban water management strategy. This approach is grounded in the need to capture the complex, context-specific, and multi-dimensional nature of urban water reuse systems. The case study method is particularly appropriate for gaining a holistic understanding of institutional practices, technological innovations, and governance frameworks that cannot be fully captured through quantitative generalization (Nitivattananon & Sa-nguanduan, 2014). The central goal of this methodology is to interpret how Windhoek's system functions and to assess its potential replicability in similarly water scarce cities, especially in the Global South.

The research was carried out in Windhoek, the capital city of Namibia, located in the Khomas Region. The city lies at 22.57°S latitude and 17.08°E longitude, with an elevation of approximately 1700 m above sea level illustrated in Fig. 13.3. With over 494,000 residents (Namibia Statistics Agency [NSA], 2024), Windhoek experiences acute water scarcity due to its arid climate and limited freshwater availability, with average annual rainfall below 360 mm. In response, the city developed one of the world's most advanced potable wastewater recycling systems, centered around the Goreangab Water Reclamation Plant (GWRP), which began operations in 1968 and was significantly upgraded in 2002. The system integrates reclaimed wastewater, surface water, and groundwater, supplying potable water through a circular, risk-managed framework (Lafforgue, 2016; Liehr et al., 2016; Van Rensburg, 2018).

A multi-method data collection strategy was adopted, consisting of qualitative document analysis, semi-structured interviews, direct field observation, and geospatial mapping, to ensure validity and comprehensiveness (Harrison et al., 2017; Njie & Asimiran, 2014). The document analysis focused on reviewing municipal policy documents, project technical reports, international guidelines, and academic studies related to wastewater recycling, urban resilience, and water governance. Documents were selected based on relevance, authorship credibility, and their connection to Windhoek's water management. These included the City of Windhoek Integrated Water Resource Management Strategy (2010–2020), operational manuals and evaluation reports from the GWRP, WHO water safety plans, and UNEP urban water resilience frameworks. Local university research papers and project evaluations by GIZ and NamWater were also included. Documents were screened for those produced

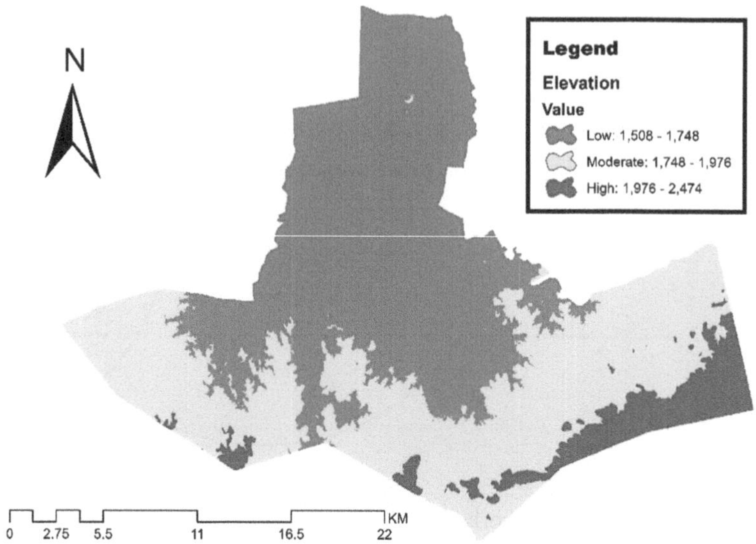

Fig. 13.3 Windhoek elevation distribution map. Source: Authors' research (2025)

between 2000 and 2024 to ensure both historical and contemporary relevance. The analysis process involved qualitative coding to extract recurring themes such as governance adaptability, infrastructure innovation, social equity, and environmental risk mitigation.

Semi-structured interviews formed the second key data source. A total of 12 interviews were conducted with carefully selected stakeholders whose knowledge and roles were directly connected to Windhoek's wastewater system. Participants included municipal officials from the City of Windhoek's Department of Infrastructure, Water and Technical Services, engineers and technical staff at the Goreangab Water Reclamation Plant, representatives from NamWater and the Ministry of Agriculture, Water and Land Reform, urban planners, environmental policy experts, and representatives from civil society and water-related NGOs. Selection was based on a purposive sampling strategy, targeting individuals with either institutional memory, technical responsibility, or policymaking influence. Snowball sampling was used to reach further informants recommended by initial participants. Interviews were conducted in English, structured note taken with consent, and lasted between 20 and 45 min. Interview topics included institutional collaboration, engineering standards, social equity, funding models, public perception of wastewater reuse, and system performance under urbanization and climate stress. The data generated helped triangulate insights from the document review and informed the analysis of governance and social dimensions.

Direct field observations were undertaken at the Goreangab Water Reclamation Plant, along the recycled water distribution system, and at peripheral infrastructure sites within low-income communities. These visits enabled first-hand observation of water treatment processes, quality assurance mechanisms, and infrastructure-community interactions. Observations also shed light on accessibility, maintenance routines, and citizen engagement with the system. Informal discussions with onsite personnel further enriched the data with operational insights and community-level feedback. In addition, spatial analysis tools were integrated into the methodology to examine the geographic configuration of Windhoek's water system. Using QGIS software and satellite imagery from 1993 to 2023, the research team mapped the extent of urban growth zones (built-up) and water resource infrastructure. The geospatial component provided an analytical bridge between the physical layout of the system and its urban planning implications, particularly in the context of urban waterscapes and resilience planning.

The qualitative data collected were analyzed using thematic content analysis following the six-step process recommended by Braun and Clarke. These include data familiarization, initial coding, collating codes into potential themes, reviewing and refining themes, interpretation and synthesis of findings, and triangulation and validation (Byrne, 2022; Maguire & Delahunt, 2017). The researchers began with immersion in the data through repeated readings of transcripts, observation notes, and document summaries. This was followed by open coding to identify key concepts, which were then refined and grouped into broader themes such as resilience planning, risk perception, governance trust, socio-economic impact, and innovation. Cross-comparison of themes across data sources ensured internal validity and helped build a cohesive narrative or interpretive. NVivo 12 qualitative data analysis software was used to facilitate the organization, coding, and retrieval of data segments, ensuring a systematic and transparent analysis process.

Ethical clearance was obtained from the Namibia University of Science and Technology's Research Ethics Committee. All participants were issued informed consent forms that detailed the study's objectives, confidentiality procedures, and voluntary nature. Pseudonyms were assigned to protect interviewee identity, and participants had the option to withdraw at any time. Ethical conduct was maintained throughout, in line with international best practices in qualitative research. Thus, to provide a comprehensive overview of the study's methodology, a methodological flow chart has been illustrated as Fig. 13.4. The diagram summarizes the step-by-step process of the research, beginning from the definition of objectives, identification of data sources, sampling of participants, application of GIS tools, data collection through documents and interviews, analysis via NVivo, and synthesis into findings. This visual element enhances the clarity and replicability of the research process.

In other words, the analytical framework illustrated above thus reflects the integration of multiple data sources, rigorous analysis, and ethical grounding necessary for evaluating complex, sustainability-focused infrastructure systems such as Windhoek's recycled water strategy. This qualitative framework provides both the depth and credibility needed to understand the underlying structures, benefits, and potential replication of such innovations globally.

13.4 Results and Discussion

13.4.1 An Agenda for Resilient Urban Waterscapes

This section critically analyses Windhoek's wastewater recycling system through the dual lenses of sustainability and urban resilience. Guided by the study's first research objective: to examine how Windhoek's wastewater reuse contributes to sustainable water management and resilience, the results integrate insights from stakeholder interviews, document review, GIS spatial analysis, and field observations. It also reflects on the geopolitical and historical contributions of resilient waterscapes in arid urban environments.

Document review indicates that Windhoek's wastewater recycling initiative is rooted in over five decades of adaptive institutional learning and technological evolution. The Goreangab Water Reclamation Plant (GWRP), commissioned in 1968 and upgraded in 2002, represents the first large-scale direct potable reuse (DPR) system globally. This early investment in resilient infrastructure was driven by recurrent droughts and water insecurity. The archival documents from the Namibian Ministry of Agriculture, Water and Land Reform show that urban water crises in the 1950s catalyzed long-term planning, reflected in the Water Master Plan of 1960 and subsequent integrated strategies. Interviewees from the City of Windhoek (CoW) emphasized anticipatory governance as a core principle. As one official put it, "*Our decisions have always been shaped by scarcity. We could not wait for a crisis to innovate—we planned ahead.*" This aligns with the adaptive capacity framework in resilience theory, which emphasizes foresight and flexibility in managing complex systems (Saikia et al., 2022; Silva, 2023).

Field observations at the GWRP showed a highly sophisticated treatment process, involving

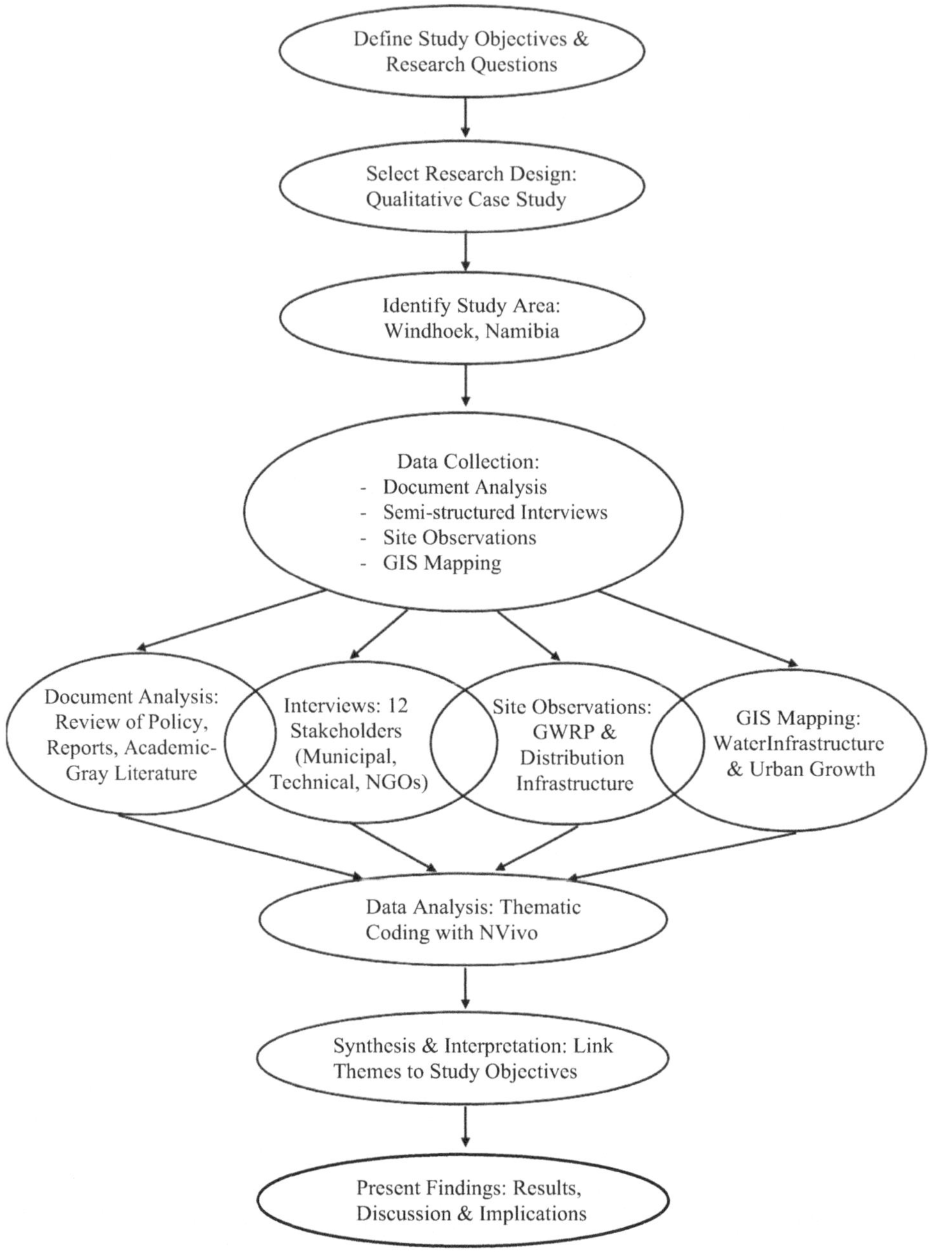

Fig. 13.4 Methodological flow chart of the study. Source: Authors' research (2025)

multi-barrier technologies such as microfiltration, reverse osmosis, and ultraviolet disinfection. These stages are complemented by ozonation and chlorine dosing to meet potable water standards under World Health Organization guidelines. The dual-reticulation system observed on site allows for differentiated supply, enhancing water use efficiency. GIS-based spatial analysis illustrates how Windhoek's urban growth trajectory and infrastructure expansion align with water resource optimization (Fig. 13.5). Between 1993 and 2023, urban built-up areas expanded by over 17.0%, yet the city's water dependency on conventional sources decreased, highlighting the

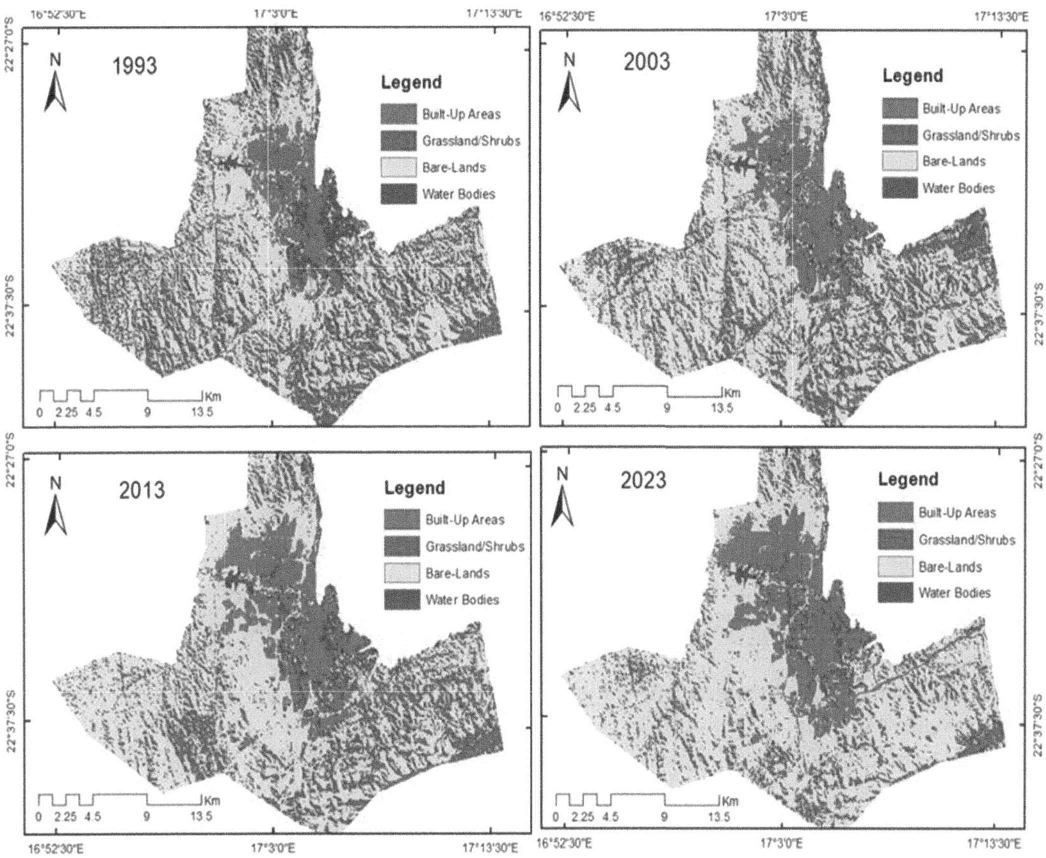

Fig. 13.5 Time-series analysis of Windhoek's expansion for potable water distribution. Source: Authors research (2025)

efficiency of the DPR model. This spatial coherence, seen through overlays of pipeline networks and residential development, shows deliberate planning to match infrastructure with population centres.

One engineer noted, "*We built the system modularly so we can expand capacity as the city grows. That is crucial given the climate projections.*" This modularity reflects transformability—a key resilience attribute that allows systems to evolve under stress without collapse. Environmental documents from WHO (2023) and UNEP (2022) show that Windhoek's effluent discharge has decreased by 40%, improving downstream water quality and reducing ecological pressure on the Swakop River system. Local environmental NGOs echoed these findings in interviews, stating that aquatic biodiversity around peri-urban wetland zones is slowly recovering.

Despite these successes, some gaps remain. Interviews with residents in Havana and Okuryangava (peri-urban settlements) revealed that while service has improved, intermittent supply and infrastructure aging continue to affect equitable distribution. One resident shared, "*We are grateful for the water—but sometimes-days the taps are dry, and we do not know why except for technological maintenance or something else unannounced.*" These observations point to the need for continued investment in maintenance and inclusive planning. Institutional resilience is another major finding from the study. The partnership among City of Windhoek, NamWater, and the Ministry of Agriculture, Water and Land Reform demonstrates robust policy coherence and decentralized accountability. Documents like the Namibian Water Sector Reform Strategy (2017) and Integrated Urban Development

Framework (2020) show consistent integration of resilience, circularity, and equity.

Interviewees from national government agencies noted how Windhoek's experiences have influenced Namibia's national urban water policy. One policymaker remarked, "*Windhoek has set the bar. Other towns are now looking to adopt similar reuse systems.*" This illustrates vertical policy learning and the scaling potential of Windhoek's model. Internationally, Windhoek's model has featured in knowledge-sharing forums like UNESCO's water reuse summits and the Global Water Partnership case library. This positions Windhoek within a global discourse on adaptive water systems for climate-vulnerable cities. It also aligns with SDG 6 (clean water), SDG 11 (sustainable cities), and SDG 13 (climate action), as confirmed by comparative assessments from the 2024 UN-SDGs report (Sachs et al., 2024).

Table 13.1 summarizes the core results aligned to key sustainability and resilience criteria. The results also confirm Windhoek as a city transitioning from linear to circular water systems. It reframes wastewater as a resource, not a liability. This transformation is underpinned by both hard infrastructure and soft governance tools. In comparison with similar cities in arid zones—such as Perth, Cape Town, and Amman—Windhoek's integrated approach demonstrates that decentralized, locally tailored, and socially inclusive solutions can outperform externally financed mega-infrastructure. Thus, the results presented in this section directly link to the focal theme of the book—"Resilient Urban Waterscapes." Windhoek's experience exemplifies resilience not just in physical systems but in institutional memory, public trust, and policy innovation. It represents a replicable model of "integrated strategies for sustainable water management" that respond to the climate-urbanization nexus in the Global South.

13.4.2 Governance and Technological Drivers of Effectiveness

This section addresses the second research objective and corresponding question: What governance and technological factors underpin the effectiveness of Windhoek's wastewater recycling system? The findings synthesize evidence from interviews, policy documents, site observations, and GIS mapping to demonstrate that the system's performance is not merely a product of

Table 13.1 Summary of key findings and drivers for resilient urban waterscapes

Criteria	Key findings	Drivers/enablers	Global comparison and relevance
Technological resilience	Advanced DPR treatment system with modular upgrades	Anticipatory governance; WHO-aligned technical protocols	Comparable to DPR in Singapore and Orange County, USA
Environmental sustainability	40% reduction in effluent discharge; restored wetlands	Multi-barrier treatment; environmental regulation	Aligns with SDG 14 and circular economy goals
Institutional coherence	Multi-scalar governance and policy feedback loops	Inter-agency collaboration; water safety plan updates	Reflects best practices from Cape Town and São Paulo
Social inclusion	85% public acceptance of recycled water; expanded service to informal settlements	Public engagement, tariff reform	Overcomes 'yuck factor' found in other reuse projects
Urban spatial planning	GIS show infrastructure aligned with urban expansion	Integrated planning; proximity of pipelines	Serves as model for climate-stressed African cities
Global contribution	Knowledge transfer through UNESCO and GWP platforms	Documented replication efforts	Strengthens South-South learning pathways

Source: Authors' research (2025)

engineering excellence but is deeply shaped by robust governance structures, technological adaptability, and institutional continuity.

The effectiveness of Windhoek's recycling system emerges from a nexus of enabling governance mechanisms. Policy coherence and continuity, evidenced in long-term strategic frameworks such as the Namibian Water Sector Reform Strategy and the Water Supply and Sanitation Policy (WASP), anchor the recycling programme within national development priorities. These documents consistently emphasize integrated water resources management (IWRM) and risk reduction, which have shaped institutional behavior at the municipal level. Interviews with officials from the Department of Infrastructure, Water and Technical Services (DIWTS) and NamWater revealed a strong tradition of multi-level collaboration. One official noted, "*We have quarterly performance reviews where the Ministry, NamWater, and the CoW jointly assess benchmarks—especially water quality, plant efficiency, and network losses.*" This collaborative model enables adaptive decision-making, reduces institutional silos, and embeds resilience within governance routines.

The city's success in implementing direct potable reuse stems from a governance model that combines top-down regulation with bottom-up community engagement. Public participation campaigns—visible through municipal outreach programmes, school-based education or green curriculum development, and regular community forums—have contributed to increasing public trust. An NGO coordinator interviewed emphasized, "*What changed minds was not just the science—it was regular dialogue. People needed to hear from trusted faces in the community.*" These soft governance tools are embedded within harder regulatory frameworks such as Water Safety Plans (WSPs), which are periodically updated and jointly reviewed by the city and national authorities. For instance, the City of Windhoek as at 2024/2025 is actively promoting water conservation also known as "save water" due to severe water scarcity. The current Water Management Plan, specifically Category D (Severe Scarcity), mandates a 15% reduction in water usage and includes restrictions on activities like garden watering, car washing, and filling swimming pools. The city is also implementing public education campaigns and penalties for excessive water consumption.

Document analysis reveals that Windhoek's regulatory landscape also reflects transboundary knowledge sharing. Namibia has consistently engaged with global institutions such as United Nations Educational, Scientific and Cultural Organization—International Hydrological Programme (UNESCO-IHP), International Council for Local Environmental Initiatives (now known as ICLEI—Local Governments for Sustainability), and the Global Water Partnership, ensuring its frameworks are aligned with international standards. Reports from these engagements document how Namibia has hosted regional learning workshops and shared operational templates for safe water reuse as well as save water/water conservation. These interactions have enhanced institutional learning, enabling the city to anticipate risks and integrate emerging technologies into legacy systems.

GIS-based spatial analysis and field observations highlight the physical design of Windhoek's system as a key contributor to its operational effectiveness. The Goreangab Water Reclamation Plant is strategically located within a short pumping radius of primary residential zones and administrative centers. This locational advantage minimizes energy use and transportation losses, a factor consistently mentioned in technical documentation and interviews. Furthermore, site observations revealed that the plant incorporates real-time sensor monitoring, with redundancy protocols to maintain water quality in the event of technical failures. Field notes confirm that control rooms operate on a continuous 24-hour basis, with back-up power and automated alerts.

The technological design itself is both redundant and modular, allowing flexibility under variable climate and demand conditions. Interviews with plant engineers identified reverse osmosis, ultrafiltration, and advanced oxidation processes as core components of the treatment train. One senior engineer explained, "*The modularity lets us upgrade one process—say the UV filters—*

without halting the entire plant. That is critical when you are operating a direct potable system." The GIS maps (Fig. 13.5) above still demonstrate an urban growth pattern increasingly oriented toward optimized water infrastructure access, a sign of effective spatial planning intertwined with technological investments.

A recurrent theme across interviews was the role of digital governance and automation in strengthening resilience. Supervisory Control and Data Acquisition (SCADA) systems enable proactive fault detection, real-time flow regulation, and quality assurance. Such innovations reflect Windhoek's maturity in transitioning from reactive to anticipatory water governance. The integration of Geographic Information Systems (GIS) into routine operations also supports monitoring infrastructure strain and spatial equity. For instance, maps show higher pipeline pressure incidents in peri-urban areas—data that have led to targeted maintenance programmes over the past 2 years.

Furthermore, cost-effectiveness emerged as a driver of system sustainability. Though initial capital investments were substantial, financial reports reviewed during the study confirmed that operational costs per cubic meter (m^3) of recycled water are competitive relative to the cost of importing freshwater. Interviews reinforced this insight, with a water economist stating, "*For every dollar spent on reuse, the city saves three in transport, pumping, and opportunity costs.*" This economic efficiency is further substantiated in Table 13.2, which compares the operational costs of recycled water and imported freshwater over a 10-year period, along with projected demand trends. As shown, recycled water consistently maintains a cost advantage of over 30% across the decade, demonstrating long-term fiscal sustainability. This efficiency is enhanced through well-calibrated tariff structures that balance affordability with cost recovery, making recycled water not only a sustainable environmental solution but a financially viable one as well.

In comparative context, Windhoek's case offers instructive insights for global water-scarce cities. Unlike many cities that depend on expensive desalination or inter-basin transfers,

Table 13.2 Estimated operational and comparative costs with water projected water demand for Windhoek, 2020–2030

Year	Recycled water (~USD/m^3)	Imported freshwater (~USD/m^3)	Relative savings (%)	Projected water demand (million m^3/year)
2020	0.58	0.88	34.0	32.0
2025	0.66	0.97	32.2	38.5
2030	0.73	1.13	34.9	46.0

Note: Exchange rate used: NAD 18.0 = 1 USD (approx.., 2025) – Cost estimates are inflation-adjusted (~2.5% annually) – Demand estimates assume 3.5%–4% urban population growth and increasing per capita usage due to industrialization and improved access
Source: Authors' computation based on Windhoek's Water Supply Master Plan (2020), NamWater tariff projections, WHO benchmarks, and expert interview insights (2025)

Table 13.3 Summary of governance and technological enablers of system effectiveness in Windhoek

Domain	Key enablers
Governance	Inter-agency coordination; policy coherence; participatory governance; memorandum of understanding
Regulatory frameworks	Updated Water Safety Plans; quality monitoring; decentralized accountability
Public engagement	Community education; multi-stakeholder forums; transparency
Technology	Multi-barrier treatment; real-time SCADA systems; modularity
Spatial integration	GIS-guided planning; strategic infrastructure siting
Financial sustainability	Cost-recovery tariff models; capital efficiency; local capacity building

Source: Authors' research (2025)

Windhoek has shown that urban resilience can be achieved by optimizing local resources through technological ingenuity and multi-scalar governance. Similar findings were observed in the Orange County DPR project (USA) and Toowoomba, Australia, but these often relied on higher financial capacities and donor aid. Windhoek's case, by contrast, reflects the power of local innovation adapted to resource-constrained environments. Table 13.3 summarizes the key governance and technological enablers identified in the study. These findings affirm that Windhoek's wastewater recycling system is effective not because of one specific intervention but due to the synergistic interaction between institutional robustness, advanced technology, and participatory planning. These insights further reinforce the central agenda of this chapter: that resilient urban waterscapes must be co-produced by technology, policy, and people. The next section presents the final implications and recommendations for policy and practice.

13.4.3 Urban Integration and Sustainability Implications

This section addresses the third research objective and question: What are the socio-economic and environmental outcomes of Windhoek's wastewater recycling system for urban populations and surrounding ecosystems? The findings indicate that the impacts of Windhoek's wastewater reuse programme transcend water conservation. The system has enabled gains in public health, environmental protection, economic participation, and spatial inclusion—core tenets of resilient urban waterscapes. The analysis is structured as a narrative synthesis of interview insights, documentary evidence, GIS mappings, and field notes.

The socio-economic outcomes are particularly visible in vulnerable communities. Interviewees from the informal settlements of Havana and Okuryangava highlighted significant improvements in water access and reduced waterborne disease incidences over the last decade. These claims are supported by municipal health reports between 2010 and 2022, which show a steady decline in reported cases of diarrhoea and cholera in areas linked to the upgraded water distribution network. Field observations confirmed the presence of new tap-off points connected to the dual reticulation system. Quotes from key stakeholder interviews offer insight into lived experiences. One community leader noted: "*Before the recycled water reached us, children often missed school due to sickness. Now, we have clean water near our homes, and it has changed everything.*" This demonstrates how access to safe water through innovative urban infrastructure significantly reduces vulnerability in low-income urban settings. The reduced disease burden contributes to improved educational attendance and productivity, reinforcing the interconnectedness between health, education, and urban resilience.

From an economic perspective, the system has enabled employment opportunities and cost savings of approximately ~0.3 USD/m^3 i.e. over 30% savings rather than importing freshwater as reference in Table 13.2. The GWRP directly employs over 150 workers and indirectly supports livelihoods through associated services such as infrastructure maintenance, sanitation outreach, and water-quality monitoring. Interviewees from the City of Windhoek's Department of Infrastructure confirmed that contracting local firms for non-specialized work, such as pipe installations and public education, has enabled wealth circulation within the local economy. Moreover, the affordability of recycled water was identified as a game-changer by residents and municipal planners alike. The average price of treated recycled water is approximately 30%–40% lower than conventional borehole or dam water sources due to local production and reduced transport costs (de Vries & Lopez, 2012; Du Pisani, 2024). This price differential has been passed down to consumers through Namibia's pro-poor water tariff framework. As one municipal engineer noted: "*We deliberately structured our water tariffs to ensure that those who need it most pay the least. Wastewater recycling allowed us to maintain low base rates for low-income users.*" This equity-focused pricing aligns with SDG 6.1, which promotes universal and equitable access to safe and affordable drinking water for all.

Environmentally, Windhoek's approach has resulted in measurable improvements to water ecosystems. Document analysis of Namibia's Ministry of Environment and Tourism reports (2015–2023) indicates a marked decline in effluent contamination in the Goreangab Dam, downstream from the recycling facility. Water quality testing data shared by NamWater demonstrates compliance with WHO standards for surface water discharge, with reduced biological oxygen demand (BOD) and nutrient levels. Satellite imagery and GIS mapping illustrate in Fig. 13.5, now translate in Table 13.4 further reveal that vegetation cover and aquatic habitat health have improved in the downstream catchment areas. This suggests that reclaimed water not only serves anthropogenic needs but also sustains natural hydrological systems.

This ecological regeneration underscores Windhoek's contribution to SDG 14 (life below water) and SDG 15 (life on land), which focus on preserving aquatic biodiversity and terrestrial ecosystems (Sachs et al., 2024). The near-zero effluent policy of the wastewater recycling system means that less untreated waste enters natural water systems, reducing environmental degradation and supporting habitat restoration. However, certain challenges persist. Interviews and site visits indicate that urban expansion is outpacing infrastructure development in some peri-urban areas. Residents in unplanned settlements continue to rely on tanker water or communal taps, which limits the reach of recycled water infrastructure. Additionally, while the Goreangab plant meets high technical standards, its capacity is finite, and projections indicate that it may be insufficient to meet Windhoek's full urban demand by 2050. This calls for future scalability strategies, such as decentralized treatment hubs, improved stormwater harvesting integration, and enhanced land-use planning that integrates water-sensitive urban design (WSUD) principles as denote in Fig. 13.6.

In sum, the socio-economic and environmental implications of Windhoek's wastewater recycling system demonstrate its multidimensional contribution to resilience. It improves equity by expanding access to safe and affordable water,

Table 13.4 Change in land surface vegetation health near the Goreangab catchment in Windhoek (1993–2023)

	Initial State (pixel to percentage count)						
		Built-up areas	Grassland/ Shrubs	Bare land	Water bodies	Row total	Class total
Final state (%)	Built-up areas	77.644	6.781	13.071	0	100	100
	Grassland/Shrubs	4.309	41.171	3.728	68.869	100	100
	Bare land	17.938	51.902	83.176	0.212	100	100
	Water bodies	0.108	0.145	0.025	30.919	100	100
	Class total	100	100	100	100	0	0
	Class changes	22.356	58.829	16.824	69.081	0	0
	Image difference	59.242	−50.755	20.518	−51.715	0	0

Source: Authors' computation based on Sentinel-2 imagery and NDVI classification (2025)

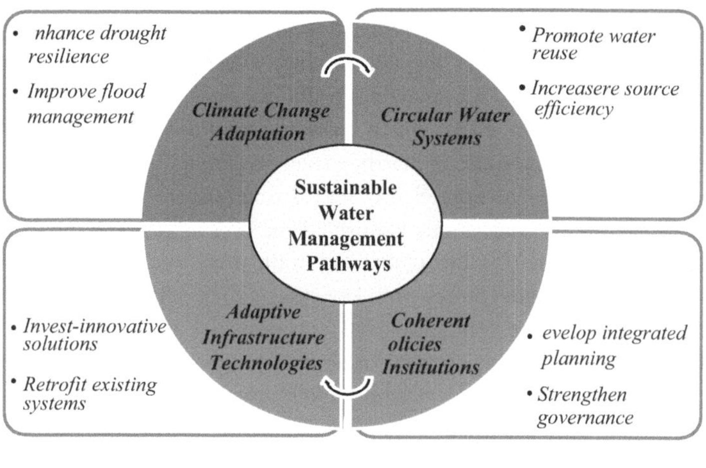

Fig. 13.6 WSUD principles for achieving resilient urban waterscapes. Source: Authors' research (2025)

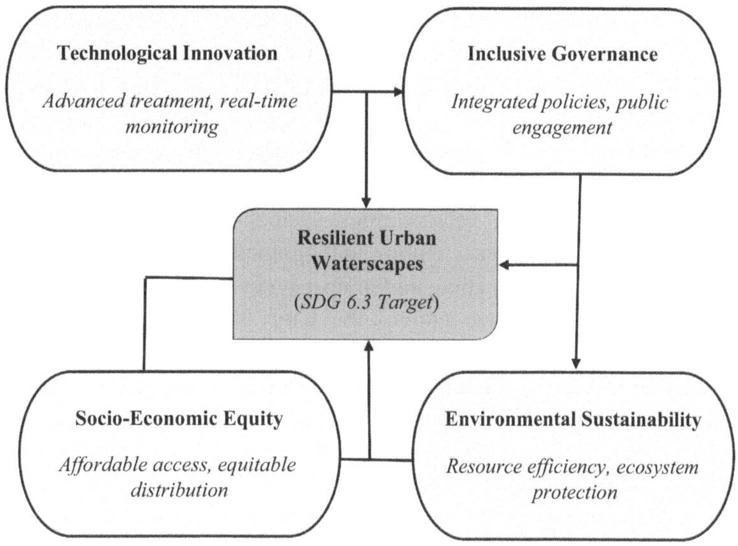

Fig. 13.7 A pathway for sustainable water management. Source: Authors' research (2025)

strengthens environmental sustainability by reducing pollution and supporting ecosystems, and reinforces economic resilience through job creation and reduced health expenditure. These findings illustrate that water infrastructure, when designed and managed inclusively, serves as a catalyst for broader urban transformation in a time of climate change and urbanization within sub-Saharan Africa. This supports the global call to integrate WSUD, as urged by the UN's New Urban Agenda and SDG Target 6.5 on IWRM. The accompanying framework in Fig. 13.7 demonstrates how local innovations like Windhoek's model can inform adaptive strategies for resilient urban waterscapes globally.

13.5 Conclusion

The analysis confirms that Windhoek's wastewater recycling system stands as a model of strategic foresight, resilience, and innovation in sustainable urban water management. A key implication emerging from this study is that sustainable urban waterscapes are achievable in even the most arid environments through the deliberate integration of technological, environmental, socio-economic, and institutional dimensions. Windhoek's approach represents a decisive shift from crisis-driven responses to circular water systems that close the loop between sanitation and supply. Enabled by early planning and long-term policy consistency, this transformation delivers potable-quality water that meets World Health Organization standards while substantially reducing dependence on depleting freshwater sources and minimizing environmental pollution. These achievements align directly with SDG Target 6.3, which calls for improving water quality, reducing contamination, and expanding recycling and safe reuse globally.

Windhoek demonstrates that while capital investment in recycling infrastructure may initially be high, the long-term economic, environmental, and social returns are substantial. Operational cost savings, reduced vulnerability to droughts, and improved water reliability position recycling as a cornerstone of circular water economies in cities worldwide. This economic resilience complements environmental sustainability by mitigating pollution and safeguarding freshwater ecosystems. On the social front, Windhoek's inclusive approach ensures equitable access to clean water, particularly benefiting vulnerable peri-urban communities. Integrating social equity with environmental protection and economic viability underscores the city's holistic commitment to resilient and sustainable development.

Framed within the lens of urban resilience, Windhoek exhibits robustness, adaptability, and foresight consistent with the City Resilience Framework (Saikia et al., 2022). The city's proactive capacity to anticipate and respond to water stress through system innovation—rather than fragmented emergency interventions—demonstrates a mature governance culture underpinned by institutional coordination and transparency. Collaborative governance among the City of Windhoek, NamWater, the Ministry of Agriculture, Water and Land Reform, and international partners has been pivotal in maintaining policy coherence, ensuring infrastructure sustainability, and strengthening accountability. These partnerships exemplify the integrated governance approach necessary to achieve SDG 6.3, where pollution control, wastewater treatment, and resource recovery are linked within coherent institutional frameworks.

Technological and policy innovations further reinforce the strength of Windhoek's model. The Goreangab Water Reclamation Plant stands as a benchmark for localized adaptation of international standards, ensuring safety and efficiency under extreme climatic conditions. Its modular design, continuous quality monitoring, and built-in redundancies enhance operational resilience, while sustained public education and transparent reporting have cultivated public trust in recycled water. This experience underscores that wastewater reuse success depends not only on technology but also on social acceptance and participatory governance.

The implications of Windhoek's model extend globally, especially for cities in Africa, Asia, and Latin America facing accelerating water insecurity due to climate change and rapid urbanization. Replicating this success, however, requires more than transferring technology—it demands adapting institutional, cultural, and governance systems to local contexts. Thus, the study underscores the importance of capacity-building, knowledge exchange, and policy learning across regions. Platforms such as UN-Habitat, ICLEI, and regional water networks can facilitate cross-city cooperation, enabling context-sensitive applications of Windhoek's model while promoting global progress toward SDG 6.3.

This study contributes to the wider policy and scientific discourse by bridging theory and practice in urban water resilience. It aligns with the overarching theme of *Resilient Urban Waterscapes: Integrated Strategies for Sustainable Water Management* by demonstrating how integrated infrastructure, governance, and community engagement can deliver measurable environmental outcomes—specifically by reducing urban water pollution, improving water quality, and promoting water reuse. Windhoek's case thus advances the implementation of SDG 6.3, offering a replicable model of how resilient urban waterscapes can emerge from sustained innovation and multi-level collaboration.

In conclusion, Windhoek's wastewater recycling initiative exemplifies how environmental protection, water quality enhancement, and recycling can converge within a unified urban governance and infrastructure model to achieve sustainable urban waterscapes. Its long-term success stems not only from technological sophistication but also from civic participation, institutional stability, and adaptive planning. By explicitly reducing pollution, improving water quality, and enhancing recycling, the Windhoek model directly operationalizes SDG 6.3, proving that resilient, equitable, and sustainable water management is attainable even under extreme scarcity.

Key lessons emerging from this study highlight the need to strengthen institutional capacity through targeted training for local authorities, promote transparent water quality monitoring to sustain public trust, and integrate water reuse policies across urban planning, health, and environmental sectors. Expanding access to green finance, public-private partnerships, and climate adaptation funds can also ensure the long-term scalability and affordability of wastewater recycling infrastructure. Moreover, fostering regional and global partnerships will be essential for accelerating innovation diffusion, technology transfer, and shared learning across cities.

By synthesizing these insights, this study affirms that Windhoek's wastewater recycling system is not merely a local success but a globally significant demonstration of how cities can mitigate pollution, safeguard freshwater ecosystems, and advance toward SDG 6.3. It reinforces the imperative of aligning science, governance, and community participation in building resilient, inclusive, and sustainable urban futures where clean water security is both a right and a shared responsibility.

Acknowledgments The corresponding author gratefully acknowledges the City of Windhoek, NamWater, the Ministry of Agriculture, Water and Land Reform, and all stakeholders and interviewees who generously contributed their time and insights to this study. Sincere thanks are also extended to academic mentors, peer reviewers, and research assistants for their valuable support throughout the research process.

Funding: The research was not supported by any external funding.

Potential Conflicts of Interest: The author declares no conflict of interest.

Informed Consent: Applicable consent was obtained from all individual participants included in the study.

Author Contributions: Conceptualization, P.M. and E.Y.; methodology, P.M.; software, P.M. and D.F.B.; validation, P.M.; writing—original draft preparation, P.M.; writing—review and editing, P.M., E.Y. and D.F.B; visualization, P.M.; supervision, E.Y.; project administration, P.M. All authors have read and agreed to the published version of the manuscript.

Data Availability Statement: Data supporting these findings are available within the chapter while subject to request from the corresponding author.

AI Disclosure Statement: The corresponding author confirms that no artificial intelligence (AI) tools were used in the conception, design, analysis, writing, editing, or revision of this chapter. All content is entirely the original work of the authors.

References

African Development Bank. (2017). Namibia Water Sector Support Program (NWSSP): Summary of the Environmental and Social Management Framework (ESMF). Rights in Development Early Warning System. Retrieved [May 15, 2024], from https://ewsdata.rightsindevelopment.org/files/documents/05/AFDB-P-NA-E00-005.pdf

African Water Facility. (2020). Integrated urban water management master plan for City of Windhoek in Namibia (Final report). https://www.africanwaterfacility.org/sites/default/files/2023-12/Integrated%20Urban%20Water%20Management%20Master%20Plan%20for%20City%20of%20Windhoek%20in%20Namibia.pdf

Akpan, V. E., Omole, D. O., & Bassey, D. E. (2020). Assessing the public perceptions of treated wastewater reuse: Opportunities and implications for urban communities in developing countries. *Heliyon, 6*(10), 1–17. https://doi.org/10.1016/j.heliyon.2020.e05246

Al-Saidi, M. (2021). From acceptance snapshots to the social acceptability process: Structuring knowledge on attitudes towards water reuse. *Frontiers in Environmental Science, 9*, 633841. https://doi.org/10.3389/fenvs.2021.633841

Banso, A. A., Olurin, J. O., Okem, E. S., & Ogunjobi, O. A. (2023). Integrated water resource management in Southwest Nigeria: A comprehensive review of strategies and outcomes. *International Journal of Applied Research in Social Sciences, 5*(8), 330–351. https://doi.org/10.51594/ijarss.v5i8.586

Byrne, D. (2022). A worked example of Braun and Clarke's approach to reflexive thematic analysis. *Quality and Quantity, 56*(3), 1391–1412. https://doi.org/10.1007/s11135-021-01182-y

Cairone, S., Hasan, S. W., Choo, K. H., Lekkas, D. F., Fortunato, L., Zorpas, A. A., et al. (2024). Revolutionizing wastewater treatment toward circular economy and carbon neutrality goals: Pioneering sustainable and efficient solutions for automation and advanced process control with smart and cutting-edge technologies. *Journal of Water Process Engineering, 63*, 105486. https://doi.org/10.1016/j.jwpe.2024.105486

Carroli, L. (2018). Planning roles in infrastructure system transitions: A review of research bridging socio-technical transitions and planning. *Environmental Innovation and Societal Transitions, 29*, 81–89. https://doi.org/10.1016/j.eist.2018.06.001

de Vries, G. E., & Lopez, A. (2012). Wastewaters are not wastes. In *Living with water: Targeting quality in a dynamic world* (pp. 101–141). Springer New York. https://doi.org/10.1007/978-1-4614-3752-9_4

Du Pisani, P. L. (2024). Direct reclamation of potable water at Windhoek's Goreangab reclamation plant. *Desalination, 188*(1–3), 79–88. https://doi.org/10.1016/j.desal.2005.04.104

Evaristo, J., Jameel, Y., Tortajada, C., Wang, R. Y., Horne, J., Neukrug, H., et al. (2023). Water woes: The institutional challenges in achieving SDG 6. *Sustainable Earth Reviews, 6*(1), 13. https://doi.org/10.1186/s42055-023-00067-2

Fuenfschilling, L., & Truffer, B. (2016). The interplay of institutions, actors and technologies in socio-technical systems—An analysis of transformations in the Australian urban water sector. *Technological Forecasting and Social Change, 103*, 298–312. https://doi.org/10.1016/j.techfore.2015.11.023

Galychyn, O. (2022). *Towards sustainable cities: A multi-criteria assessment framework for studying urban metabolism*. PhD thesis, Parthenope University of Naples. https://hdl.handle.net/2158/1362132

Ghernaout, D., Elboughdiri, N., & Ghareba, S. (2019). Drinking water reuse: One-step closer to overpassing the "yuck factor". *Open Access Library Journal, 6*(11), 1–12. https://doi.org/10.4236/oalib.1105895

Gondhalekar, D., Hu, H. Y., Chen, Z., Tayal, S., Bekchanov, M., Sauer, J., et al. (2021). The emerging environmental economic implications of the urban water-energy-food (WEF) nexus: Water reclamation with resource recovery in China, India, and Europe. In *Oxford research encyclopedia of environmental science*. https://doi.org/10.1093/acrefore/9780199389414.013.590

Gul, S., Govender, I. G., Bux, F., & Baba, S. H. (2024). Role of public attitude, knowledge, and trust in shaping the acceptance of recycled wastewater for potable and non-potable uses. *AQUA - Water Infrastructure, Ecosystems and Society, 73*(4), 722–736. https://doi.org/10.2166/aqua.2024.225

Hariram, N. P., Mekha, K. B., Suganthan, V., & Sudhakar, K. (2023). Sustainalism: An integrated socio-economic-environmental model to address sustainable development and sustainability. *Sustainability, 15*(13), 10682. https://doi.org/10.3390/su151310682

Harrison, H., Birks, M., Franklin, R., & Mills, J. (2017). Case study research: Foundations and methodological orientations. *Forum: Qualitative Social Research, 18*(1), 1–17. https://doi.org/10.17169/fqs-18.1.2655

He, C., Liu, Z., Wu, J., Pan, X., Fang, Z., Li, J., & Bryan, B. A. (2021). Future global urban water scarcity and potential solutions. *Nature Communications, 12*(1), 4667. https://doi.org/10.1038/s41467-021-25026-3

Heenetigala, U., Kapetas, L., & Fenner, R. (2021). Building resilience in water supply infrastructure in the face of future uncertainties: Insight from Cape Town. In *Water-wise cities and sustainable water systems: Concepts, technologies, and applications* (pp. 201–233). https://doi.org/10.2166/9781789060768_0201

Herrera, V. (2019). Reconciling global aspirations and local realities: Challenges facing the Sustainable Development Goals for water and sanitation. *World Development, 118*, 106–117. https://doi.org/10.1016/j.worlddev.2019.02.009

Johannessen, Å., & Wamsler, C. (2017). What does resilience mean for urban water services? *Ecology and Society, 22*(1) https://www.jstor.org/stable/26270048

Jones, E. R., Van Vliet, M. T., Qadir, M., & Bierkens, M. F. (2021). Country-level and gridded estimates of wastewater production, collection, treatment and reuse. *Earth System Science Data, 13*(2), 237–254. https://doi.org/10.5194/essd-13-237-2021

Khair, S. M., Mushtaq, S., Reardon-Smith, K., & Ostini, J. (2019). Diverse drivers of unsustainable groundwater extraction behaviour operate in an unregulated water scarce region. *Journal of Environmental Management, 236*, 340–350. https://doi.org/10.1016/j.jenvman.2018.12.077

Kwembeya, E. G., & Shikangalah, R. N. (2023). Towards understanding rainfall variability in Namibia: An analysis of spatial and temporal variations of rainfall from 2010 to 2019. *International Science and Technology Journal of Namibia, 16*, 26–39. https://journals.unam.edu.na/index.php/ISTJN/article/view/1844

Lafforgue, M. (2016). Supplying water to a water-stressed city: Lessons from Windhoek. *La Houille Blanche, 4*, 40–47. https://doi.org/10.1051/lhb/2016038

Liehr, S., Brenda, M., Cornel, P., Deffner, J., Felmeden, J., Jokisch, A., et al. (2016). From the concept to the tap—Integrated water resources management in Northern Namibia. In *Integrated water resources management: Concept, research and implementation* (pp. 683–717). https://doi.org/10.1007/978-3-319-25071-7_26

Maguire, M., & Delahunt, B. (2017). Doing a thematic analysis: A practical, step-by-step guide for learning and teaching scholars. *All Ireland Journal of Higher Education, 9*(3), 33510–33514. https://doi.org/10.62707/aishej.v9i3.335

Mishra, B. K., Kumar, P., Saraswat, C., Chakraborty, S., & Gautam, A. (2021). Water security in a changing environment: Concept, challenges and solutions. *Water, 13*(4), 490. https://doi.org/10.3390/w13040490

Mukherjee, M., & Jensen, O. (2020). Making water reuse safe: A comparative analysis of the development of regulation and technology uptake in the US and Australia. *Safety Science, 121*, 5–14. https://doi.org/10.1016/j.ssci.2019.08.039

Namibia Statistics Agency [NSA] (2024). Annual national accounts, 2024. [accessed on 3 May 2025]. Available from: https://nsa.org.na/document/annual-national-accounts-2024/

Narayan, A. S., Marks, S. J., Meierhofer, R., Strande, L., Tilley, E., Zurbrügg, C., & Lüthi, C. (2021). Advancements in and integration of water, sanitation, and solid waste for low- and middle-income countries. *Annual Review of Environment and Resources, 46*(1), 193–219.

Ngene, B. U., Nwafor, C. O., Bamigboye, G. O., Ogbiye, A. S., Ogundare, J. O., & Akpan, V. E. (2021). Assessment of water resources development and exploitation in Nigeria: A review of integrated water resources management approach. *Heliyon, 7*(1). https://doi.org/10.1016/j.heliyon.2021.e05955

Nitivattananon, V., & Sa-nguanduan, N. (2014). Domestic water reuse situation in the context of middle-income countries: A case analysis. *Frontiers in Environmental Engineering, 2*(1) http://www.dpi-journals.com/index.php/FIEE/article/view/611/479

Njie, B., & Asimiran, S. (2014). Case study as a choice in qualitative methodology. *Journal of Research and Method in Education, 4*(3), 35–40. http://www.iosr-journals.org/

Nyamwanza, A. M., & Kujinga, K. K. (2017). Climate change, sustainable water management and institutional adaptation in rural sub-Saharan Africa. *Environment, Development and Sustainability, 19*, 693–706. https://doi.org/10.1007/s10668-016-9762-2

Obaideen, K., Shehata, N., Sayed, E. T., Abdelkareem, M. A., Mahmoud, M. S., & Olabi, A. G. (2022). The role of wastewater treatment in achieving sustainable development goals (SDGs) and sustainability guideline. *Energy Nexus, 7*, 100112. https://doi.org/10.1016/j.nexus.2022.100112

Olivieri, F., Koop, S. H. A., Van Leeuwen, K., & Hofman, J. (2022). Enhancing governance capacity to ensure a long-term water supply: The case of Windhoek, Namibia. *Sustainability, 14*(4), 2387. https://doi.org/10.3390/su14042387

Pisano, U. (2012). Resilience and sustainable development: Theory of resilience, systems thinking. *European Sustainable Development Network (SN), 26*, 50. https://www.cjwalsh.ie/wp-content/uploads/2013/01/ESDN_Resilience-and-Sustainable-Development_September-2012.pdf

Prouty, C., Mohebbi, S., & Zhang, Q. (2018). Socio-technical strategies and behavior change to increase the adoption and sustainability of wastewater resource recovery systems. *Water Research, 137*, 107–119. https://doi.org/10.1016/j.watres.2018.03.009

Russo, T., Alfredo, K., & Fisher, J. (2014). Sustainable water management in urban, agricultural, and natural systems. *Water, 6*(12), 3934–3956. https://doi.org/10.3390/w6123934

Sachs, J. D., Lafortune, G., & Fuller, G. (2024). *The SDGs and the UN summit of the future. Sustainable development report*. SDSN, Dublin University Press. https://doi.org/10.25546/108572

Saikia, P., Beane, G., Garriga, R. G., Avello, P., Ellis, L., Fisher, S., et al. (2022). City water resilience framework: A governance-based planning tool to enhance urban water resilience. *Sustainable Cities and Society, 77*, 103497. https://doi.org/10.1016/j.scs.2021.103497

Saketa, Y. (2022). Assessment of future urban water demand and supply under socioeconomic scenarios: A case of Assosa town. *Water Supply, 22*(10), 7405–7415. https://doi.org/10.2166/ws.2022.329

Samaraweera, S. A. P. T., & Najim, M. M. M. (2024, February). Optimizing wastewater treatment: Harnessing non-potable treated wastewater reuse to foster sustainable development goals. *Proceedings of International Forestry and Environment Symposium, 28*(1), 73. https://doi.org/10.31357/fesympo.v28.6947

Santos, E., Carvalho, M., & Martins, S. (2023). Sustainable water management: Understanding the socioeconomic and cultural dimensions. *Sustainability, 15*(17), 13074. https://doi.org/10.3390/su151713074

Scruggs, C. E., Pratesi, C. B., & Fleck, J. R. (2020). Direct potable water reuse in five arid inland communities: An analysis of factors influencing public acceptance. *Journal of Environmental Planning and Management, 63*(8), 1470–1500. https://doi.org/10.1080/09640568.2019.1671815

Shemer, H., Wald, S., & Semiat, R. (2023). Challenges and solutions for global water scarcity. *Membranes, 13*(6), 612. https://doi.org/10.3390/membranes13060612

Silva, J. A. (2023). Wastewater treatment and reuse for sustainable water resources management: A systematic literature review. *Sustainability, 15*(14), 10940. https://doi.org/10.3390/su151410940

Suman, A. B., & Toscano, A. (2021). Public acceptance of water reuse for agriculture in the wake of the new EU regulation: Early reflections. *Journal for European Environmental & Planning Law, 18*(3), 225–255. https://doi.org/10.1163/18760104-18030001

Truelove, Y. (2019). Rethinking water insecurity, inequality and infrastructure through an embodied urban political ecology. *Wiley Interdisciplinary Reviews: Water, 6*(3), e1342. https://ui.adsabs.harvard.edu/link_gateway/2019WIRWa...6E1342T/doi:10.1002/wat2.1342

United Nations. (2022). Republic of Namibia: Namibia Water and the UN 2023 Water Conference preparatory meeting, New York. United Nations Sustainable Development Goals. https://sdgs.un.org/sites/default/files/2022-11/NAMIBIA%20inputs.pdf

United Nations Environment Programme [UNEP] (2022). Chapter 2: The state of the environment—measuring progress: Water-related ecosystems and the SDGs. United Nations Environment Programme. https://wedocs.unep.org/handle/20.500.11822/42097

Van Rensburg, P. (2018). Overcoming global water reuse barriers: The Windhoek experience. In *Water reuse policies for potable use* (pp. 123–137). Routledge. https://doi.org/10.4324/9781315205489

Visseren-Hamakers, I. J. (2015). Integrative environmental governance: Enhancing governance in the era of synergies. *Current Opinion in Environment Sustainability, 14*, 136–143. https://doi.org/10.1016/j.cosust.2015.05.008

Wester, J., Timpano, K. R., Çek, D., Lieberman, D., Fieldstone, S. C., & Broad, K. (2015). Psychological and social factors associated with wastewater reuse emotional discomfort. *Journal of Environmental Psychology, 42*, 16–23. https://doi.org/10.1016/j.jenvp.2015.01.003

Weststrate, J., Dijkstra, G., Eshuis, J., Gianoli, A., & Rusca, M. (2019). The sustainable development goal on water and sanitation: Learning from the millennium development goals. *Social Indicators Research, 143*, 795–810. https://doi.org/10.1007/s11205-018-1965-5

World Health Organization [WHO] Regional Office for Africa. (2023). WHO biennial report 2022–2023. Available from: https://www.afro.who.int/sites/default/files/2024-06/WHO23%20BIENNIAL%20REPORT%20ONLINE.pdf

Zellner, M., Massey, D., Rozhkov, A., & Murphy, J. T. (2023). Exploring the barriers to and potential for sustainable transitions in urban-rural systems through participatory causal loop diagramming of the food-energy-water nexus. *Land, 12*(3), 551. https://doi.org/10.3390/land12030551

Zhang, M., Liu, R., & Li, Y. (2022). Diversifying water sources with atmospheric water harvesting to enhance water supply resilience. *Sustainability, 14*(13), 7783. https://doi.org/10.3390/su14137783

14 Forecasting the Pollution and Eutrophic Status of the Upper Crocodile West Catchment, South Africa by a Random Forest Regression-Based Machine Learning Approach

Ernestine Atangana, Paul J. Oberholster, Heba Bedair, Ashkan Khalife, and Soumya Ghosh

Abstract

Contaminant levels in water quality downstream of four wastewater treatment works in the Crocodile West Catchment, as well as their spatial- and seasonal variations were examined. A total of 24 samples were collected during dry and wet seasons from four sampling site locations: Lethabile, Losperfontein, Rietfontein, and Brits. Physicochemical parameters such as pH, EC, temperature, suspended solids, chemical oxygen demand, dissolved oxygen, sulphates, phosphates, nitrogen, and ammonia were determined. Spatial and seasonal variations in water quality parameter concentrations were evalauted using the Kruskal-Wallis test and Mann-Whitney U test respectively, at a significance level of $p < 0.05$). The results indicated that pollution levels followed the order; Lathabile > Losperfontein > Rietfontein > Brits. Values of R^2 greater than 0.6 obtained from random forest models of predicting pollutant concentrations from environmental variables indicates strong predictive performance and low prediction error. Urban and industrial wastewater treatment plants were identify as the main sources of contamination, while pollution indices, including total organic pollution and eutrophication, reached their maximum level during the rainy season. All sampling sites were classify as heavily polluted and eutrophic. These results suggest that machine learning models, such as random forest, are effective tools for predicting eutrophication and detecting water pollution in complex aquatic systems.

E. Atangana (✉) · P. J. Oberholster
Centre for Environmental Management, University of the Free State, Bloemfontein, South Africa
e-mail: AtanganaE@ufs.ac.za; OberholsterPJ@ufs.ac.za

H. Bedair
Department of Botany, Faculty of Science, Tanta University, Tanta, Egypt
e-mail: heba.taha@science.tanta.edu.eg

A. Khalife
Mathematical and Physical Sciences, College of Arts and Sciences, University of Nizwa, Nizwa, Oman
e-mail: a.khalifeh@unizwa.edu.om

S. Ghosh
Natural and Medical Sciences Research Center, University of Nizwa, Nizwa, Oman
e-mail: soumyaghosh@yahoo.com

U. Chatterjee, A. I. Almulhim (eds.), *Urban Waterscapes*, Sustainable Development Goals Series,
https://doi.org/10.1007/978-3-032-20649-7_14

Keywords

Wastewater treatment · Eutrophication · Pollution indices · Crocodile West Catchment · Random forest model · Hartbeespoort Dam

14.1 Introduction and Background Chapter

The overabundance of nutrients in water bodies, leading to elevated rates of primary production, is known as eutrophication. (Njagi et al., 2022). Eutrophication happens naturally, but human activity has the potential to increase the quantity of nutrients in a water body, raising the possibility of eutrophication (Njagi et al., 2022). According to a chapter in 2022 (Sirunda et al., 2022), the main excessive nutrient enrichment of water causes oligotrophic water bodies to transition to mesotrophic, eutrophic, and hypertrophic conditions. The primary sources of nutrients for aquatic body enrichment include rainfall, runoff from agricultural regions, household detergents, and inadequately treated industrial and sewage effluents. One of the biggest threats to water quality is significant impoundment eutrophication, and South Africa is no exception to this general trend (Oberholster & Ashton, 2008). As a result of improperly treated sewage effluents, 41% and 76% of South Africa's reservoirs were respectively categorized as eutrophic or hypertrophic in earlier research (Njagi et al., 2022).

In an analysis of the Hartbeespoort Dam conducted in 2012 to assess the introduction of the waste discharge charge system, it was found that sewage system leaks, which account for about 5% of the flow, contributes approximately 50% of the phosphorus load (Department of Water Affairs, 2012). Poor farming practices, inefficient waste management procedures, diffuse sources in rural and urban areas, poorly treated industrial effluent, and inadequately treated sewage effluent have all been implicated in South Africa's severe eutrophic terrible eutrophic conditions (Oberholster & Ashton, 2008). Some of the most affected water bodies in South Africa includes the Roodeplaat and Hartbeespoort Dams (Makia, 2021). Earlier studies on eutrophication focused on specific affected areas. One such example is in the Buffalo River Situation Analysis by Van Ginkel et al. (1996), which highlighted eutrophication, its impacts and management strategies to adress it within a single catchment. Rainfall is a major factor driving eutrophication, as it increases both the volume of freshwater inflow and nutrients loading in the Crocodile catchment (Liu et al., 2011). Rainfall also alters water column stratification and can increase dissolved oxygen (DO) levels, thereby limiting nutrient release from sediments. There are many issues regarding water quality in South Africa and around the globe, especially in urban water resources, including eutrophication, which results from the enrichment of water supplies with nutrients.

Indicators used to determine the eutrophication status of the water body facilitate comparisons of water quality across space and time among geographically diverse trophic conditions. These indicators includes; eutrophication index (EI), the integrated pollution index, and the organic pollution index (OPI) (Sirunda et al., 2022). Previous studies have focussed on assessing the concentrations of selected biological and/or physicochemical variables to evaluate the effects of aeration on impoundment water quality (Masawi & Oberholster, 2022). Some researchers have also emphasized the importance of developing tools, such as hydrological indices, that can be used to detect and mitigate environmental risks (Masawi & Oberholster, 2022; Oberholster et al., 2021). Various water quality monitoring parameters have been used by various researchers across the globe in the development of surface water quality indices. These includes; the Water Quality Index, comprehensive, modified pollution index, and eutrophication (Atangana & Oberholster, 2021; Chaturvedi & Bassin, 2010; Gupta et al., 2003; Liu et al., 2011; Matta et al., 2017; Oberholster et al., 2021; Rocha et al., 2015; Şener et al., 2017). The Indices used in this chapter to assess eutrophication status are intended to facilitate uniform comparisons of water quality across different spatial and seasonal trophic conditions in the

Crocodile West Catchment (Barnwal et al., 2015; Son et al., 2020).

Although numerous studies assessing eutrophication and pollution in coastal waters have been reported across the world, there is a still a lack of literature on the use of water quality indicators to assess eutrophic surface water quality in rivers and streams and to forecast pollution level in South Africa using machine learning technique.

This chapter aims to assess the water quality of the Crocodile West Catchment using pollution indices, namely the eutrophication index (EI), comprehensive pollution index (CPI), and organic pollution index (OPI). A random forest artificial intelligence model was also used to estimate pollutant levels in relation to environmental variables. The selected parameters used in the evaluation of EI, CPI and OPI include temperature (°C), pH, sulphates (SO_4^{2-}), dissolve inorganic nitrogen (DIN), ammonium (NH_4^+), suspended solids (SS), electrical conductivity (EC), dissolve phosphate (DIP), dissolve oxygen (DO) and chemical oxygen demand (COD). The random forest model was further applied at the catchment level to predict eutrophic state and general patterns of pollution.

The hightest contaminated values identified across the indices were used to assess the eutrophic status of the Crocodile West Catchment, particularly for COD, DO, DIN, and DIPcompared to conventional periodic water quality assessments, the use of machine learning algorithms to estimate the eutrophic status of rivers offers several advantages. Machine learning models can process large volumes of historical and real-time data and continuously provide accurate predictions of water quality indicators such as chlorophyll-a concentrations, which are essential for evaluating eutrophication (Karimian et al., 2023). To improve prediction accuracy and spatial resolution, these algorithms can integarte multiple data sources including in-situ measurements and remote sensing data (CCME, 2012; Karimian et al., 2023). In contrast to periodic surveys, which may overlook short term variations and require considerable time and financial resources, machine learning algorithms can provide rapid insights and early warnings, enabling more proactive water management. Furthermore, the ability of machine techniques to capture complex, non linear relationships among environmental variables makes them powerful tools for analyzing and predicting the effect of various environmental factors on river health (Kaddoura, 2022). The random forest model has proven particularly effective for forecasting and assessing the eutrophic condition of water bodies due to its robustness and high predictive performance. Random forest, an ensemble learning method, effectively handles complex non linear relationships and interactions among multiple water quality variables, such as nutrient concentrations and chlorophyll-a levels (Bedair et al., 2023; Karimian et al., 2023). This method is less prone to overfitting, a common limitation of many machine learning models, such as decision Trees1. In addition, random forest models can efficiently handle large and high dimensional datasets, often producing more reliable predictions than methods such as support vector machines and neural networks (Jimeno-Sáez et al., 2020). Their ability to manage missing data while maintaining high prediction accuracy makes them particularly suitable for environmental monitoring, where incomplete data set are common (Jimeno-Sáez et al., 2020). By leveraging random forest, researchers and policymakers can obtain timely insights into the eutrophic state of rivers, thereby supporting more proactive and informed water management decisions (Ha et al., 2020).

This chapter focuses on the upper Crocodile West Catchment, including the Hartbeespoort Dam, which supplies potable water and serves as an impoundment within the Historically, the region was surrounded by mining and and agricultural activities. Since 1923, mining towns and cities such as Johannesburg and Pretoria have expanded due to the growth of the mining industry (Makia, 2021). The need for water for agricultural usage has therefore risen over time as population growth has increased (Makia, 2021). Consequently, the demand for water for agricultural use has increased alongside population growth (Makia, 2021). The increased population has also led to higher contaminant loads entering the dam, particularly through sewage discharge (Son et al., 2020). Nutrient influx is one of the most significant sources of contamination and has intensified problems, such as algal blooms and accelerated eutrophication rates (Sirunda et al., 2022). The water quality of the Hartbeespoort Dam has a significant influence on the aquatic environment of the Crocodile West Catchment (Oberholster & Ashton, 2008). However, eutrophication remains a major concern in the catchment and requires further investigation, highlighting the need for continued research in the region.

Furthermore, numerous machine learning techniques have been widely used in previous studies to predict water quality metrics. For example, several researchers have applied algorithms such as Support Vector Machines, decision trees, and neural networks (Akhlaq et al., 2024; Yan et al., 2024). These studies demonstrated the effectiveness of machine learning in handling complex datasets and generating accurate predictions. However, the application of random forests is particularly notable due to its stability and ability to manage high dimensional data with a low risk of overfitting (Akhlaq et al., 2024). Random forest is especially well-suited for environmental monitoring because, unlike many other algorithms, it can efficiently handle missing data while maintaining high prediction accuracy (Azrour et al., 2022). This present chapter employs a random forest model and is novel in its comprehensive integration of diverse data sources and its focus on reducing model uncertainty, thereby addressing several limitations identified in previous studies.

The objectives of the chapter contributed significantly to understanding and addressing eutrophication issues in the upper Crocodile West Catchment in the following ways:

1. **Evaluation of water quality**: By assessing water quality across various sub-catchments discharging into the Hartbeespoort Dam over a period of 6 months, the chapter identified critical sources of nutrient enrichment. This evaluation highlighted anthropogenic influences and natural factors contributing to eutrophication, particularly during different seasons.
2. **Eutrophication status and nutrient distribution**: Determining the temporal distribution of key nutrients, such as nitrogen and phosphorus, provides insight into the dynamics of nutrient loading. This analysis is essential for understanding how these nutrients contribute to eutrophication in the Hartbeespoort Dam and other water bodies in the catchment. It also helps in identify hotspots of nutrient pollution and periods of increase eutrophic risk.
3. **Predictive modelling with random forest regression**: The use of machine learning to predict pollutant levels enhanced the ability to monitor and forecast eutrophication trends effectively. The random forest regression model not only tracked pollutant concentrations but also helped in identifying critical environmental variables that influenced water quality. This predictive capability was invaluable for implementing proactive management strategies and mitigating the adverse effects of eutrophication. Indeed, the novelty of this chapter lies in employing Random Forest Regression to estimate the eutrophic status of the Upper Crocodile West Catchment by incorporating PO_4-P, SO_4, NH_4-H, NO_3, and SS as objective variables and T, PH, EC, and COD as explanatory variables. The choice of these explanatory variables is justified due to their strong correlations with the target variables, ease of measurement, availability of consistent and high-quality data, and their significant impact on predictions. Measuring temperature, pH, EC, and COD is more straightforward and cost-effective compared to directly measuring the objective variables, allowing for continuous and real-time monitoring. This approach not only enhances the model's accuracy and reliability but also provides a proactive and effective method for water quality management. By leveraging the strengths of Random Forest Regression, this chapter offers a robust and innovative solution to forecasting environmental contaminants, thereby contributing to the advancement of water quality manage-

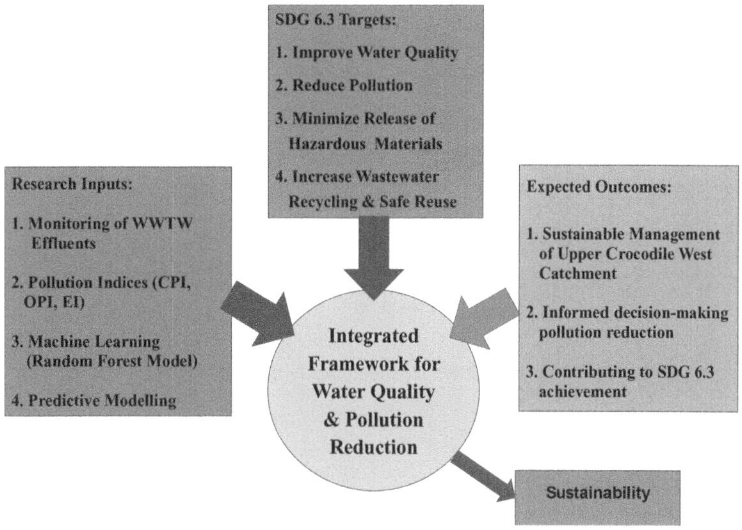

Fig. 14.1 Conceptual framework linking the Research Methodology to the targets of SDG 6.3 for Sustainable water quality management in the Upper Crocodile west Catchment

ment practices. Overall, these objectives collectively addressed the critical need for comprehensive water quality management and targeted interventions to control eutrophication, ensuring sustainable water resource use and ecosystem health in the region. Eutrophication and water pollution are global concerns addressed by SDG 6.3, which calls for improved water quality through pollution reduction and safe reuse. The Upper Crocodile West Catchment, impacted by wastewater treatment works, serves as a key case study. Using a random forest model to forecast pollutants (PO_4-P, NH_4-H, NO_3, SS) from simple variables (T, pH, EC, COD), this research provides water managers with a tool to curb nutrient and chemical discharges. Fig. 14.1 illustrates the framework linking this approach to SDG 6.3.

14.2 Materials and Methods

14.2.1 Description of the Chapter Area

The chapter consisted of the upper Crocodile West Catchment. The upper Crocodile River and its catchment areas that drain into the Hartbeespoort Dam act as a sink for all the inflowing pollutants. The dam is situated between the towns of Pretoria and Rustenburg on the eastern boundary of the Northwest Province of South Africa. The nearby Magaliesburg quartzite mountains produced a valley and lake. Commercial, residential, agricultural, and recreational uses are all present in this mixed-use region. The town of Hartbeespoort is next to the dam wall. Sampling surface water is challenging because most sites near the margins of the dam are privately owned, commercial, and residential properties. The surface area of the dam is 20 km^2, it has a 195,000,000 m^3 maximum storage capacity, and a mean depth of 9.6 m. It is 45 m deep at its deepest point. High nutrient loads, primarily from point sources (16 sewage treatment facilities), consisting of 80–300 metric tons of phosphate (P), end up in the dam annually (Carroll & Curtis, 2021). This area comprises 16 sewage works including Lethabile, Brits, Losperfontein, and Rietfontein WWTWs, which release annually into the Crocodile-Jukskei River system, which feeds the reservoir, resulting in dam hypertrophication (Fig. 14.2). The general geological structure is of the Chuniespoort Group, overlain discordantly by the Pretoria Group, and invaded by the Bushveld Complex. The karstic dolomites, which constitute the rocks, would have moved between 2643 and 2520 million years ago (Leyland, 2008).

The region has a two season climate, consisting of a wet season and dry season. Like the rest

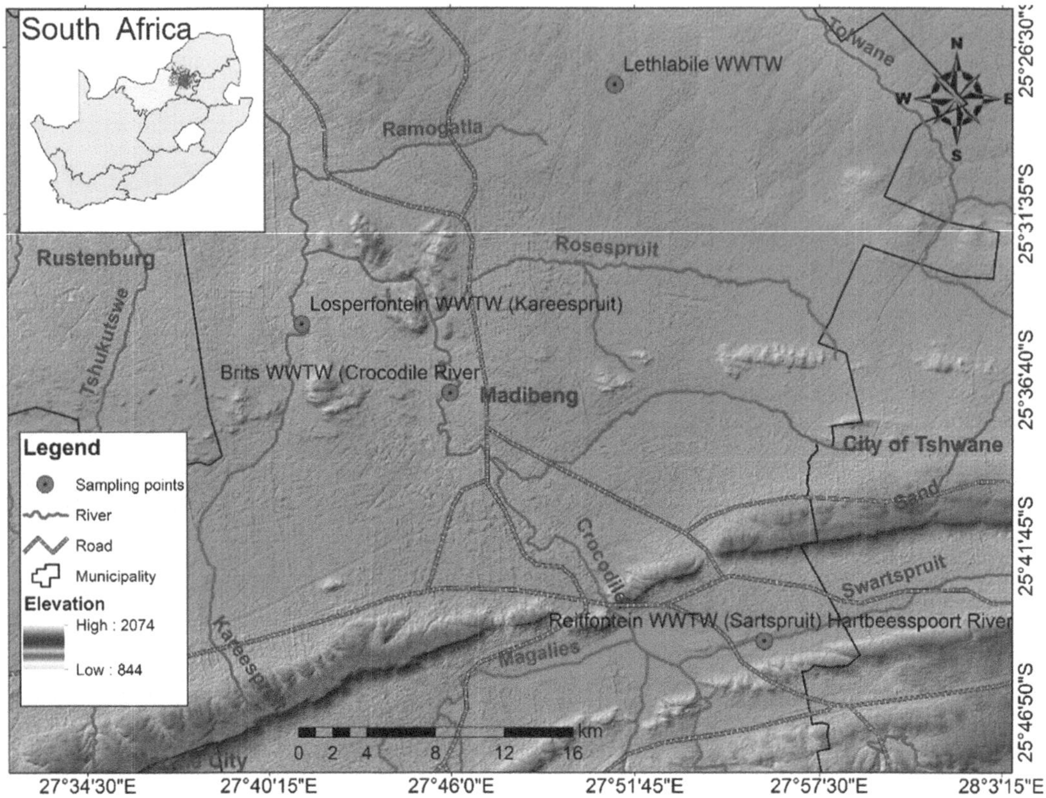

Fig. 14.2 Upper Crocodile West Catchment depicting the individual sampling sites, including Hartbeespoort Dam

of South Africa, the area experiences hot, and rainy summers between September/October and March, and cool, dry winters during the remaining months (Department of Water Affairs and Forestry, 2009). The orographic and microclimate effects resulting from the dam and local thunderstorms result in 559–678 mm rainfall here each year due to the unique semi-arid climate (Department of Water Affairs and Forestry, 2009). Average summer temperatures vary from 15 °C to 30 °C, and average winter temperatures vary from 5 °C to 24 °C in the afternoon. The predominant geographical wind direction is from the northeast. There are various thermoclines that develop in the water column of Hartbeespoort Dam during the summer, the deeper thermal water layer becoming anaerobic. Under these conditions, phosphorus is released from the sediment, leading to the optimal cyanobacterial growth in the lake. However, for unknown reasons, hyperscums are produced in more significant quantities and at a faster rate in the Hartbeespoort Dam than anywhere else in South Africa (Oberholster et al., 2021).

14.2.2 Field Experiment and Sampling Assessment

A field research was conducted in the upper Crocodile West Catchment near the town of Rustenburg in the North West province, South Africa. The study area was selected based on the predominant land use activities within the catchment, which has the potential to influence water quality. Four sampling sites were establish along the mainstream of the upper Crocodile West Catchment and its tributaries to represent the different activities and sections of the river. Site selection also considered accessibility and safety to ensure consistent sampling. For the purpose of this research, the upper reaches of the river were divided into four sampling location designated as Sites (sites 1–4).

This research was carried out over 6 months between June and November 2021 to investigate Water Quality parameters in the upper Crocodile West Catchment. Water samples were collected from the outlets of the WWTPs. Testing was done to find out the degree of pollution of the surface water and eventually to provide a suggestion on how to ensure improvement in the condition of the water. During the 24 sampling periods of the research work, each sampling site was visited monthly. Water samples for physicochemical analysis were taken from the site using the grab sampling technique into sterilized 2-L polyethylene bottles filled to the brim and labelled accordingly.

To ensure sample integrity, the bottles were kept in a cooler box, packed with ice packs for cold storage to keep the samples on lower temperature while transferring. All samples were transported to the Mpumamanzi Laboratory (Nelspruit) and Water Lab (Pretoria, Gauteng Province, South Africa) within 24h. Samples were filtered on arrival (0.45 μm filter) before analysis. Objective water quality parameters were chosen taking into account requirements for instream water users as well as the need for beneficial return flows to the environment. The objective of this method was to reveal the driving forces, results and indicators of water degradation in the area.

Monitoring sites were optimally selected to determine the status of water quality because they were located near upstream pollution sources and water users downstream (Department of Water Affairs and Forestry, 1996b). The suitability of the site was determined in order to take into account the objectives of the research, the local context validity, and the water user features that could be useful to impact by the data and the possible limitations of the available dataset. The main water quality indicators were therefore chosen as the pH, chemical oxygen demand (COD), inorganic phosphate as total phosphorus (PO_4-P/TP), electrical conductivity (EC), dissolved oxygen (DO), inorganic phosphate as total phosphorus (PO_4-P/TP), total nitrogen (TN) as nitrate and nitrite, nitrate (NO_3-N), nitrite (NO_2-N), ammonium (NH_3-N), water temperature (T), suspended solid (SS), and sulphate (SO_4^{2-}). These variables were monitored and studied in this chapter in order to determine variations in quality of the water according to season and space.

The inherent properties of the data on land use activities were also considered. Only the dry season (June-August) and wet season (September-November) were studied for the selected sampling points.

For the physical parameters, the pH, EC, temperature, and suspended solids of the water samples were analysed using a multimeter as described by APHA methods (1998: part 4500 H B; 1998: 2510 B). The other parameters, such as chemical oxygen demand (COD), were also analyzed by using the guidelines set forth by the Standard Methods and the method of the American Public Health Association (APHA, 1998). Furthermore, other parameters such as total phosphate and nitrogen, suspended solids, sulphate, and modified ammonia were measured as follows;

Total phosphate water samples analyses were carried out using the Ascorbic Acid method according to (APHA, 1998, Par 5400-P E). A colorimetric analysis was performed using a spectrophotometer with infrared phototubes. Total nitrogen was measured by the Cadmium Reduction method (APHA, 1998, Part 4500-NO_3^- E) using a reduction column with a spectrophotometer. Sulphate concentration was determined gravimetrically according to APHA (1998, Parts 4500-SO_4^{2-} C and D). Ammonium was also determined by a titrimetric method as described by APHA (1998, Part 4500-NH_3-HC). Data analysis.

14.2.3 Water Quality Standard and Pollution Indices Parameters

The concentrations of the selected water quality variables were determined through laboratory analysis. The results were subsequently compared to target water quality objectives as follows: (a) South African Target Water Quality

Objectives (TWQOs) for the Crocodile River (Department of Water Affairs and Forestry [DWAF], 1996c), both on a regional scale and based on recreational use, (b) the South African Water Quality Guidelines for Aquatic Ecosystems (DWAF, 1996c), (c) the Australian and New Zealand Environment and Conservation Council guidelines (ANZECC, 2000), and (d) the Canadian Council of Ministers of the Environment guidelines (CCME, 2012). These laboratory derived indices were later used to calculate three indices: eutrophication index (EI), comprehensive pollution index (CPI), and organic pollution index (OPI) for evaluation of the pollution status and general water quality in the upper Crocodile West Catchment. The use of these three international and national standards resulted in assessing the contents in order to determine whether the concentrations of other parameters were above the recommended limits.

Initailly the South African Water quality guidelines was apply. The quality water in a stream or dam is asssessed in terms of its suitability for the intended use. Water is not good or bad just because scientific tests indicate it so, not knowing the end purpose of it. Given this, the question must be addressed whether the water satisfies the quality for industrial, farming or potable performance, and if it is a good resource for the functioning of an healthy ecosystem. In South Africa, DWAF (1996b) has produced a set of primary foundations for assessing the quality of water, which includes the Water Quality. The water quality parameters are compared to the generic South African water quality guidelines. Adopted water quality standards. Background water quality standards are modelled upon international literature. However, the adopted water quality standards contain more specific technical and scientific details for each of the constituents of the water quality (Department of Water Affairs and Forestry, 1996a, 1996b, 1996c). The content guidance for each water quality constituent, if available, is considered in terms of numerical information or a narrative description of its impact on the acceptability of a water use. In this chapter, the recreational (DWAF, 1996a) and aquatic ecosystems values (Department of Water Affairs and Forestry, 1996b), were the quality guidelines used.

The second WDWC value applied was that of the Australian and New Zealand Environment and Conservation Council (ANZECC, 2000). Methods to derive the physical and chemical water quality guidelines for ecosystem management (referred to as guidelines trigger values) have also been updated, given advances in ecosystem knowledge and technology. There is now more emphasis on problem-oriented water quality management than on managing one parameter at a time.

In a practical sense, this means having monitoring programs that enable managers to measure biological parameters and related physical and chemical factors associated with them both in water and in the sediment. However, guidelines have been developed for these other indicator types (e.g., biological assessment, sediment quality and environmental flows).

The guidelines highlight key elements of water quality monitoring and assessment programs, with extensive reference to the recent Monitoring and Reporting Guidelines of the National Water Quality Management Strategy (ANZECC, 2000). The Canadian Council of the Ministry of the Environment (CCME) is one of the water quality guidelines used in this research. A universal communication tool that summarizes complex water quality data into a simple index format (CCME, 1999). In 1990, CCME WQI was created for locations in Australia and Canada. It will enable effective utilization of monitoring data, provide consistency among beneficial uses, and facilitate reporting at various diversity of temporal and spatial scales. The index can include at least four and up to 47 water quality parameters (Kachroud et al., 2019; Poonam et al., 2013). The CCME guidelines have been applied by many scientists in the assessment of surface water and other water bodies (rivers, lakes, reservoirs) around the world (Lumb et al., 2006; Ruiz-Medina & Espejo, 2012; Uddin et al., 2021). The water quality parameters in this chapter were also compared with four international and local water quality standards and three pollution indices (EI, CPI, and OPI) in this chapter to determine whether the water had a higher quality status.

14.2.3.1 Comprehensive Pollution Index

This technique has been extensively applied as a qualitative evaluation of the quality of water (Njagi et al., 2022). Many researchers have used the CPI to assess surface water quality. (Matta et al., 2017; Mishra et al., 2016; Son et al., 2020; Tang et al., 2011). The CPI was chosen because it offers insightful data on how pollution affects the water quality of rivers. The comprehensive pollution index (CPI) offers an overall assessment of dam water quality by incorporating the individual pollution indices (PIs) of each measured parameter. By using data from water quality parameters, the CPI for the four sites was calculated and applied to determine the river's pollution level (Liu et al., 2011). This is calculated as follows:

$$CPI = \frac{1}{n}\sum_{i=1}^{n} PI \tag{14.1}$$

The CPI is the average of the PI and of the monitored pollutants. Here, n is the number of parameters of the water quality selected, and PI is the pollution index of the ith parameter. And the PIs of each parameter were calculated as follows:

$$PI = \frac{C_i}{S_i} \tag{14.2}$$

In this sense, C_i Represent the concentration measured of the parameter, and the corresponding water quality standard is S_i or the permissible limit of the parameter. The CPI consists of five groups. (Son et al., 2020):

- Clean: CPI between 0 and 0.20 (Category 1).
- Sub-clean: CPI between 0.21 and 0.40 (Category 2).
- Slightly polluted: CPI between 0.41 and 1.00 (Category 3).
- Moderately polluted: CPI between 1.01 and 2.00 (Category 4)
- The heavily polluted CPI is greater than 2.01 in Category 5.

14.2.3.2 Eutrophication Index

The EI suggested by the literature. Chapter et al. (Mnyango et al., 2022), was used to determine the trophic status of the Crocodile West Catchment. EI is useful because it provides information on how pollution levels impact the risk of eutrophication and assesses the risk of eutrophication based on the concentrations of TN (DIN), PO_4-H (DIP), and COD in the river. Using the four communities in the sub catchment region of the main upper Crocodile West Catchment, the EI of the river was determined during the 6-month chapter for the upper Crocodile West Catchment as follows:

$$EI = \frac{COD \times DIP \times DIN}{4500} \times 10^6 \tag{14.3}$$

where COD is the concentration of chemical oxygen demand; DIN is dissolved inorganic nitrogen; and DIP is dissolved inorganic phosphorus (as PO_4-H). the threshold value of DIP is an important index to distinguish the eutrophication level of water bodies. The EI can then be categorized into two groups (Son et al., 2020):

EI < 0 (zero eutrophication in the body), and
EI > 0 indicates eutrophication (there is eutrophication in the water body).

14.2.3.3 Organic Pollution Index

The organic pollutant index (OPI) measures pollution in rivers and lakes. (Liu et al., 2011). It is considered a reliable method for evaluating pollution in the water quality. chemical water quality. In this chapter, the organic pollution levels in the Crocodile West Catchment were evaluated using the OPI, based on the average values relative to established standard limits. (Liu et al., 2011). The standard limit for evaluating the organic pollution types is shown in Equation 14.3.

$$OPI = \frac{COD}{COD_s} + \frac{DIN}{DIN_s} + \frac{DIP}{DIP_s} + \frac{DO}{DO_s} \tag{14.4}$$

Dissolved inorganic nitrogen (DIN, mg/L), dissolved inorganic phosphorus (DIP, mg/L), dissolved oxygen (DO, mg/L), and chemical oxygen demand (COD, mg/L) were also assessed in the Crocodile West Catchment.

In the Crocodile West Catchment Resource Water Quality Objectives as specified by the

South African Department of Water and Sanitation, the following benchmark concentrations were applied.

CODs = 15 mg/L; DINs = 10 mg/L; DIPs = 0.125 mg/L; and DOs = 4 mg/L (Department of Water Affairs and Forestry, 2003).

Thus, OPI was classified as follows (Guasmi et al., 2010; Son et al., 2020):

1. Category 1: $OPI < 0$ (excellent).
2. Category 2: $0 \leq OPI < 1$ (good).
3. Category 3: $1 \leq OPI < 2$ (beginning to be contaminated).
4. Category 4: $2 \leq OPI < 3$ (lightly polluted).
5. Category 5: $3 \leq OPI < 4$ (moderately polluted).
6. Category 6: $OPI \geq 4$ (heavily polluted).

14.2.4 Statistical Analysis

14.2.4.1 Mann-Whitney U-Test and Kruskal-Wallis Test

The Mann-Whitney U-test is a non-parametric statistical method used to compare two independent groups. The Mann-Whitney U-test focuses on comparing the medians of two independent groups rather than their means, making it particularly robust when the data contain outliers or do not follow a normal distribution. If the distributions of the two groups are symmetric and have a similar shape, the mean and median for each group will be the same. In such cases, the Mann-Whitney U-test can give results similar to the t-test, as both tests examine differences in central tendency.

Due to the small sample size in our data (three observations per group for each variable) and the absence of normal distribution, the non-parametric Mann-Whitney U-test is an appropriate method for assessing significant seasonal differences. By focusing on the medians of these parameters across seasons, the test provides insights into whether seasonal variations have a meaningful impact on water quality, helping to better understand environmental trends and dynamics.

According to the Mann-Whitney U test (Table 14.1), the season has a sole impact on temperature, SS, and $NO_2 + NO_3$ at the 0.05 significance level.

The Kruskal-Wallis test is used here is a non-parametric test that considers whether several samples come from a common distribution, was developed by Kruskal and Wallis in 1952. It is a generalization of the Mann-Whitney U-test for two groups. The null hypothesis of the test states that the mean ranks of the groups are equal. Known as a "one-way Kruskal-Wallis Test on ranks", this method serves as the nonparametric counterpart to the one-way Kruskal-Wallis Test but does not require the data to follow a normal distribution. This made the Kruskal-Wallis test particularly suitable for analyzing our data, exhibiting non-normal distribution patterns and including significant outliers. By focusing on rank data rather than raw values, the test minimizes the influence of outliers and non-normal distributions.

In the current chapter, the Kruskal-Wallis test was employed to analyze the effects of four sites on the dependent variables [COD, EC, T, PO_4, SO_4, NH_4, pH, SS, $NO_2 + NO_3$ (N dissolved in water) and DOAs] our data did not satisfy the normality assumptions required for parametric tests, the Kruskal-Wallis test provided a robust non-parametric alternative. This test allowed us to evaluate whether there are statistically significant differences in water quality parameters across the different sites without assuming a normal distribution. Each dependent variable [COD, EC, T, PO_4, SO_4, NH_4, pH, SS, $NO_2 + NO_3$ (N dissolved in water), and DO] was analyzed separately to isolate its specific impact on water quality.

Based on the results of the Kruskal-Wallis test (Table 14.2), there was no significant effect of the site on pH and temperature at the 0.05 significance level. The p-values for both variables exceeded the alpha level of 0.05, indicating that we failed to reject the null hypothesis. Therefore, we conclude that there is no statistically significant difference in pH and temperature across the different sites.

Table 14.1 Results of the Mann-Whitney U test, which compare the DO, EC, COD, NH_4, pH, $NO_2 + NO_3$, SO_4, SS, T and PO_4 levels between dry and wet seasons

Variable	COD	DO	EC	NH_4	$NO_2 + NO_3$	pH	PO_4	SO_4	SS	Temperature
Mann-Whitney U test statistics	50	57	63	51	28	44	45	46	34	1
Wilcoxon W test statistics	128	135	141	**129**	**106**	**122**	**123**	**124**	112	79
Z test statistics	−1.271	−0.866	−0.52	−1.212	−2.54	−1.617	−1.559	−1.501	−2.194	−4.1
p-Value (2-tailed)	0.204	0.386	0.603	0.225	0.011*	0.106	0.119	0.133	0.028*	0*

*p-Value < 0.05

Table 14.2 Kruskal-Wallis test results for COD, DO, EC, NH_4, $NO_2 + NO_3$, pH, PO_4, SO_4, SS, and temperature across four sampling sites (degree of freedom = 3 for all tests)

Variable	COD	DO	EC	NH4	$NO_2 + NO_3$	pH	PO_4	SO_4
Kruskal-Wallis test statistics (H)	10.644	18.28	14.4	9.553	13.027	13.027	8.667	18.207
p-Value	0.014*	0*	0.002*	0.023*	0.005*	0.155	0.034*	0*

*p-Value < 0.05

14.2.4.2 Machine Learning - Model Building and Performance Metrics

The Random Forest Regression approach was implemented using the R programming language environment to forecast the contaminants. Choosing random forest as a regression method over simpler options such as linear regression or decision trees can be justified for several reasons: 1. Handling non-linearity; 2. Robustness to overfitting; 3. Providing feature importance; 4. Flexibility in analyzing all types of variables without requiring preprocessing; 5. High accuracy performance; and 6. Parallelization (Antoniadis et al., 2024; Couronné et al., 2018). PO_4-P, SO_4, NH_4-H, NO_3, and SS in the Upper Crocodile West Catchment based on the explanatory variables (T, PH, EC, and COD) were forecasted. These variables were specifically chosen to be forecasted for certain reasons: 1. These variables have shown strong correlations with the environmental variables; 2. These variables are available with consistent and high-quality data; 3. Diversity and redundancy of these variables to reduce overfitting of the model; 4. These variables can significantly impact the prediction and are likely to influence the target outcome of the paper. Random forest is a powerful machine learning technique in which a large number of individual decision trees (n = 500 in the present chapter) are randomly sampled to create this ensemble supervised learning model, which then reports the total response of all the trees. The model was constructed using the *random forest* package in the R environment, version 3.4.2 (Liaw & Wiener, 2014). The number of trees, the number of input variables, and the size of the nodes are, in fact, some of the critical characteristics that control the performance of the model. The random forest regression model was further improved by optimizing these parameters.

Indeed, training a random forest regression model involved several steps to ensure robust and accurate predictions. First, the data was prepared through collection, cleaning, feature engineering, and splitting into training and testing sets, where each subset is known as a bootstrap. Multiple decision trees (n = 500 in this chapter) were subsequently trained for different bootstrap samples of the training events, with each tree examining a random subset of features at each split in order to decorate the trees. They are allowed to grow to their maximum depth (or stop due to a predefined criterion).

The final model aggregates the predictions from all trees, typically by averaging, to form the ensembleHyperparameters including the number of trees, maximal depth, and maximum number of features per split were optimized with methods such as grid-search and cross validation.

The performance of the generated model was assessed using statistical parameters Feature importance, Mean Squared Error (MSE) and R-squared (R^2) were extracted from random forest package in R programming language. R-squared (R^2) explains how good of a fit the regression line is to the data; MSE indicates how far the average data point is from the fitted line. Afterward, the optimized model was trained on the full dataset and tested in new data.

This approach ensured robustness to overfitting, handled non-linear relationships, and provided insights into feature importance. We used the "Increase in Node Purity" (IncNodePurity) to evaluate the importance of predictor variables for our model. The IncNodePurity expresses the change in the homogeneity of the groups created by the trees (using the Gini coefficient as a measure). Additionally, the "ggplot2", "reshape2", "R0", and "gridExtra" packages were used to build the graphical visualization of the results (Fig. 14.3).

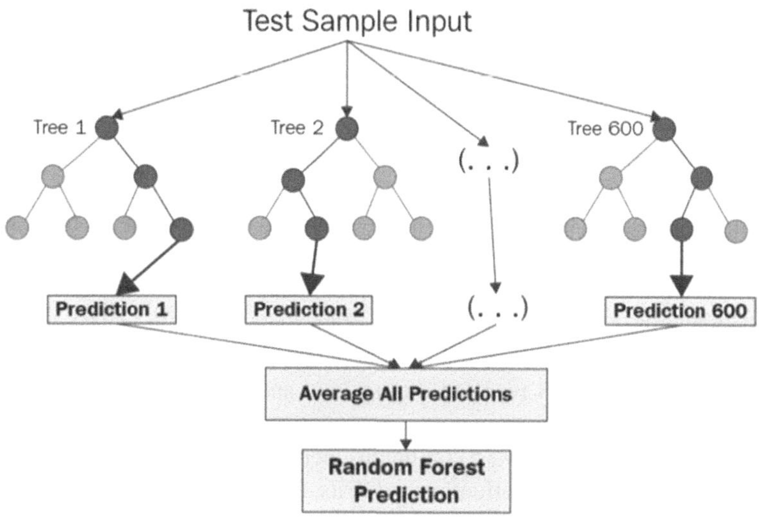

Fig. 14.3 Illustration of the random forest regression model algorithm

14.3 Results

14.3.1 Chemistry Parameters

The water quality status of the water quality parameters for sampling Sites 1–4 (Lethabile WWTW, Brits WWTW, Loose-fitting WWTW, Rietfontein WTW), separated by seasons. The Crocodile West Catchment has an unacceptable amount of nutrient loading, according to the Crocodile Catchment's compliance status. TP and TN concentrations were above the levels of the Klip River Water Quality objectives (South Africa, DWAF, 2003) at their maximum allowable levels. The high concentration of TN (>1.3 mg/L) and total phosphate (>0.1 mg/L) at Sampling Sites 1–4 provided evidence for the degree of eutrophication. The average phosphate level was above 0.125 and between 0.144 and 0.58 mg/L, the average total nitrogen level, screened at the 1–4 sites, was above the TWQOs.

The ammonium levels in the Crocodile West Catchment exceeded the allowed limits. Ninety per cent (90%) of the ammonium nitrogen complied with the recommended 0.044 mg/L of the Klip River Water Quality objectives (Department of Water Affairs and Forestry, 2003). The NH_4-N concentration varied between the sampling sites (Sites 1–4), ranging from 0.36 to 0.76 mg/L, with higher values occurring at Sampling Site Lopserfontein (0.76 mg/L), which does not comply with the Klip River Water Quality guidelines (Department of Water Affairs and Forestry, 2003). After the junction of the Crocodile and Hennops rivers, below the town of Brits, there was a modest increase in the TSS from the sampling sites. The findings show that the upper Crocodile West Catchment had average TSS concentrations that ranged between 22.64 and 33.03 mg/L, which were over the guidelines of the Klip River Water Quality objectives (Department of Water Affairs and Forestry, 2003). At sampling Sites 1–4, the average pH ranges were between 7.81 and 8.34 mg/L which were within the guidelines of the Klip River Water Quality objectives (Department of Water Affairs and Forestry, 2003).

The COD average over the chapter period ranged between 46.38 and 50.63 mg/L, which was above the Klip River Water Quality objectives. The highest average concentration of COD was recorded at the sampling site point (81.40 mg/L), close to Brits. The average EC ranged between 95.09 and 612.04 mS/m, below the Klip River Water Quality limits of 80 mS/m. The average DO was between 4.04 and 8.13 mg/L, which was above the Klip River Water Quality limit for aquatic ecosystems (<0.4 mg/L). However, most other parameters, such as temperature (25.1–25.2°C) and sulfate (46.46–114.68 mg/L), were within the recommended Klip River Water Quality limits at all sampling sites.

14.3.2 Trend Analysis

The results of the trend analysis of the four sampling sites (Site 1—Lethabile; Site 2—Brits; Site 3—Losperfontein; Site 4—Rietfontein) during the 6 months of chapter. All three sampling sites have shown a common TN trend, with the lowest concentrations of TN reported during June and the highest in November.

The trend for TN shows a continuous increase within sampling sites Rietfontein and Brits, from July (0.457) to August (4.5426), whereas the sampling points were at an equilibrium state (TN = 1.61) and later slowly increased to another constant state. TN was 1.952, except for Site 3 (Losperfontein WWTW), which was higher above the same TN strength (TN = 2.71) than the other sites (Sites 2 and 4; Brits and Rietfontein). The only exception was Lethabile WWTW (Site 1), which showed a continued increase in concentrations from TN at 2.85–2.65 mg/L, and a later increase at a contact state of TN was 6.83. The TN ranges remained within the TWQOs. The trend analysis showed that PO_4^{3-} PO_4^{3-} concentrations followed a similar trend as pH, which goes in a zigzag (Brownian movement) during the 6-month season (June-November) for Sampling Sites 1–4. The PO_4-P concentrations were highest at 0.73 mg/L (Lethabile WWTW) during August and later at their lowest at 0.022 mg/L (Brits WWTW) in October, compared to the other sampling sites, the lowest concentrations of PO_4-P (Rietfontein WWTW, Brits WWTW, and Losperfontein WWTW) at a range of <0.07. The data revealed that the TN concentrations fluctuated above and below the Klip River Water Quality limits at the four sampling sites. There was a general increase in the trend of ammonium concentrations at the four sampling sites (Sites 1–4), with the maximum concentration measured in August at Site 3, Losperfontein (0.97 mg/L), while the minimum concentration (0.14 mg/L) occurred in September during the 6 months of sampling. In September, an increase and decrease were observed at Sampling Sites 2 and 3, with a decrease (0.14 and 0.71 mg/L) and an increase (0.48 and 0.78 mg/L) in the NH_4 concentrations. However, the ammonium ranges remained above the Klip River water quality values. There was a slight increase in the DO trend from June to July (4–8 mg/L) for the four sampling sites (Brits WWTW, Rietfontein WWTW, Losperfontein WWTW, and Lethabile WWTW). From July, DO decreased to August and dropped further to October (3.50–5.10 mg/L), while increasing towards November. The exception was observed downstream of the Losperfontein WWTW, where the DO concentration increased from June to October and later decreased to November (DO was 8.94 mg/L). The lowest concentration of DO was measured downstream of the Brits WWTW (3.11 mg/L), which was within Klip River water quality value range, while the highest concentration was measured downstream of the Losperfontein WWTW (9.65 mg/L). The latter value was above the Klip River Water Quality standards. The pH values did show an increase and decrease from June to November during the sampling of the four sampling sites (1–4). Three sampling sites (Rietfontein WWTW, Lethabile WWTW, and Brits WWTW) slightly increased from 8.732 to 7.633 and decreased to about 7.633, while a sharp increase up to a pH value of 8.61 was observed. The trend in SS shows a slow increase from Sites 3 and 4 from June to July (21–26 mg/L), then a decrease to August (21–27 mg/L), which later increases to November (30–37 mg/L).

On the other hand, a contrary observation was observed at sites 1 and 2 with similar concentration ranges, 27 mg/L (July), and decreasing to 22–25 mg/L in August, and finally increasing to November (21–28 mg/L). The spike in the SS highest value (Site 1) was recorded in September, where the concentration was 41.54 mg/L. This indicates that the rainy/wet season (October-November) of the chapter area may have resulted from contamination from runoff containing SS flowing into the main Crocodile River catchment, where the WWTW is being discharged. The concentrations in all sampling sites remained above the Klip River Water Quality objectives (South Africa, DWAF, 2003).

The COD trend analysis shows an increase with sharp spikes at Sites 2 and 4 (70–120 mg/L), which falls in September (23–69 mg/L) and later increases at a constant rate (68–86 mg/L). Sites 1 and 2 increased throughout the research period of June-November. The trend in EC remained above the limit at all sampling sites (Sites 1–4). This was expected because EC is an indication of salts in the water and vice versa for sulfate concentrations. The EC range of Site 1 was higher with two peaks at 898 and 915 mg/L (August and October) in comparison to the other sampling sites (Sites 2, 3, and 4), which were all below the concentration of 350 mg/L from June to November. The concentrations in all sampling sites remained above the Klip River Water Quality limits (Department of Water Affairs and Forestry, 2003). The sulphate concentration gradually increased, peaking at 40–105 mg/L and the lowest at 55–122 mg/L. On the other hand, from June to November, the temperature concentrations grew rapidly across the sampling sites (1–4), with Site 4 recording the highest temperature (29.21°C) and the lowest temperature (16.34°C). These concentrations observed remained below the Klip River Water Quality limits (Department of Water Affairs and Forestry, 2003).

14.4 Water Quality Indices

14.4.1 Comprehensive Pollution Index

The CPI results was calculated for the four sampling sites (Sites 1–4) with regard to their seasonal variation. The findings demonstrated that the standard deviation (minimum-maximum) of the average concentration among Sites 1–4 for all watrere samles ($n = 4$) were 3.60 ± 0.6, 2.35 ± 0.4, 3.56 ± 0.5 and 3.64 ± 0.3 for wet season and 2.88 ± 1.1, 2.49 ± 0.9, 3.02 ± 1.2, 2.65 ± 0.75 for the dry season, respectively. The CPI for all sampling sites ranged from 3.87 to 2.93, 1.79 to 3.94 (Site 1); 2.08 to 2.81, 1.47 to 2.79 (Site 2); 3.11 to 4.12, 1.67 to 3.13 (Site 3); 3.11 to 3.53, 1.79 to 3.11 (Site 4) for both the dry and wet season resptively. This results demonstrate that the hightest concentration was observed in Novemever a Site 1 (4.115 mg/L), wet season, whereas the lowest CPI of 1.477 was observed in June at Site 2 (dry season). These sites were close to Rietfontein and Brits.

14.4.2 Eutrophication Index

The EI calculations showed that at Site 1, Wet season (September to November) had three times increase in eutrophication in comparison to the dry season (June to August) during the chapter period of 6 months' sampling in 2021. During the sampling period of Site 2, the wet season experienced eutrophication conditions of more than one and a half times (>1.5 times) from September to November, in comparison to the dry season (June to August). A similar pattern was observed at Sites 3 and 4, where eutrophication levels were higher during the wet season compared to the dry season. The results of the eutrophication index (EI) calculations for all four sampling sites across both seasons are discussed.

14.4.3 Organic Pollution Index

Monthly variations in the OPI showed fluctuating values at the four sampling sites. The highest value (11.58) was recorded at Site 1 (Lethabile WWTW) in November (wet season), while the lowest value (4.49) was recorded in June (dry season) at the same sampling site. The OPI disparity in values was as follows: 11.58–4.49 during November and June at Site 1 (Lethabile WWTW), while it ranged from 11.24 and 6.70 in July and June at Site 2 (Brits WWTW). At the third site (Losperfontein WWTW), the values ranged from 35 and 11.41 in June and November, respectively. The values of the OPI that were recorded at Site 4 (Rietfontein WWTW) ranged from 4.97 and 10.35 in September and November (wet season). The results from all four sites (1–4) indicated that pollution levels were significantly higher during the wet season (OPI = 11.31, 10.18,

9.15, and 7.80) in comparison to the dry season (OPI = 7.88, 9.67, 7.37, and 8.48).

14.4.4 Seasonal Distribution of the Common Pollution Parameters

The rainfall in June, July, August, September, October, and November was 2.0 mm (0.1 in.), 7.0 mm (0.03 in.), 21 mm (0.8 in.), 57 mm (2.2 in.), 8.9 mm (3.5 in.), 9.1 mm (4.3 in.), respectively. The mean and standard deviation for the chemical parameter at all four sites were calculated from 24 water samples collected during this sampling period (Lethabile WWTW, Brits WWTW, Losperfontein WWTW, and Rietfontein WWTW). COD, NH_4-H, DO, TN, PO_4-P), SO_4^{2-}, pH, T, and SS were all examined using the Kruskal-Wallis Test and the Mann-Whitney U-test.

14.4.5 Kruskal-Wallis Test and Paired Mann-Whitney U-Test

Twenty-four surface water samples were collected in June, July, August, September, October and November 2021. During the sampling period, the mean value for the four chemical parameters was COD, NH_4-H, DO, TN, PO_4-P, SO_4^{2-}, pH, T, and SS for the 6 months (June to November 2021), as well as the Kruskal-Wallis Test for all four sampling sites (Lethabile WWTW, Brits WWTW, Losperfontein WWTW, Rietfontein WWTW). From these results, it can be observed that NH_4-H (0.178), PO_4-P (0.051), pH (0.0643), and SS (0.616) did not show any statistical differences ($p > 0.05$), whereas TN (0.00027), EC (0.997), SO_4^{2-} (0.959) and DO (0.984) were the only variables that showed significant variation ($p < 0.01$ and $p < 0.05$) across the 6-month sampling period between the four sampling sites.

According to the geographical sampling location, the maximum concentration of the water temperature, pH, SS, NH_4, SO_4, DIN, PO_4-P (DIP), COD, PO_4, ranged from 3.062 to 310 mg/L, occur during wet season and DO and EC (5–310 mg/L) during the dry season. For the analysis of mean difference (Kruskal-Wallis Test), the only variables that showed statistically significant changes were SS, T ($p < 0.01$), NH_4-H, PO_4-P, SO_4^{2-}, pH, and TN ($p < 0.05$). COD, EC, and DO did not exhibit any statistical differences ($p > 0.05$). The Mann-Whitney U-test findings indicated that the following variables were statistically significant for PO_4-P, pH, SO_4^{2-}, $\alpha = 0.05$ ($p = 0.02, 0.03, 0.68$); whereas the opposite was true for COD, EC, T, NH_4-H and DO, $\alpha = 0.05$, ($p = 0.69, 0.34, 0.69, 0.69, 2.84$), respectively. However, this difference was only significantly closer to the towns of Brits, Lethabile, Losperfontein and Rietfontein. To evaluate the ability of a water body to purify itself and to ascertain its pollution level, the COD and DO levels can be utilized as pollution indicators. The higher the COD, the closer the degree of organic pollution of water will be. COD is usually measured to determine the discharge and treatment of wastewater (Saral & Goncaloglu, 2008).

14.4.5.1 Regression Model Approach

All measures showed that our model performed better in terms of prediction (Table 14.3). The Random Forest Regression model's prediction accuracy varies according to the Mean Squared Error (MSE) values for the various environmental variables. The MSE for PO_4-P is the lowest at 0.01 mg/L, suggesting highly accurate predictions for this variable. In contrast, SO_4 has the highest MSE at 75.35 mg/L, indicating less precise predictions. NH_4-H and NO_3 have relatively low MSE values of 0.02 and 0.39 mg/L, respec-

Table 14.3 Metrics for measuring the performance of the optimized random forest regression model

	Environmental variables				
Parameters of statistics	PO_4-P (mg/L)	SO_4 (mg/L)	NH_4-H (mg/L)	NO_3 (mg/L) (mg/L)	SS (mg/L)
MSE	0.01	75.35	0.02	0.39	4.61
R^2	0.70	0.89	0.68	0.76	0.62

Table 14.4 Feature importance of the predictor variables of the random forest model according to the IncNodePurity

	PO_4 (mg/L)	SO_4 (mg/L)	NH_4 (mg/L)	NO_3 (mg/L)	SS (mg/L)
Temperature (°C)	0.09	801.2	0.13	2.97	18.21
pH	0.07	1597.8	0.16	2.24	23.20
EC (µg/L)	0.12	2430.1	0.15	4.03	25.45
COD (mg/L)	0.06	1033.5	0.17	3.05	24.02

tively, reflecting good prediction accuracy. The MSE for SS is 4.61 mg/L, which is higher than PO_4-P, NH_4-H, and NO_3 but significantly lower than SO_4, suggesting moderate prediction accuracy. These differences highlight the model's varying performance across different environmental variables, likely due to the inherent complexity and variability of each variable. According to Table 14.4, the IncNodePurity results revealed that EC was the most highly contributed variable for all the models. Moreover, Fig. 14.4 compares the actual pollutant values and the forecasted ones using the proposed random forest model for some time, and it also demonstrates how close the forecasted readings are to the actual readings, besides the ability of our random forest model to follow the actual pollutant level accurately. This model's exceptional performance can be attributed to its Machine learning architecture which gives the random forest model the privilege of extracting the most essential features from the input time series data of environmental predictors, allowing the model to understand complex patterns and trends within the dataset (Fig. 14.4).

14.4.5.2 Assessment of CPI, EI, and OPI in the Crocodile River

Evaluation levels of pollution are highest at distances upstream of the Crocodile River. The water quality aggregates depended on the location and the sampling time. Water quality at the four sampling sites were higher during both dry and wet seasons compared to the rest of the river. The river was seriously polluted, OPI > 4 and CPI > 2.0 at most of the river sections with organic pollution. River water was at a high risk of eutrophication, since EI > 0. Pollution indices calculation. Water pollution indices calculations results (Table 14.1) indicated that the pollution levels in the rainy season recorded higher values than in the dry season for Sites 1–4 for both CPI, OPI, and EI. As a result, the water of the Crocodile River was found to be highly contaminated and eutrophic according to CPI, EI, and OPI estimates.

The river catchment was most contaminated at the Rietfontein WWTW (wet season), followed by the Losperfontein WWTW, according to the estimated CPI, which gives a general assessment of the river water quality situation (dry season).

The CPI had declared this river as one of the most polluted. The general rating with regard to the river was also estimated using the CPI, with the results indicating potable water quality during the dry season. The difference in measurement elements at each site was attributed to the ambient nutrient concentrations for EI and OPI. For example, sites located closer to the Lethabile WWTW near Brits town had higher levels of DIN and DIP, implying that nutrient loading may have been because of water runoff. Furthermore, the hydrodynamics of each water system (faster flow rates, more tidal exchange, or wave activity) could affect the nutrient content (Park et al., 2005). Considering the relatively high DIP levels from the upper Crocodile West Catchment, which is draining into the Hartbeespoort Dam, as compared to the lower DO concentrations after the discharge of nutrient-contaminated river water, high nutrient loading will have a profound effect on the dam.

Additionally, we found that COD levels above the baseline standard concentration (15 mg/L) were used to calculate the value of OPI at all four sampling sites. The water in the Upper Crocodile West Catchment was eutrophic and included organic pollution due to the EI and OPI assessments.

The mathematical selection of standard concentrations is essential for determining the quality of the water, based on CPI, EI, and OPI calculations. There is no doubt, however, that

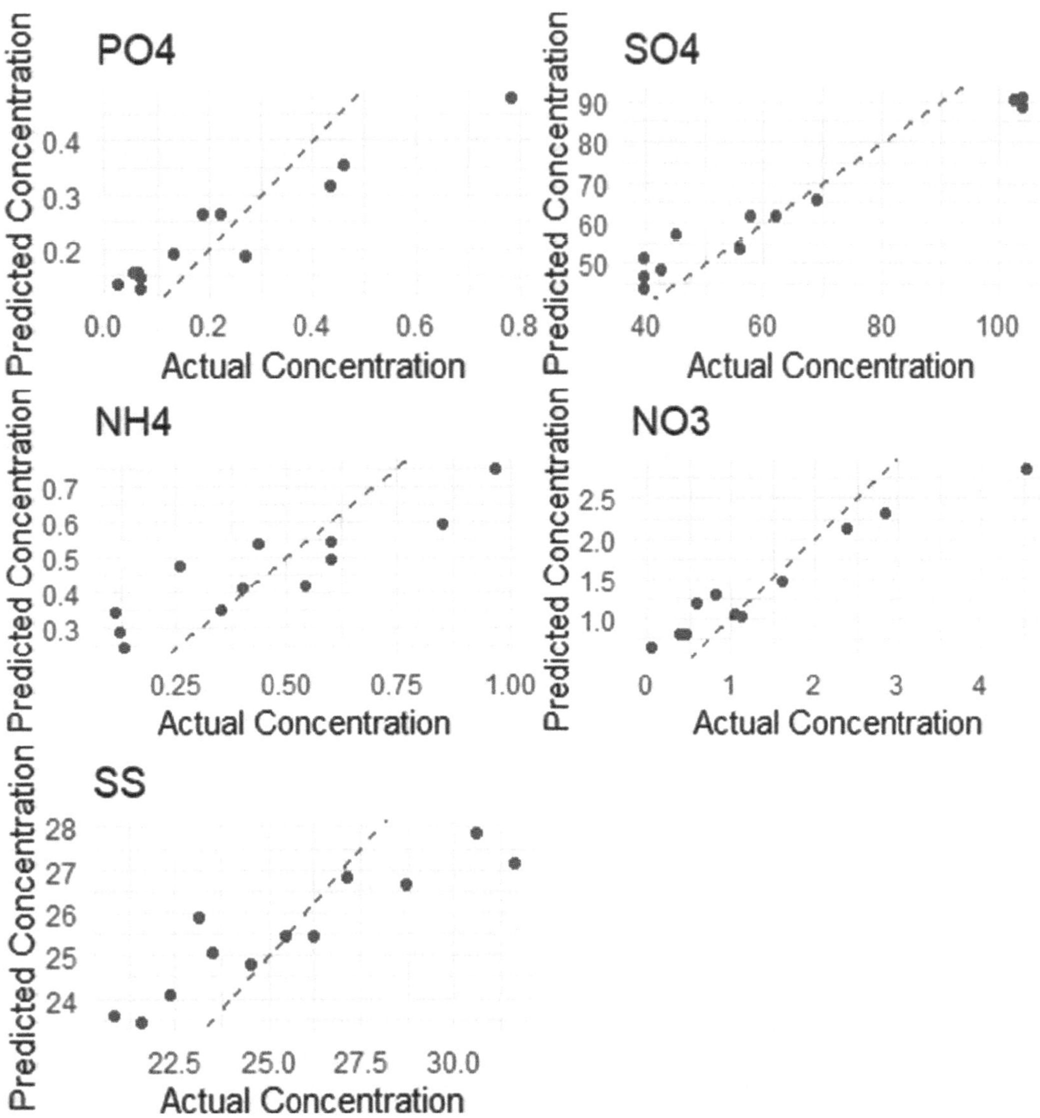

Fig. 14.4 The actual pollutants compared to the forecasted values using the proposed random forest model (unit for variables is mg/L)

four Crocodile River catchments were found to be severely polluted. The average CPI and OPI were 3.025 and 8.35, respectively. The comprehensive and organic pollution degrees were less than 4 and 0.2, indicating that the upper Crocodile West Catchment survey area had high overall and organic pollution. With an average EI value of 14,776.03, it is noticeably higher than 0. The eutrophication of the rivers is abnormally high. This area was negatively impacted by organic pollution and nutrient enrichment. Variations in DIP had the most significant impact on the EI and OPI values of the Crocodile River. The usage of fertilizers and the discharge of industrial and municipal waste may have caused high P concentration.

Moreover, low dissolved oxygen (DO) levels and high salinity in rivers and dams can elevate dissolved inorganic phosphate (DIP) concentrations, as DIP provides sulphate that supports microbial reduction in sediment, resulting in the issue of phosphorus in eutrophic and hypertrophic environments (Venter, 2012). The results demonstrate that eutrophication in rivers and dams can be prevented by managing urban and industrial sewage and effluent flows, as well as by removing N and P from water bodies through agricultural processes (Venter, 2012).

14.5 Discussion

Most parameters that reflect eutrophication processes and their driving forces exhibit pronounced seasonal variation, similar to the nutrient loads in surface waters. Parameters such as temperature, ammonium, phosphate (DIP), total nitrogen (DIN), DO, pH, SS, COD, EC, and SO_4 concentrations downstream of the water sample body change due to seasonal variation. This chapter examined the role of eutrophication in the seasonal variation of key water quality parameters and compared the comprehensive and organic pollution indices between the dry and wet seasons and further employed a random forest machine learning algorithm model to predict pollutants based on environmental contaminants. In this chapter, the average concentration of the individual variables DIN, DIP, DO, COD, SS, pH, NH_4, EC, SO_4, and T for all water samples in the Crocodile River was 2.26, 0.305, 5.69, 57.14, 0.52, 28.67, 8.08, 294.15, 71.02, 25.14 mg/L, respectively.

A previous chapter conducted by Masawi & Oberholster (2022) reported that DIN and DIP concentrations in water samples were the collected from Boksburg Lake in South Africa, ranged from 0.05–0.12 mg/L and 0.417–1.10 mg/L, respectively, during pre-and post-aeration periods. DIN and DIP concentrations of the WWTW sample concentrations in the Crocodile River were more than two times higher than those collected from the Boksburg Lake. The measured nutrient concentrations are certainly large numbers, indicative of severe eutrophication in the area. Bunce et al. (2018) observed that nutrient loading may cause seasonality shifts in aquatic systems. During the wet and dry seasons of 2021, high phosphate and ammonium concentrations were measured at Site 2 (Lethabile) and Site 3 (Losperfontein). The ammonia concentration was significantly increased by effluent from wastewater treatment works. This implies that further treatment, especially the removal of residual ammonia, might be needed for the anammox conversion method and the discharge criterion selected. In a conventional treatment process, ammonium-nitrogen is entirely or partly converted to nitrate by the action of microorganisms, which may produce high levels of nitrate in the effluent. These results are also in agreement with other South African rivers such as Mhlathuze, Vaal, and Klip (Bezuidenhout et al., 2002; Phungela et al., 2022). High ammonium levels can have an impact on both the aquatic habitat and human health (Department of Water Affairs, 2012; Van Ginkel et al., 1996). Municipal wastewater treatment facilities, fertilizer use, and extensive livestock raising are the main sources of ammonium. Industries have the potential to leak some ammonium into water resources.

Phosphate is a crucial limiting nutrient for phytoplankton proliferation. As such, the eutrophication and algal blooms were caused by excessive amounts of water resources (Welch & Cooke, 2005). High concentrations of phosphorus were detected at Site 3, located downstream of a WWTW works with activated sludge systems. These systems are based on the enrichment of PAOs for efficient biological phosphorus removal. As outlined in chapter (Bunce et al., 2018), operating conditions, especially in terms of presence of metabolic requirements including carbon, glycogen and proper electron acceptors, are very important for growth and activity of PAOs. Thus, the efficient utilization of these factors is necessary to maintain the proliferation of PAOs and enhance the phosphorus removal capability of wastewater-based systems. Results obtained from Bunce's (2018) chapter show that the macrophyte effectively removes phosphorus from wastewater, but it does not always produce effluent with phosphate levels lower than the regulatory standard of 1 mg/L from the water outflow license. These results are also confirmed in the results of another chapter (Cai et al., 2020), which stated that BNR systems typically only remove approximately 60% of the influent total phosphorus and cannot achieve the complete elimination of phosphorus.

Although high concentrations of phosphate, nitrate, and ammonia were expected at Site 1, it was Site 5 that exhibited relatively lower levels of these nutrients. This anomaly suggests that anthropogenic activities, including potential sewage discharge, may be influencing nutrient levels in certain areas. The concentration of nutrients in the medium was a crucial determinant responsible for the level of variation of the measured parameters within the sites. High levels of DIP and NH_4^+ were found near the Lethabile WWTW, which indicates that municipal wastewater and agricultural return flows contribute predominantly to the nutrient loadings of the Crocodile River.

It is reported that the sewage treatment works discharge approximately 650 million L/day of treated effluent to the Hartbeespoort Dam catchment, which is subsequently conveyed into the Crocodile River (Annandale et al., 2011; Creamer Media, 2012). As well as from municipal sources, the dam also receives effluent from informal settlements, agricultural runoff, and storm water from industrial areas, discharges from the South African Nuclear Energy Corporation, and acid mine drainage in the form of mine water decanting from 18 Winze in the Western Basin, starting in 2002. The aggregate discharge of sewage, industrial waste, and storm water adds large amounts of dissolved nutrients, especially phosphorus and nitrogen, to the water system, which promotes the growth of algae and pollutes the catchment area (Atta et al., 2020).

The River Health Programme (2005) identified urban return flows, sewage spills and poultry and dairy farming activities as the major sources of pollution in the Crocodile West Catchement has also indicated that urban return flows, sewage spills, and chicken and dairy farming activities were the major sources of pollution in Crocodile West Catchments. In addition, the present chapter observed that COD and SS concnetrations at all sampling sites exceeded the reference values of 15 and 20 mg/L respectively, used in the calculation of the organic pollution index (OPI). This result was supported by a previous chapter in Boksburg Lake, where COD samples were recorded during pre and post-aeration. The continuing discharge of contaminants, such as sewage, into the Hartbeespoort Dam by inflowing rivers may have caused the high SS and COD values at sampling Site 2, which is near Brits, the WWTW. COD readings suggest that organic contamination may impact the water at this site.

Furthermore, DO was detected at a higher level in August during the dry season in Site 3, close to the Losperfontein WWTW. According to a chapter on the Hartbeespoort Dam by Venter (2012), the author measured high DO concentrations in the wet season in March (17.60 mg/L), which was twice the value obtained in the current chapter. Research studies have identified that the Hartbeespoort Dam in South Africa and the Von Bach Dam in Namibia both experienced excessive nutrient enrichment, with parts of the dams already in a chronic eutrophic state (Sirunda et al., 2022).

More so, the other variables, such as DIN, T, pH, EC, and SO_4, were detected with reasonable concentrations since they stayed below the standard used in the chapter.

Large scale surveys and national assessments of knowledge produce large amounts of data that require suitable statistical techniques for analysis and precise interpretation. The body of literature currently in publication advises researchers to employ machine learning methods, like the random forest regression analysis, to better understand the variables that may affect pollution status on various assessments (Osborne, 2019; Yoo et al., 2014). Because of their high accuracy and robustness, random forests, a well-liked machine learning algorithm, provide several advantages in terms of accuracy, generalization, and robustness for regression tasks. Since random forests are ensemble-based, they yield accurate predictions. When there are enough trees in the forest, they manage overfitting effectively.

Scalability: Random forests are able to handle big datasets with more dimensions efficiently.

Importance of features: They offer feature importance that sheds light on the most pertinent predictors for the regression task (Osborne, 2019; Yoo et al., 2014). The results presented in this chapter demonstrate the strong predictive performance of our random forest model. Its effectiveness stems from an architecture well suited to extracting key features from time-series data. Our model could be applied across all metrics, showcasing its robustness and reliability in forecasting environmental data. This performance underscores the effectiveness of the architecture of our model and its ability to handle complex datasets. Furthermore, the findings suggest that EC plays a crucial role in predicting pollutant levels, and its accurate measurement is vital for reliable forecasts. The close alignment between the forecasted and actual readings demonstrates the precision of the model and its capability to track pollutant levels accurately. This visual evidence supports the quantitative metrics, reinforcing the model's reliability. In summary, the combination of superior predictive performance, the critical role of the EC variable, and the model's adeptness at feature extraction and pattern recognition collectively contributes to the outstanding results achieved by our random forest model. These results also demonstrate the capability of high-level machine learning methodologies in the realm of environmental forecasting and the necessity of correct modelling to infer consistent forecast variables.

The results obtained from the indices (CPI, OPI, EI) in the present chapter are consistent with findings reported in the previous research. For instance, an assessment of the water quality of the Cau River employed a range of indicators, including the comprehensive pollution index (CPI) and organic pollution index (OPI), and identified significant organic pollution and eutrophic conditions in downstream sections. Similarly, studies using a combination of some indices (CPI, OPI and EI) have concluded that the water in affected system ranges from eutrophic to very polluted (Son et al., 2020). Comparable results were reported for Von Bach and Swakoppoort dams in sub-Saharan Africa, where the CPI indicated severe pollution, driven by most measured variables, except fluoride (F), sulfate (SO_4^{2-}), and manganese (Mn) (Sirunda et al., 2022). In contrast, a chapter conducted in the Chambal River in India, using similar indices and parameters, reported no evidence of eutrophication and classified the water quality as excellent in terms of pollution level (Yadav et al., 2018).

Eutrophication, caused by nutrient enrichment of water resources, is one of the primary challenges to water quality in South Africa and is facing the rest of the world, particularly regarding water resources in urban areas (Liu et al., 2011). According to a previous chapter (Griffin et al., 2014), poorly maintained sewerage systems are one of the main causes of water resource contamination, contributing to low oxygen and excessive nutrient levels in South Africa's water supplies. Carroll and Curtis (2021) also highlighted that, during the development of the South African National Water and Sanitation Plan, more than 40% of the country's water treatment works and almost 60% of sewage treatment works were identified to be in poor or in a very poor condition. In 1999, the amount of water identified as in poor ecological condition was five times what it was in 2011.

14.6 Conclusion and Recommendations

This chapter assessed the water quality and eutrophication status of the Upper Crocodile West Catchment using pollution indices and machine learning model. Results showed that nutrients enrichment and organic pollution are the main environmental concern in the catchment, mainly caused by wastewater treatment works, urban runoff and, and agricultural activities. The calculated comprehensive pollution index (CPI), the Organic pollution index (OPI), and Eutrophication index (EI)n showed most sampling sites were heavily polluted and exhibited clear eutrophic conditions. Seasonal variation reveled that organic pollution and eutrophication levels were higher during the wet season due to increase runoff and nutrient transport into the river system. High concentrations of dissolve inorganic nitrogen (DIN) and dissolve inorganic phosphorus

(DIP) were detected near the sampling site at Lathabile wastewater treament works, likely due to the transport of organic materials and sediments within the watershed. Statistical analysis using the Kruskal Wallis and Paired Mann Whitney U tests indicated that sampling sites did not significantly influence pH and temperature, while seasonal variation significantly affected temperature, suspended solids (SS), and nitrate concentrations. Predicting pollution using a machine learning, random forest regression model demonstrated strong predictive performance, highlighting the usefulness of machine learning in forecasting pollution concentrations and supporting proactive water quality managements due to excessive nutrients inputs into freshwater system. Eutrophication still remains a growing challenge in many South African catchments due to excessive nutrients inputs ino freshwater system. Addressing this problem,

Water resources management on eutrophication should include of following:

- The surveillance program will continue.
- Continued initiatives to decrease phosphorus emissions into water resources, such as introducing detergents low in phosphorus.
- The establishment of a knowledge base containing every practical and efficient eutrophication management technique for the climate and sociological requirements of South Africa.
- Options for managing hypertrophic areas inside lakes.
- Consider extending the use of efficient eutrophication management techniques to further impacted locations.

Safeguarding adequate provision of funds to address the problem of nutrient enrichment of fresh waters in a sustainable manner.

The novelty of this research lies in its multifaceted approach to assessing water quality downstream of WWTWs in the Crocodile West Catchment. The application of a random forest machine learning algorithm to predict pollutant levels based on environmental variables is a significant advancement. Its high predictive accuracy ($R^2 > 0.6$) and ability to track pollution trends provide a robust tool for environmental management and decision-making. Furthermore, the chapter evaluated a wide range of physicochemical parameters and indices (CPI, OPI, and EI) to provide a detailed pollution profile. This holistic assessment captures seasonal variations and identifies the impact of urban and industrial wastewater sources. By comparing dry and wet season samples, the research highlighted how seasonal dynamics influence pollution levels and eutrophication, offering critical insights into temporal water quality changes. Through advanced statistical analysis (Kruskal-Wallis test, Mann-Whitney U-tests) and indices calculations, the research categorizes the water samples as heavily polluted and eutrophic, emphasizing the environmental and health risks. This chapter integration of machine learning with traditional water quality assessment provides a cutting-edge framework for monitoring and managing pollution in aquatic ecosystems, particularly under the pressures of urbanization and industrialization. This work shows that, by reducing sewage discharges and managing agricultural runoff are essential for sustainable urban waterscapes. Combining pollution indices with machine learning offers a practical toolkit for proactive water quality management, directly supporting SDG 6.3 in South Africa and beyond.

Acknowledgments Our special thanks go to the University of the Free State for all technical equipment and software used for this research. Thanks to Dr. Yolandi Schoeman and Shaquille Barnes for assisting with the maps.

Funding: The authors declare that no grants or funds, or other support, were received during the preparation of this manuscript.

Author Contributions: Ernestine Atangana: Writing, preparing the first draft, conceptualization, technique, software, validation, formal analysis, research, resources, data curation, and investigation. Paul J. Oberholster: writing, including project management, editing, review, and visualization. Heba Bedair: writing, review editing, visualization, supervision, project administration. Ashkan Khalife: Statistical analysis, Writing, review, editing, project administration. Soumya Ghosh: Review, visualization, Supervision, project administration.

Corresponding Author: Correspondence sent to Dr. Ernestine Atangana.

Ethical Approval: The authors listed in the manuscripts have granted their agreement to be listed as co-authors, have reviewed and approved the work, and have given their approval for submission and publishing.

Consent to Participate: All authors have approved the material of the text for publication.

Consent to Publish: All listed authors have agreed to submit the work.

Statements and Declarations

Competing Interests: The authors have no relevant financial or non-financial interests to disclose. The authors did not receive support from any organisation for the submitted work.

Institutional Review Board Statement: Not applicable.

Informed Consent Statement: Not applicable.

Data Availability: The corresponding author will disclose the data used in the current work upon reasonable request.

AI Discloser Statement: No AI tool has been used in the chapter.

References

Akhlaq, M., Ellahi, A., Niaz, R., Khan, M., Sammen, S. S., & Scholz, M. (2024). Comparative analysis of machine learning algorithms for water quality prediction. *Tellus A, 76*(1), 177–192. https://a.tellusjournals.se/articles/10.16993/tellusa.4069

Annandale, J., Beletse, Y., Stirzaker, R., & Bristow, K. (2011). Managing poor quality coal-mine water: Is irrigation part of the solution? In *Observations on environmental change in South Africa* (pp. 174–177). Sun Press.

Antoniadis, A., Cugliari, J., Fasiolo, M., Goude, Y., & Poggi, J.-M. (2024). Random forests. In *Statistical learning tools for electricity load forecasting* (pp. 99–111). Springer. https://doi.org/10.1007/978-3-031-60339-6_5

ANZECC (Australian and New Zealand Environment and Conservation Council). (2000). *Australian and New Zealand Guidelines for fresh and marine water quality* (Vol. 1). The Guidelines, https://www.waterquality.gov.au/sites/default/files/documents/anzecc-armcanz-2000-guidelines-vol1.pdf

APHA. (1998). *Standard methods for the examination of water and wastewater* (20th ed.). American Public Health Association, American Water Works Association and Water Environmental Federation. https://books.google.co.za/books?hl=en&lr=&id=ErAbUHJhfR8C&oi=fnd&pg=PR7&dq=APHA+(1998)+Standard+Methods+for+the+Examination+of+Water+and+Wastewater.+20th+Edition,+American+Public+Health+Association,+American+Water+Works+Association+and+Water+Environmental+Federation,+Washington+DC.&ots=2ORyz6lr9P&sig=qiZwFS_5dvDFKvGBrmDMbFlluCk&redir_esc=y#v=onepage&q&f=false

Atangana, E., & Oberholster, P. J. (2021). Using heavy metal pollution indices to assess water quality of surface and groundwater on catchment levels in South Africa. *Journal of African Earth Sciences, 182*, 104254. https://doi.org/10.1016/j.jafrearsci.2021.104254

Atta, K. P. T., Maree, J. P., Onyango, M. S., Mpenyana-Monyatsi, L., & Mujuru, M. (2020). Chemical phosphate removal from Hartbeespoort Dam water, South Africa. *Water SA, 46*(4), 610–614. https://doi.org/10.17159/wsa/2020.v46.i4.9074

Azrour, M., Mabrouki, J., Fattah, G., Guezzaz, A., & Aziz, F. (2022). Machine learning algorithms for efficient water quality prediction. *Modeling Earth Systems and Environment, 8*, 2793–2801. https://doi.org/10.1007/s40808-021-01266-6

Barnwal, P., Mishra, S., & Singhal, S. K. (2015). Risk assessment and analysis of water quality in Ramgarh Lake, India. *Journal of Integrated Science and Technology, 3*, 22–27. http://www.thesciencein.org/isciencepubs/journal/index.php/jist/article/view/266

Bedair, H., Shaltout, K., & Halmy, M. W. A. (2023). Stacked machine learning models for predicting species richness and endemism for Mediterranean endemic plants in the Mareotis subsector in Egypt. *Plant Ecology, 224*, 1113–1126. https://doi.org/10.1007/s11258-023-01366-6

Bezuidenhout, C. C., Mthembu, N., Puckree, T., & Lin, J. (2002). Microbiological evaluation of the Mhlathuze River, KwaZulu-Natal. *Water SA, 28*(3), 281–286. https://doi.org/10.4314/wsa.v28i3.4895

Bunce, J. T., Ndam, E., Ofiteru, I. D., Moore, A., & Graham, D. W. (2018). A review of phosphorus removal technologies and their applicability to small-scale domestic wastewater treatment systems. *Frontiers in Environmental Science, 6*(8), 6. https://doi.org/10.3389/fenvs.2018.00008

Cai, H., Liang, J., Ning, X.-A., Lai, X., & Li, Y. (2020). Algal toxicity induced by effluents from textile-dyeing wastewater treatment plants. *Journal of Environmental Sciences, 91*, 199–208. https://doi.org/10.1016/j.jes.2020.01.004

Canadian Council of Ministers of the Environment (2012). Canadian water quality guidelines for the protection of aquatic life. In Canadian Environmental Quality Guidelines; Canadian Council of Ministers of the Environment: Winnipeg, MB, Canada, Available online: https://ccme.ca/en/resources/water-aquatic-life.

Carroll, A. S. D., & Curtis, C. J. (2021). Increasing nutrient influx trends and remediation options at Hartbeespoort Dam, South Africa: A mass-balance approach. *Water SA, 47*, 210–220. https://doi.org/10.17159/wsa/2021.v47.i2.10917

Chaturvedi, M. K., & Bassin, J. K. (2010). Assessing the water quality index of water treatment plant and bore wells, in Delhi, India. *Environmental Monitoring and Assessment, 163*, 449–453. https://doi.org/10.1007/s10661-009-0848-2

Couronné, R., Probst, P., & Boulesteix, A.-L. (2018). Random forest versus logistic regression: A large-scale benchmark experiment. *BMC Bioinformatics, 19*, 270. https://doi.org/10.1186/s12859-018-2264-5

Creamer Media. (2012). *Water 2012: A review of South Africa's water sector. Water report*. Creamer Media. https://static.pmg.org.za/docs/120911sector.pdf

Department of Water Affairs. (2012). Implementation of the waste discharge charge system: Crocodile river situation assessment and resource quality management plan report. Pretoria, South Africa.

Department of Water Affairs and Forestry. (1996a). *South African water quality guidelines. Recreational use. Pretoria, South Africa* (2nd ed., Vol. 2), https://www.dws.gov.za/iwqs/wq_guide/edited/Pol_saWQguideFRESHRecreationalUsevol2.pdf

Department of Water Affairs and Forestry. (1996b). *South African water quality guidelines. Agricultural use: Irrigation. Pretoria, South Africa* (2nd ed., Vol. 4), https://www.dws.gov.za/iwqs/wq_guide/edited/Pol_saWQguideFRESHIrrigationvol4.pdf

Department of Water Affairs and Forestry. (1996c). *South African water quality guidelines. Aquatic ecosystems. Pretoria, South Africa* (2nd ed., Vol. 7), https://www.dws.gov.za/iwqs/wq_guide/edited/Pol_saWQguideFRESH_vol7_Aquaticecosystems.pdf

Department of Water Affairs and Forestry. (2003). *Klip River water quality objectives. Pretoria, South Africa*

Department of Water Affairs and Forestry. (2009). *Hartbeespoort Dam resource management plan, South Africa*. https://www.gov.za/news/water-affairs-and-forestry-resources-management-plan-hartebeesport-dam-26-nov-2007

Griffin, N. J., Palmer, C. G., & Scherman, P. A. (2014). *Critical analysis of environmental water quality in South Africa: Historic and current trends. WRC Rep 2184/1/14*. Water Research Communication. https://wrcwebsite.azurewebsites.net/wp-content/uploads/mdocs/2184-1-14.pdf

Guasmi, I., Kherici-Bousnoubra, H., Kherici, N. et al. Assessing the organic pollution of surface water of Medjerda watershed (NE Algeria). *Environ Earth Sci 60*, 985–992 (2010). https://doi.org/10.1007/s12665-009-0237-8

Gupta, A. K., Gupta, S. K., & Patil, R. S. (2003). A comparison of water quality indices for coastal water. *Journal of Environmental Science and Health, Part A, 38*, 2711–2725. https://doi.org/10.1081/ESE-120024458

Ha, N. T., Nguyen, H. Q., Truong, N. C. Q., Le, T. L., Thai, V. N., & Pham, T. L. (2020). Estimation of nitrogen and phosphorus concentrations from water quality surrogates using machine learning in the Tri An Reservoir, Vietnam. *Environmental Monitoring and Assessment, 192*, 1–20. https://doi.org/10.1007/s10661-020-08731-2

Jimeno-Sáez, P., Senent-Aparicio, J., Cecilia, J. M., & Pérez-Sánchez, J. (2020). Using machine-learning algorithms for eutrophication modeling: A case chapter of Mar Menor Lagoon (Spain). *International Journal of Environmental Research and Public Health, 17*, 1189. https://www.mdpi.com/1660-4601/17/4/1189

Kachroud, M., Trolard, F., Kefi, M., Jebari, S., & Bourrié, G. (2019). Water quality indices: Challenges and application limits in the literature. *Water, 11*(2), 1–26. https://www.mdpi.com/2073-4441/11/2/361

Kaddoura, S. (2022). Evaluation of machine learning algorithm on drinking water quality for better sustainability. *Sustainability, 14*(18), 11478. https://doi.org/10.3390/su141811478

Karimian, H., Huang, J., Chen, Y., Wang, Z., & Huang, J. (2023). A novel framework to predict chlorophyll-a concentrations in water bodies through multi-source big data and machine learning algorithms. *Environmental Science and Pollution Research, 30*, 79402–79422. https://doi.org/10.1007/s11356-023-27886-2

Leyland, R. C. (2008). *Vulnerability mapping in karst terrains, exemplified in the wider Cradle of Humankind WHS*. Univ Pretoria. https://www.wrc.org.za/wp-content/uploads/mdocs/KV%20208-Groundwater.pdf

Liaw, A., & Wiener, M. (2014). *Package 'randomForest': Breiman and Cutler's random forests for classification and regression*. Univ California. https://www.stat.berkeley.edu/~breiman/RandomForests/

Liu, S., Lou, S., Kuang, C., Huang, W., Chen, W., Zhang, J., & Zhong, G. (2011). Water quality assessment by pollution-index method in the coastal waters of Hebei Province in western Bohai Sea, China. *Marine Pollution Bulletin, 62*(10), 2220–2229. https://www.sciencedirect.com/science/article/pii/S0025326X11003523

Lumb, A., Halliwell, D., & Sharma, T. (2006). Application of CCME water quality index to monitor water quality: A case of the Mackenzie River Basin, Canada. *Environmental Monitoring and Assessment, 113*, 411–429. https://doi.org/10.1007/s10661-005-9092-6

Makia, M. S. (2021). *The impact of the Hartbeespoort Dam on the water quality of the Crocodile River*. MSc thesis. University of South Africa

Masawi, N. & Oberholster, P.J. (2022). Mini-dissertation by publication submitted in partial fulfilment of the requirements in respect of the master's degree, Master of Science majoring in Integrated Water Resource Management in the Centre for Environmental Management in the Faculty of Natural and Agricultural Sciences at the University of the Free State, April 2022.

Matta, G., Kumar, A., Uniyal, D. P., Sing, P., Kumar, A., Dhingra, G. K., Naik, A. K., Pradeep, K., & Shrivastva, N. G. (2017). Temporal assessment using WQI of River Henwal, a tributary of River Ganga in Himalayan region. *International Journal for Environmental Rehabilitation and Conservation, 8*(1), 187–204. https://www.academia.edu/38167392/Temporal_assessment_using_WQI_of_River_Henwal_a_Tributary_of_River_Ganga_in_Himalayan_Region_pdf

Mishra, S., Sharma, M. P., & Kumar, A. (2016). Assessment of surface water quality in Surha Lake using pollution index, India. *Journal of Materials and Environmental Science, 7*(3), 713–719. http://www.pubs.iscience.in/journal/index.php/ira/article/view/732/414

Mnyango, S. S., Thwala, M., Oberholster, P. J., & Truter, C. J. (2022). Using multiple indices for the water resource management of a monomictic man-made dam in Southern Africa. *Water, 14*(21), 3366. https://doi.org/10.3390/w14213366

Njagi, D. M., Routh, J., Odhiambo, M., Luo, C., Basapuram, L. G., Olago, D., Klump, V., & Stager, C. (2022). A century of human-induced environmental changes and the combined roles of nutrients and land use in Lake Victoria catchment on eutrophication. *Science of the Total Environment, 835*, 155425. https://www.sciencedirect.com/science/article/pii/S0048969722025189

Oberholster PJ, Ashton PJ (2008) An overview of the current status of water quality and eutrophication in South African rivers and reservoirs. CSIR Deliverable, Pretoria. https://www.anthonyturton.com/assets/my_documents/my_files/44B_Eutrophication__Human_Health_in_South_Africa_-_Final_-_5Mar200.pdf

Oberholster, P. F., Goldin, J., Xu, Y., Kanyerere, T., Oberholster, P. J., & Botha, A.-M. (2021). Assessing the adverse effects of a mixture of AMD and sewage effluent on a sub-tropical dam situated in a nature conservation area using a modified pollution index. *International Journal of Environmental Research, 15*, 321–333. https://link.springer.com/article/10.1007/s41742-021-00315-3

Osborne, J. W. (2019). Prediction in multiple regression. *Practical Assessment, Research and Evaluation, 7*(1), 2. https://doi.org/10.7275/7j20-gg86

Park, K., Jung, H.-S., Kim, H.-S., & Ahn, S.-M. (2005). Three-dimensional hydrodynamic-eutrophication model (HEM-3D): Application to Kwang-Yang Bay, Korea. *Marine Environmental Research, 60*, 171–193. https://doi.org/10.1016/j.marenvres.2004.10.003

Phungela, T. T., Maphanga, T., Chidi, B. S., Madonsela, B. S., & Shale, K. (2022). The impact of wastewater treatment effluent on Crocodile River quality in Ehlanzeni District, Mpumalanga Province, South Africa. *South African Journal of Science, 118*. https://doi.org/10.17159/sajs.2022/12575

Poonam, T., Tanushree, B., & Sukalyan, C. (2013). Water quality indices – important tools for water quality assessment. *International Journal of Advanced Chemistry, 1*(1), 15–28. https://www.researchgate.net/publication/262730848_Water_quality_indices-_important_tools_for_water_quality_assessment_a_review

River Health Programme. (2005). *State of river report: Monitoring and managing the ecological state of the rivers in the Crocodile (West) Marico Water Management Area*. Department of Environmental Affairs and Tourism. https://www.dws.gov.za/iwqs/rhp/state_of_rivers/crocwestmarico_05/crocmarico_eng&tswana_05.pdf

Rocha, F. C., Andrade, E. M., & Lopes, F. B. (2015). Water quality index calculated from biological, physical and chemical attributes. *Environmental Monitoring and Assessment, 187*, 1–15. https://doi.org/10.1007/s10661-014-4163-1

Ruiz-Medina, M. D., & Espejo, R. M. (2012). Spatial autoregressive functional plug-in prediction of ocean surface temperature. *Stochastic Environmental Research and Risk Assessment, 26*, 335–344. https://doi.org/10.1007/s00477-012-0559-z

Saral, A., & Goncaloglu, B. I. (2008). Determination of real COD in highly chlorinated wastewaters. *Clean—Soil, Air, Water. Scientific Research, 36*, 996–1000. https://doi.org/10.1002/clen.200800073

Şener, Ş., Şener, E., & Davraz, A. (2017). Evaluation of water quality using water quality index (WQI) method and GIS in Aksu River (SW-Turkey). *Science of the Total Environment, 584*, 131–144. https://doi.org/10.1016/j.scitotenv.2017.01.102

Sirunda, J., Oberholster, P., & Wolfaardt, G. (2022). Assessing the adverse effects of land use activities on the water quality of selected Sub-Saharan Africa reservoirs using a combination of water quality indices. *Water, Air, and Soil Pollution, 233*, 267. https://doi.org/10.1007/s11270-022-05703-9

Son, C. T., Giang, N. T. H., Thao, T. P., Nui, N. H., Lam, N. T., & Cong, V. H. (2020). Assessment of Cau River water quality using a combination of water quality and pollution indices. *Journal of Water Supply: Research and Technology-AQUA, 69*, 160–172. https://doi.org/10.2166/aqua.2020.122

Tang, T., Zhai, Y., & Huang, K. (2011). Water quality analysis and recommendations through comprehensive pollution index method. *Management Science and Engineering, 5*(2), 95–100. https://doi.org/10.3968/j.mse.1913035X20110502.011

Uddin, M. G., Nash, S., & Olbert, A. I. (2021). A review of water quality index models and their use for assessing surface water quality. *Ecological Indicators, 122*, 107218. https://doi.org/10.1016/j.ecolind.2020.107218

Van Ginkel, C., O'Keeffe, J., Hughes, D., Herald, J., & Ashton, P. (1996). *A situation analysis of water quality in the catchment of the Buffalo River, Eastern Cape with special emphasis on the impacts of low-cost, high-density urban development on water quality. WRC Rep 405/2/96*. Water Research Communication. https://www.wrc.org.za/wp-content/uploads/mdocs/405-2-96.pdf

Venter, P. (2012). Overview of the programme: Hartbeespoort Dam remediation. *Civil Engineering, 2012*(7) https://hdl.handle.net/10520/EJC125575

Welch, E. B., & Cooke, G. D. (2005). Internal phosphorus loading in shallow lakes: Importance and control. *Lake and Reservoir Management, 21*(2), 209–217. https://doi.org/10.1080/07438140509354430

Yadav, N. S., Kumar, A., Mishra, S., & Singhal, S. (2018). Assessment of water quality using pollution-index in the chapter stretch of River Chambal, India. *Integrated Research Advances, 5*, 20–25. http://www.thescience.org/isciencepubs/journal/index.php/ira/article/view/732

Yan, X., Zhang, T., Du, W., Meng, Q., Xu, X., & Zhao, X. (2024). A comprehensive review of machine learning for water quality prediction over the past five years. *Journal of Marine Science and Engineering, 12*(1), 159. https://www.mdpi.com/2077-1312/12/1/159

Yoo, C., Ramirez, L., & Liuzzi, J. (2014). Big data analysis using modern statistical and machine learning methods in medicine. *International Neurourology Journal, 18*(2), 50–57. https://www.ncbi.nlm.nih.gov/pmc/articles/PMC4076480/

Conclusion & Way Forward 15

Gouri Sankar Bhunia, Susanta Patra, Soumen Bramha, S. R. Kamalesh, and Uday Chatterjee

Abstract

Urban waterscapes are at the temperament of sustainable urban design but face significant threat from heightened urban growth, climate change, growing population levels, and unmaintainable consumption of resources. These are manifested in the form of heightened scarcity of water, pollution, ecosystem degradation, unequal distribution, environmental exposure to weather extremes such as floods and droughts. While that is happening, the interconnection of water with energy, food, health, and urban ecosystems makes governance and management very complex. Traditional water infrastructure, ideally planned with a focus on proficiency, falls short in developing resilient and adaptive cities. New methods such as integrated urban water management (IUWM), water-sensitive urban design (WSUD), citizen science, nature-based solutions, and application of remote sensing, GIS, and AI offer new possibility to boost pliability. This book scrutinizes the intricate encounters of urban waterscapes critically and offers directions for future elasticity, highlighting innovation in governance, technology, community, and ecological refurbishment. The dialogue emphasizes the necessity for transdisciplinary thinking, circular economy, and participatory water management to safeguard that waterscapes in cities evolve as forces of sustainability, equity, and suppleness in emerging cities.

Keywords

Urban waterscapes · Sustainable cities · Climate resilience · Water-energy-food nexus · Water-sensitive urban design · Remote sensing and GIS · Citizen science · Water governance · Circular economy · Urban river restoration · Water reuse and recycling

G. S. Bhunia
Independent Researcher, Kharagpur, West Bengal, India

S. Patra · S. Bramha
Department of Geography, Subarnarekha Mahavidyalaya, Jhargram, West Bengal, India

S. R. Kamalesh
Principal Govt. Bilasa Girls Autonomous P.G. College, Bilaspur, Chhattisgarh, India

U. Chatterjee (✉)
Department of Geography, Bhatter College Dantan (Autonomous), Kharagpur, West Bengal, India

With growing city sizes and densities being forced by population dynamics, climate change, or economic shifts, water infrastructure rivers, wetlands, lakes, canals, floodplains, coastal areas come into prominence as crucial eco-technical infrastructure shaping the future of cities

U. Chatterjee, A. I. Almulhim (eds.), *Urban Waterscapes*, Sustainable Development Goals Series,
https://doi.org/10.1007/978-3-032-20649-7_15

(Mabrouk et al., 2024). Urban water infrastructure provides immense potential for developing resilience if it can be protected by integrated, nature-based, and inclusive approaches. Blue-green infrastructure, ecosystem-based adaptation, water-sensitive design, or participative governance approaches demonstrate how cities can transcend the reactive paradigm dominated by traditional engineering responses for developing resilient water infrastructure (Becker et al., 2024). This book symbolizes at once the highest risk factor as well as the biggest opportunity for future cities. As climate change becomes more extreme and the size of city populations swells, the relationship between cities and water becomes more crucial for the sustainability of the city.

Accelerated urbanization and population growth expand impervious surfaces, strain water supply and sanitation infrastructure, and amplify flood and pollution risks, particularly in rapidly growing cities of the Global South (UN-Habitat, 2020; World Bank, 2019). Economic transitions, including industrialization and service-sector expansion, further alter water demand patterns and generate complex pollution loads, necessitating more integrated and adaptive water governance frameworks (OECD, 2018). Climate change acts as a systemic stress multiplier, increasing the frequency and intensity of droughts, floods, heatwaves, and sea-level rise, thereby exposing the vulnerability of conventional, supply-driven water infrastructure (IPCC, 2022). In this context, resilient urban waterscapes align strongly with the Sustainable Development Goals: SDG 6 (Clean Water and Sanitation) through improved water quality, efficiency, reuse, and equitable access; SDG 11 (Sustainable Cities and Communities) by integrating blue-green infrastructure, risk reduction, and inclusive planning into urban development; and SDG 13 (Climate Action) by enhancing adaptive capacity, reducing climate-related water risks, and promoting nature-based and low-carbon solutions (UN, 2015; WWAP, 2023). Together, these drivers and SDG linkages underscore that resilient urban waterscapes are not only environmental necessities but also strategic instruments for achieving climate-resilient, inclusive, and sustainable urban futures (Fig. 15.1).

Chapter 1 presents a critical analysis of how water insecurity in expanding cities across the

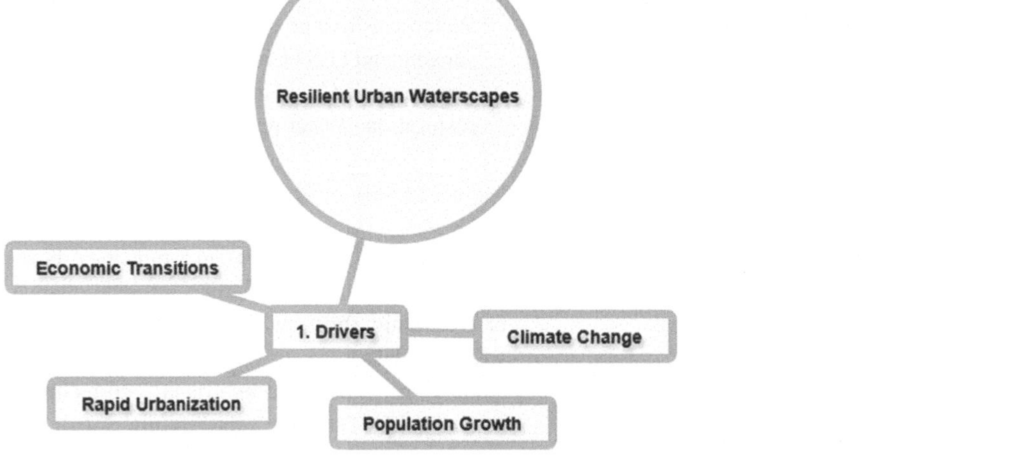

Fig. 15.1 Drivers of resilient urban waterscapes and SDG (Sustainable Development Goals) alignment

world today is as much a consequence of governance flaws as it is a result of climate change and scarcity reality. Through comparative analysis of Cape Town, Riyadh, Medellín, Chennai, and Jakarta, Chap. 1 presents how water crises exist as a reality when it comes to inefficient designs of institutions, infrastructure, as well as a lack of social engagement, as opposed to the physical reality of water scarcity itself. Examples of effective implementation of these methodologies exist through Cape Town's social demand management tactics utilized during the Day Zero crisis, Medellín's efficacious social engagement-based designs, Chennai's civil society-driven rebooting of traditional water tanks, as well as Riyadh's data-based designs for how the city utilizes its water resources (Aiyetan, 2025). Perhaps the most extreme example against which these efficacious tactics exist would be through the reality of Jakarta itself, as its centrality-driven, data-based designs often leave its communities behind through more efficacious or smarter approaches when it comes to water management practices. Together, the cases evidence a paradigm shift towards a multi-level, participatory form of governance in which the state's role is that of a facilitator-regulator, and in which the provision of services cannot remain the monopoly of the state. This chapter stresses that the pursuit of climate-resilience, equity, and sustainable urban water resources in the South can only be realized through comprehensive planning and the strategic use of information technology and strong laws, in partnership with local stakeholders who must have a stronger role in this endeavor. While acknowledging the representational imbalance, this chapter links the reform of urban water sector governance with the pursuit of SDG 6 and SDG 11, acknowledging the need to overcome top-down approaches and adopt a more bottom-up perspective in order to achieve a water-resilient future in cities of the South.

In Chap. 2, the issue of the Water Crisis in Darjeeling has been discussed as an intricate socio-ecological issue framed by the sensitive mountain environment, rapid urbanization, tourism, aging infrastructure, and the inability of policies to address the ever-increasing gaps between the demand and supply for water. Rather than viewing the issue through the lens of the distant environment, the study views the issue from a more holistic environment by considering the nearby environment through the means of the hydro-geological environment, springs, the ecological environment, the environment through infrastructure, and the institutional environment (Riaz et al., 2025). It focuses on the importance of hill-specific policy frameworks regarding water within the context of missions such as National Mission for Sustaining the Himalayan Ecosystem and Integrated Watershed Management Programme for ensuring conservation within existing sources as well as identifying options for sustainable sources in the future within hills states (Fennell et al., 2023). This chapter focuses on the need for urgent action like minimizing distribution losses by maintaining/replacing old piping networks, preventing wastage of water by people, and Harvesting rainwater on a massive scale during monsoon months with treatment for utilization during summer months when tourism is at its peak. The chapter elevates the need for conservation of natural springs as well as management of Senchal Lake. In addition to that, it is important to consider that effective and sustainable water resource management in hill townships is impossible without proper governance structures in place along with community involvement and awareness about conservation of water resources. In this regard, to address the concerns and apprehensions about the health of potable water in these hill townships, the conclusion of this chapter is in support of comprehensive hydrogeological studies that are based on the analysis of spring discharge measurements, microbial parameters of water quality assessments, and ecological studies.

Chapter 3 provides a critical and complex analysis of the interplay between rapid urbanization and climate change in exacerbating the threat of flooding in modern urban spaces, specifically through the case study of Mumbai. Chapter 3 locates the city of Mumbai in the dual capacity of being at once a cause for climate change through rapid urbanization, land change, and being an affected city through annual instances of extreme flooding, infrastructure breakdown, and enhanced eco-social vulnerabilities. This chapter also reveals the complexity in urban spaces in which economic growth associated with rapid urbanization unleashes prosperity, it also unleashes vulnerabilities in being unplanned for in terms of unregulated real-estate developments, natural drainage invasion by unregulated buildings, destruction of wetlands, and lack of infrastructure. This chapter rests upon an interdisciplinary approach in defining the analysis in terms of technological rationale associated with flooding in being political-socioeconomic-legal-legal in sharply locating the inequity in climate change vulnerabilities. The analysis highlights the need for a convergent approach to environmental sustainability and equity, minimizing disparities in the vulnerability of marginalized populations to climate change. The implications and recommendations highlighted by this analysis outline upgrading existing drainage and stormwater systems, and protecting and restoring natural systems, building and strengthening capacity, and implementing equity-focused planning approaches that prioritize the requirements of at-risk populations. Broadly related to, and specifically building from, the Sustainable Development Goals adopted by the United Nations, specifically collectively focusing and building from SDG 6: Clean Water and Sanitation, SDG 11: Sustainable Cities and Communities, and SDG13: Climate Action, this analysis finds a broader application and significance than within the context of Mumbai, and finds applicability within rapidly urbanizing cities globally. This analysis can lead to further research studies and advances related to hot islands and flooding, nature and climate-resilient infrastructure, and mapping and AI-based early warning systems.

In Chap. 4, there is an analysis of inequities with regards to access to water in Ghana, specifically in Greater Accra, and it highlights how issues of access and inequity cannot be seen from an infrastructural or technical point of view, but have long-standing socio-political and economic underpinnings that have cumulatively led to an extremely stratified water landscape in cities, with informal areas deliberately left outside of access. The chapter contends that indeed water policies and practices in Ghana have derailed in their redistributive functions and how this has cumulatively led to an incredibly under-regulated private sector delivery sector that remains uneven and prejudicial on several fronts (Mangulama, 2024). With the increasing commodification of water, access to it has come to be dependent on land tenure security and purchasing capacity, leading to the establishment of a dual system in which better-off and formally planned parts of the city are provided subsidized piped water services while marginalized sections are left with expensive, erratic, and often unsafe alternative sources of water. This chapter points out the irony of the fact that it is the group of people who can least afford it who are required to pay the highest prices for it and are thereby exposed to the greatest risks to health, thereby firmly situating unequal access to water as the negation of urban citizenship rather than merely being an issue of service delivery. It empirically embeds urban water inequity within wider political economies of infrastructure, planning, and governance, putting local realities into conversation with global development agendas such as SDG 6. Drawing on comparative experiences from the Global South, this chapter calls for a more paradigmatic shift toward a transformative urban water governance that merges accountability at the public level and community participationist approaches with regulatory reform that embeds affordability, public investment in consistently neglected areas, and formal recognition of informal settlements. Historicizing contemporary reforms through integrating perspectives from political ecology and governance, the chapter concludes that any hope for resilient, equitable, and sustainable urban waterscapes requires dismantling entrenched privilege and fragmentation of political authority in water governance.

Chapter 5 proposes an innovative, exploratory argument for wetlands integration within urban planning as a foundation for resilient, regenerative cities by placing Integrated Water Resources Management (IWRM), as formulated within SDG 6.5, and protection and restoration of water-related ecosystems under SDG 6.6 at the core of sustainable urban development. The chapter further illustrates partial feasibility of the Urban Wetland Concept by an interdisciplinary synthesis of urban design, water engineering, and Nature-based Solutions (NbS) and by introducing a multicriteria GIS-based screening tool to identify suitable locations for blue infrastructure within dense urban environments. Central to this chapter is the concept of urban autopoiesis, here understood as a capacity of cities to develop self-reconstructing and regenerative systems, notably through wetlands that simultaneously provide technical water management functions, ecological services, and social benefits. Whilst acknowledging that full autopoiesis cannot be achieved due to functional limits of diverse urban construction layers and material systems, this chapter puts forward a methodological framework for linking architectural design with hydrological performance, supported by iterative tools for optimization, evaluation, and design refinement. The produced urban design strategy is primarily based on the principles of multi-functionality; that is, a balance between social functionality, biodiversity gain, and technical functionality, while highlighting the complexity and relationships among spatial variables, biodiversity variables, and infrastructure variables. More specifically, the chapter states that "the value of the experiment resides less in its result—its final design—than in its ability to illuminate the difficulties, trade-offs, and hidden possibilities within regenerative NbS—and specifically their relationships among land use effectiveness, biodiversity benefits, biomass production." Moreover, the analysis underlines that a process aimed at a complete autopoietic process will necessitate a co-evolution process among traditional buildings and urban morphology; this is characterized by a need for a completely different paradigm with regard to planning, governance structures, and business models that could incorporate biomass production, carbon reduction, fire resilience, and the everyday use of cities. Despite the focus on the specific case in Denmark, the importance of the chapter lies in the clarification that autopoietic wetland design in urban environments is highly context-dependent, particularly in the Global South. This, instead of offering a solution that can be universally applied, helps the chapter serve as a conversational starter that helps to develop methods for further research on autopoietic wetlands in urban settings.

Chapter 6 provides an integrated assessment of the vulnerability of waterlogging in the Chaktai Commercial Area by integrating scientific analysis, policy views, and local knowledge to understand the complexities involved. Focusing on Chattogram, the chapter uses a combination approach that blends Participatory Rural Appraisal (PRA) with GIS analysis to understand exposure to waterlogging as well as the experiences of the inhabitants affected by waterlogging. Findings indicate that whereas lack of drainage infrastructure is an essential condition, waterlogging is further exacerbated by socio-institutional factors such as improper management of solid waste, encroachment of canals/waterbodies, unplanned land use change, and lack of coordination between concerned organizations. This study fills an important knowledge gap in scientific research by delving into specifics of vulnerabilities often missed in technically focused investigations. Locating the case study within the wider Global South, the chapter sees the problem of waterlogging within the context of climate change, the pace of urbanization, and the issue of governance with a focus on the argument for the adoption of Integrated Water Resources Management (IWRM) as the tool that envisages the integration of the infrastructure dimension with the aspect of institutions and finance to ensure synergistic movement on SDG 6.5. Based on community responses and the overall thrust of the Chittagong Storm Water & Drainage Master Plan of 1995, the chapter suggests intervention strategies including the right implementation of drain maintenance and the use of proper coverage of drains; better systems of solid waste; the reinstatement of the filled-up canals and low-lying retention; a total ban on construction within water bodies or open spaces; and appropriate designs of roads as well as drains to not have a build-up of water run-offs.

Chapter 7 investigates the adaptive use of urban blue-green voids in Chennai to understand the importance of BGI in enhancing mental wellbeing, positioning BGI as a critical factor in resilient and health-oriented urban planning. The chapter highlights a strong positive relation between BGI attributes and community happiness in general, with visit frequency, purpose of use, and perceived urban-nature balance acting as important variables within a structured methodological framework combining spatial analysis with perception-based surveys across selected urban parks. The correlation analyses bring out that frequent and purposeful use of BGI contributes significantly to improved mental health outcomes, visual access, aesthetic quality, and ecological integration of green and blue elements contribute positively even when parks are in close proximity. Findings from the case-specific analysis reveal that the design elements of canalized shade walks, seating areas, and water features stimulate physical activity, recreational fun, and psychological relaxation, thus reinforcing mental wellbeing. The study, however, identifies some inequities in the distribution level and neighbourhood accessibility to BGI, suggesting that an unbalanced distribution and the reduced accessibility to the neighbourhoods may mitigate the positive mental health effects even for high-quality designed parks. The parks that were centrally located, multitask, and intensively incorporated within the environment showed the positive effectiveness outcome results for mental wellbeing benefits, as opposed to the others. A key contribution of the chapter is the introduction of the Community Happiness Index (CHI), a replicable, multi-criteria tool that links psychological responses with spatial equity, enabling planners to systematically embed mental well-being into land-use and blue-green planning strategies. While acknowledging limitations related to geographic scope, reliance on self-reported data, and limited demographic and temporal coverage, the chapter calls for future research to expand longitudinal and comparative analyses, integrate physiological and environmental performance data, and connect well-being indicators with hydrological functions.

Chapter 8 critically evaluates the relevance of geospatial evidence towards improving the management of urban water through a critical assessment of surface water dynamics within the city of Dhaka, where it depicts the alarming effects of unbridled urbanization and the lack of enforcement of regulations on the urban water landscape. With a focus on the Dhaka Metropolitan area, it demonstrates how surface water features essential for micro-climate modulation, biodiversity conservation, and maintenance of ecological balance have drastically deteriorated, even when these features formally received recognition through plans like the Dhaka Metropolitan Development Plan and the Detailed Area Plan. Employing multi-temporal remote sensing analysis of the Landsat images (2000, 2010, and 2020) along with the Modified Normalized Difference Water Index (MNDWI), it represents a drastic reduction of 77.9% of surface water area over a period of two decades, along with a significant increase of urban non-water surface area. These findings expose the critical disconnect between policy intent and on-ground implementation, where continued encroachment and infilling of ponds, canals, and wetlands have undermined urban ecological resilience. Beyond documenting hydrological change, this chapter makes a significant methodological contribution through the presentation of a replicable, data-efficient geospatial framework for the long-term surface water monitoring of data-scarce urban contexts and via reinforcing GIS and remote sensing as a basis for evidence-based planning. Situating Dhaka within a broader Global South narrative, the chapter draws parallels with international experiences of resilient and adaptive water governance, underscoring the need for integrated, flexible, and participatory management approaches. it contends that urbanization, although inevitable, has to be managed on the basis of full hydrological analysis and strong legal instruments prioritized for the protection and restoration of public and private surface water bodies and their surroundings, and sees sustainability, resilience, and liveability in the appropriate management of urban surface water, highlighting areas of improvement for similar cities in developing nations across the globe that face similar challenges in their own fast-growing cities, such as

Dhaka, and learn appropriate strategies from this successful case study.

Chapter 9 gives a geographically informed analysis of the effects of unchecked urbanization upon the groundwater in the historical city of Agra, which finds itself under pressure due to population growth, land use change, and an escalating water crisis. With the application of geoinformatic tools, it highlights the changing demographic patterns over time from 188,022 in 1901 to over 1.76 million in 2011, in addition to a substantial increase in the number of households, impervious surfaces, and administrative wards, which denote the profound change in the hydrological cycle of the city caused by the rapid rate of urbanization. It emerges from this study that an increase in the extent of urbanized land, industries, and infrastructure projects has substantially lowered the rate of natural groundwater recharge in the city simultaneously making it more reliant upon groundwater extraction, which in turn makes it more prone to contaminants through leakage, wastewater, and waste. With the description of data procurement and geospatial analysis tools and techniques in the chapter, the importance of evidence-based decision-making for the management of groundwater has also hit the spotlight. This has created an imperative for a holistic approach towards improving the efficiency of recharge and demand management and aligning land use planning efforts with water sustainability targets. These have also included the managed aquifer recharge plans of the Central Ground Water Board, Jal Shakti Abhiyan, Mission Amrit Sarovar, National Aquifer Mapping (NAQUIM), and community resource person efforts through the MGNREGA plan. This chapter has also established how these schemes and initiatives have enabled the achievement of SDG Target 6.4 as part of Integrated Urban Water Management (IUWM) by improving the efficiency of water use and reducing the overexploitation of groundwaters through the implementation of tools and techniques such as permeable pavement and GIS monitoring of urban expansion and recharge areas for the control of the "unregulated growth of concrete jungles."

Chapter 10 examines the relationships between the concepts of resilience and spatial justice within urban waterscaping through a Geographical Information System (GIS) case analysis on Sylhet, a monsoon-affected city with a fire-storm transformation rate in blue space masses at a breakneck pace. In this chapter, Geographically Weighted Regression (GWR) analysis is employed to show that accessibility and resilience levels for blue spaces within those cities are largely affected by their spatial heterogeneity rather than the citywide processes that affect other cities and their blue spaces as reservoirs for precipitation augmentation purposes. From the research findings in Chap. 10, both high population density and blue space degradation are common within Sylhet city; however, there is a prospect for reviving blue space within cities with high population densities once the major blue spaces within those cities are preserved for such purposes. On the other hand, the regions where there is landfilling as well as hard engineering practices along the floodplains show non-uniform hydrologic systems, while the surrounding areas maintain stable waterbodies that remain unutilized due to a lack of accessibility. Extreme non-stationarity as shown by the GWR analysis of Sylhet's Strategic Zones of Planning emerges as highly explanatory (Local R^2 of 0.954) even when dealing with the complexities associated with monsoon hydrology as well as the geographical variations, implying the dominant role of site-specific factors on the stability of waterbodies, with identity-fixed parameters supporting higher density as a stability-reducing factor but with area as a positive ingredient of stability as well as its relevance to urban waterbody scales and conservation practices. The chapter argues that equitable and resilient urban waterscapes require context-responsive, decentralized planning strategies that transform density pressures into opportunities for stewardship, prevent landfilling of both public and floodplain waterbodies, and integrate nature-based development alongside expanding grey infrastructure. By foregrounding spatial justice, the study emphasizes that protecting waterbodies, ensuring public accessibility, and halting encroachment especially on government-owned blue spaces are essential for replicating best practices and achieving sustainable, locally grounded urban waterscape resilience in rapidly urbanizing cities.

Chapter 11 presents a detailed review of advanced SWMS, prepared using various integrations of IoT, AI, and renewable energy technologies that provide transformative solutions for emerging global water scarcity problems. Chapter 11 therefore has been prepared against the dramatic backdrop of increasing water stress affecting over 2.4 billion people around the world today, with the situation bound to deteriorate considerably, according to all projections. This document outlines the transition from traditional and reactive water management methods to an intelligent and data-driven system by monitoring, prediction, and optimization in real-time (Vogel et al., 2015). It outlines the architecture of SWMS arranged in layers: sensor networks, protocols of communication, data processing platforms, AI-driven intelligence layers, and mechanisms for automatic control, which together enable holistic management of water supply, quality, distribution, and reuse. Drawing on recent empirical work, this chapter provides evidence of significant performance improvements in a wide range of applications: deep learning-based leak detection with accuracies up to 95.4%, smart irrigation systems reducing water consumption by 30%–50% while increasing agricultural yields by up to 70%, and machine learning models predicting water quality indices with accuracies up to 99.8%. Beyond the beneficial effects on technical efficiency, this chapter shows good prospects for economic viability, as the cost-benefit analyses using high returns on investment show that certain projects on the recycling of wastewater yield Net Present Values of almost US$160 million and Benefit-Cost Ratios > 4, besides a range of low-cost deployment options, such as IoT-based irrigation systems developed at ~US$215. The use of renewable energy sources also increases the sustainability of the system, thus cutting operating costs. Third, the chapter conceives SWMS as a paradigm shift in the governance of global water/resources. This paradigm shift enables design universality but adapted to the settings. Furthermore, it outlines future research topics in sensor resilience, protocol standardization, affordable modular systems, and AI hybrid models for improved SWMS capabilities for future use. Notably, it becomes apparent that SWMS can facilitate achievement of the SDG Target 6.3 in measurable terms, such as by eliminating pollution by real-time detection, improving water quality by predictive analytics, and improving efficiency in recycling beyond global means. This chapter affirms that smart, integrated technological innovations are central to creating sustainable, resilient urban waterscapes and advancing equitable access to clean water and sanitation in an era of climate uncertainty.

Chapter 12 integrates lessons from the world experience in urban water management by analyzing some illustrative cases between 2000 and 2024 to show how different cities are coping with increasing pressures arising from the pace of urbanization, climate variability, old infrastructure, and increasing water demand. Framed within five broad thematic areas of water supply, waste water management, storm water management, protection of water quality, and management of groundwater resources and maintenance of infrastructure, the chapter shows that effective management of urban water resources is necessarily integrative (Wheeler et al., 2025). By taking examples from Stuttgart and Melbourne, this book explores the impact of state-of-the-art wastewater treatment and reuse technologies, whereas examples from Portland, Chicago, and Sydney will reveal the significance of green stormwater infrastructure in addressing flooding issues and making a city livable (Ragusa, 2021). The book will explore the effect of a diversified supply mix, such as rainwater harvesting, desalination, and reuse, introduced by water-scarce cities such as Singapore and Cape Town in addressing water-related issues. Similarly, proactive groundwater management practices, including artificial recharge, monitoring, and conservation, are highlighted through cases in Delhi, Nairobi, and Bangalore, where aquifer stress necessitated integrated policy and technical responses. Cities with legacy systems such as New York and Toronto illustrate the importance of infrastructure modernization to reduce losses and improve efficiency. Collectively, these cases reinforce that resilient urban water systems depend on strong governance, public participation, cross-sector collaboration, and data-driven decision-making.

Chapter 13 provides a thorough discussion on the case study of Windhoek, known for its resilience across the world, which portrays the effectiveness of wastewater recycling in overcoming the challenge of extreme water scarcity in arid climates. The chapter uses a qualitative method of case study research, wherein the technicality, governance, socio-economics, and equity components of windhoek's wastewater recycling plant for making potable water would be discussed, keeping in mind its relevance within the context of a closed-loop water economy that is integral to sustainable development discussions across the world. It exemplifies the foresight of Namibia's water policies that over the years over came the crisis water management strategy for making potable water by adopting a closed-loop water economy system that meets international standards laid by the World Health Organization. The key element of this strategy is the Goreangab Water Reclamation Plant, which is a precursor for the Goreangab Water Reclamation Plant's model of context-sensitive technological innovations using modularity, quality monitoring, or redundancy (Hemmati et al., 2020). The results have made it clear that the Water System of Windhoek functions efficiently with high resource use efficiency, cost-effectiveness, and equitable water distribution; this helps decrease the reliance on the depleting freshwater resources base significantly, the extent of pollution of the environment, as well as the impact on the disadvantageous peri-urban area populations. This chapter is presented within the conceptual framework of windhoek Water System resilience/sustainability. Economically, while initial capital investments were substantial, long-term returns include operational cost savings, reduced drought vulnerability, and enhanced water security, reinforcing recycling as a cornerstone of resilient urban water economies. Socially, sustained public education and participatory governance have been critical in fostering acceptance of potable reuse, illustrating that technological success must be matched by civic engagement and trust-building. This chapter places the Windhoek case in the context of target 6.3 of the SDG 6, in terms of the meaningful progress achieved in combatting pollution, enhancing the quality of water, as well as the promotion of the safe reuse of water, indirectly contributing to the conservation of ecosystems under the SDG 14. In terms of the wider implications of the case, it is submitted that while the Windhoek case provides an adaptive model to be adopted by similar cities in Africa, Asia, and Latin America that are increasingly being subjected to exacerbated levels of water stress, implementation of the technology solutions presented in the book cannot be transferred.

Chapter 14 presents a comprehensive and methodologically advanced assessment of pollution and eutrophication dynamics in the Upper Crocodile West catchment, demonstrating the value of integrating machine learning with conventional water quality analysis for resilient urban and peri-urban waterscape management. Based on 24 water samples collected during dry and wet seasons from four strategically important locations Lethabile, Losperfontein, Rietfontein, and Brits, the chapter evaluates a wide suite of physicochemical parameters, including nutrients, organic load indicators, and suspended solids, to characterize water quality conditions downstream of wastewater treatment works (Mnyango et al., 2025). Using robust non-parametric statistical tests, the study identifies clear spatial and seasonal patterns, with the wet season exhibiting higher nutrient loads and eutrophication indices due to intensified runoff and sediment transport toward downstream systems such as the Hartbeespoort Dam (Acuña-Alonso et al., 2020). All sampling sites were classified as heavily polluted and eutrophic based on Complementary Pollution Index (CPI), Organic Pollution Index (OPI), and Eutrophication Index (EI), highlighting severe ecological stress driven primarily by municipal and industrial effluent discharges and nutrient-rich agricultural runoff. A key contribution of the chapter lies in the application of a random forest ensemble regression model optimized through cross-validation and grid search which achieved strong predictive performance ($R^2 > 0.6$) in forecasting pollutant concentrations and eutrophic status from environmental variables. This machine learning approach successfully captured complex, non-linear relationships within the catchment, offering a powerful decision-support tool for real-time monitoring and proactive management. By combining advanced statistical analysis, pollution indices, and machine learning, the chapter delivers a holistic pollution profile that captures both spatial heterogeneity and temporal variability, addressing critical knowledge gaps in conventional monitoring frameworks. The findings reinforce the urgency of reducing nutrient inputs through improved wastewater treatment, stricter control of sewage discharges, and better management of agricultural practices, while also emphasizing the need for sustained surveillance, targeted phosphorus reduction strategies, and long-term funding commitments.

In this chapter, urban waterscapes are presented as essential geo-socio systems and resources for sustainable, equitable, and climate-resilient urban development, which, however, are facing increased endangerment because of rapid urbanization, climate change, population growth, and unsustainable commodity extraction and utilization rates. In this chapter, urban water challenges, with their complexity and multifaceted symptoms of water scarcity, pollution, degradation, inaccessibility, floods, droughts, and increasing vulnerability to climate change, are presented as inextricably intertwined with energy, food, and health, and hence complex to manage and govern. The chapter is critical of 'conventional, efficiency-oriented, and water infrastructure and water planning' approaches to urban water, and finds them unsuitable for and unprepared to deal with uncertainty, complexity, and risk. Other innovative approaches, such as Integrated Urban Water Management, Water Sensitive Urban Design, nature-based solutions, circular water economy thinking, citizens' science, and the strategic use of geo-technologies, AI, and decision support systems, are also mentioned as potential approaches to improve urban resilience to change. This chapter brings together insights from the book to highlight the necessity for transdisciplinary thinking across environmental, social, and economic goals. These are seen as especially pertinent in understanding that urban water problems are both multiscale and historically embedded within wider water-energy-food-climate nexuses. This will require restoring and protecting blue-green infrastructure, mainstreaming water reuse and recycling, strengthening the governance of groundwater, improving stormwater management, and embedding digital monitoring to enable anticipatory and adaptive responses. Of equal importance, there will be governance reforms aimed at multi-level coordination, inclusive participation, and equitable access, which will ensure that water management becomes a driver of social justice rather than exclusion. The chapter concludes that the transition toward sustainable urban water futures is not a one-time technical fix but an iterative process of learning, monitoring, and adaptation, where cities that successfully align technological innovation, ecological restoration, robust governance, and community engagement can transform urban waterscapes into engines of resilience, livability, and sustainability for present and future generations.

Across its four thematic sections, the book demonstrates that urban waterscapes are not merely technical infrastructures, but complex socio-ecological systems produced through interactions among climate processes, urban growth, governance structures, technology, and everyday human practices. Section I establishes the foundational challenges facing urban water systems under rapid urbanization and climate stress, revealing how historical planning legacies, fragmented governance, and increasing climate extremes exacerbate water scarcity, flooding, and systemic vulnerability. Case studies on hill towns, megacities, and flood-prone metros illustrate that conventional, efficiency-driven water management models are increasingly inadequate, calling instead for adaptive, multi-level governance and integrated planning frameworks. Section II deepens this analysis by foregrounding equity, justice, and resilience, showing that unequal access to water, exposure to hazards, and degradation of wetlands and blue-green spaces are expressions of deeper socio-political and spatial inequalities. Through participatory approaches, nature-based solutions, and attention to social well-being, this section demonstrates that resilience must be co-produced with communities and embedded in inclusive planning processes that recognize water as both a right and a shared urban common. Section III advances the discussion by highlighting the transformative potential of geospatial technologies, GIS-based modelling, machine learning, IoT, and AI in diagnosing urban water challenges and enabling evidence-based, anticipatory decision-making. These chapters show how smart technologies, when aligned with integrated urban water management principles, can enhance efficiency, transparency, and adaptability, while also supporting Sustainable Development Goal targets related to water quality, efficiency, and governance. Section IV broadens the perspective through global case studies, illustrating how diverse cities have responded to water stress through infrastructure modernization, wastewater recycling, pollution control, groundwater management, and circular water economy approaches. Together, these experiences demonstrate that resilience emerges not from isolated technical fixes but from sustained investment, strong institutions, participatory governance, and the integration of ecological restoration with technological innovation.

The book argues that achieving resilient, just, and sustainable urban waterscapes requires a fundamental paradigm shift from sectoral, supply-driven water management toward integrated, adaptive, and people-centred systems. Urban water futures must be shaped through transdisciplinary thinking that bridges planning, ecology, engineering, data science, governance, and social justice, while embracing nature-based solutions, digital intelligence, and circular economy principles. The collective insights of the chapters affirm that resilience is an ongoing process of learning, monitoring, and adaptation, not a fixed endpoint. Cities that successfully align governance reform, technological innovation, ecological rehabilitation, and community engagement can transform urban waterscapes into engines of sustainability, equity, and climate resilience. As such, this volume offers both a critical reflection on current urban water crises and a forward-looking roadmap for scholars, planners, policymakers, and practitioners seeking to build water-secure, livable, and resilient cities for present and future generations.

References

Acuña-Alonso, C., Álvarez, X., Lorenzo, O., Cancela, A., Valero, E., & Sánchez, A. (2020). Assessment of water quality in eutrophized water bodies through the application of indexes and toxicity. *Science of the Total Environment, 728*, 138775.

Aiyetan, A. O. (2025). Enhancing sustainable and resilient water infrastructure in South Africa in the face of climate change: Smart and resilient infrastructure in the wake of climate change. *Urban Planning, 10*. https://doi.org/10.17645/up.10096

Becker, M., Seeger, K., Paszkowski, A., Marcos, M., Papa, F., Almar, R., et al. (2024). Coastal flooding in Asian megadeltas: Recent advances, persistent challenges, and call for actions amidst local and global changes. *Reviews of Geophysics, 62*, e2024RG000846. https://doi.org/10.1029/2024RG000846

Fennell, J., Soulsby, C., Wilkinson, M. E., Daalmans, R., & Geris, J. (2023). Assessing the role of location and scale of Nature Based Solutions for the enhancement of low flows. *International Journal of River Basin Management, 21*(4), 743–758. https://doi.org/10.1080/15715124.2022.2092490

Hemmati, M., Ellingwood, B. R., & Mahmoud, H. N. (2020). The role of urban growth in resilience of communities under flood risk. *Earths Future, 8*(3), e2019EF001382. https://doi.org/10.1029/2019EF001382. Epub 2020 Mar 20. PMID: 32715013; PMCID: PMC7375139.

IPCC. (2022). Climate change 2022: Impacts, adaptation and vulnerability. In *Contribution of working group II to the sixth assessment report of the intergovernmental panel on climate change*. Cambridge University Press.

Mabrouk, M., Han, H., Mahran, M. G. N., Abdrabo, K. I., & Yousry, A. (2024). Revisiting urban resilience: A systematic review of multiple-scale urban form indicators in flood resilience assessment. *Sustainability, 16*(12), 5076. https://doi.org/10.3390/su16125076

Mangulama, J. A. (2024). Water resource grabbing: A focus on rural Malawi. *Cogent Social Science, 10*(1). https://doi.org/10.1080/23311886.2024.2423857

Mnyango, S. S., Thwala, M., Truter, C., Xulu, N. G., Schoeman, Y., & Oberholster, P. J. (2025). Drivers and Management of Nutrient Overload in Dams: Insights from Roodeplaat Dam. South Africa. *Hydrology, 12*(3), 57. https://doi.org/10.3390/hydrology12030057

OECD. (2018). *Water governance in cities*. Organisation for Economic Co-operation and Development (OECD) Publishing.

Ragusa, A. T. (2021). We have plenty of water, don't we? Social norms, practices, and contentions in a drought-ridden country. *Hydrology, 8*(4), 161. https://doi.org/10.3390/hydrology8040161

Riaz, A., Nijhuis, S., & Bobbink, I. (2025). The role of spatial planning in landscape-based groundwater recharge: A systematic literature review. *Water, 17*(6), 862. https://doi.org/10.3390/w17060862

UN-Habitat. (2020). *World cities report 2020: The value of sustainable urbanization*. United Nations Human Settlements Programme (UN-Habitat).

United Nations. (2015). *Transforming our world: The 2030 agenda for sustainable development*. United Nations General Assembly.

Vogel, R. M., Lall, U., Cai, X., Rajagopalan, B., Weiskel, P. K., Hooper, R. P., & Matalas, N. C. (2015). Hydrology: The interdisciplinary science of water. *Water Resources Research, 51*, 4409–4430. https://doi.org/10.1002/2015WR017049

Wheeler, S. A., Nauges, C., & Quentin Grafton, R. (2025). Water pricing and markets: Principles, practices and proposals. *Applied Economic Perspectives and Policy, 47*(2), 487–514. https://doi.org/10.1002/aepp.13505

World Bank. (2019). *Quality unknown: The invisible water crisis*. World Bank.

WWAP (United Nations World Water Assessment Programme). (2023). *United Nations World Water development report 2023: Partnerships and cooperation for water*. UNESCO.

The manufacturer's authorised representative in the EU is Springer
Nature Customer Service Centre GmbH, Europaplatz 3, 69115 Heidelberg,
Germany. If you have any concerns regarding our products, please
contact ProductSafety@springernature.com

Printed and bound by CPI Group (UK) Ltd, Croydon, CR0 4YY
21/09/2026
02219989-0020